European Data Protection: In Good Health?

Serge Gutwirth · Ronald Leenes · Paul De Hert
Yves Poullet
Editors

European Data Protection: In Good Health?

Springer

Editors
Serge Gutwirth
Center for Law, Science, Technology
and Society Studies (LSTS)
Vrije Universiteit Brussel (VUB)
Pleinlaan 2, Brussels
Belgium

Paul De Hert
Center for Law, Science, Technology
and Society Studies (LSTS)
Vrije Universiteit Brussel (VUB)
Pleinlaan 2, Brussels
Belgium

Ronald Leenes
Tilburg Institute for Law, Technology,
and Society (TILT)
Tilburg University
Warandelaan 2, AB Tilburg
The Netherlands

Yves Poullet
Research Centre for Information
Technology & Law
University of Namur
Rempart de la Vierge 5, Namur
Belgium

ISBN 978-94-007-2902-5 e-ISBN 978-94-007-2903-2
DOI 10.1007/978-94-007-2903-2
Springer Dordrecht Heidelberg London New York

Library of Congress Control Number: 2012931001

Printed on acid-free paper

Springer is part of Springer Science+Business Media (www.springer.com)

Preface

The informational society is in a state of constant flux. After the adoption of the Internet as a prominent channel of information (websites) and communication (e-mail, chat, IM, VOIP), we are now witnessing a transition whereby internet infrastructure is also used for storing and processing data. Cloud computing is replacing direct control of data on local devices with flexibility, scalability and accessibility from anywhere. Cloud computing however also complicates the privacy and data protection landscape because crucial concepts such as the 'data controller' and consequently their responsibilities, liabilities, duties, and the 'purpose of the processing' (which indicates what a processing is), are (further) blurred.

Next to this, we face an enormous growth of tracking, monitoring and surveillance applications. Automatic number plate recognition is not only being used to detect passing cars that are already on black-lists, but increasingly as a blanket method of collecting the number plates of all passing cars, only to be analysed afterwards in order to detect interesting or pertinent correlations. This shift from targeted to all-round monitoring is significant because it is at odds with and undermines the constitutional principle of the presumption of innocence, by actually turning it upside down. In the domain of commerce, internet users are increasingly taking for granted the free services that the internet offers, whilst ignoring the manner in which it works from the perspective of the service providers and the webmasters. The bottom line is however that if you do not pay for a service, you are not the customer, but rather the product that is actually being sold (to advertisers). The monitoring and profiling of online behaviour is the driving force in advertising, even though it may be to the detriment of human rights such as autonomy, privacy, data protection, non-discrimination, due process and dignity.

Although Europe has a significant legal data protection framework, built up around EU Directive 95/46/EC and the Charter of Fundamental Rights, the question of whether data protection and its legal framework are 'in good health' is increasingly being posed. Advanced technologies raise fundamental issues regarding key concepts of data protection and especially the relationship between the various stakeholders. Falling storage prices, increasing chips performance, the fact that technology is becoming increasingly embedded and ubiquitous, the convergence of technologies and other technological developments are broadening the scope and possibilities of applications rapidly. Society however, is also changing, affecting the privacy and data protection landscape. The 'demand' for free services, security, convenience, governance, etc., changes the mindsets of all the stakeholders involved. Privacy is being proclaimed dead or at least worthy of dying by the captains of industry; governments and policy makers having to manoeuvre between competing and incompatible aims; and citizens and customers are considered to be indifferent.

In the year in which the plans for the revision of the Data Protection Directive will be revealed, the current volume brings together a number of chapters highlighting issues, describing and discussing practices, and offering conceptual analysis of core concepts within the domain of privacy and data protection. The chapters were written for and following up on the 4th international *Computers, Privacy and Data Protection* (CPDP2011) Conference: *In good health*?[1] The CPDP-conferences are annually held in Brussels. In 2011 the venue has been *Les Halles*, a prestigious cultural location offering facilities for both large plenary sessions, smaller interactive sessions and also small get-togethers. The conferences offer a unique format bringing together academics, (legal) practitioners, policy-makers, business representatives, data protection authorities, civil society representatives, activists and artists. They represent a multidisciplinary forum for participants with backgrounds in law, social sciences, technology, and humanities where the participants can exchange ideas, discuss current trends and issues regarding privacy and data protection, and (initiate) work towards solutions. The conference is composed of panels, side tracks, and side events, such as artistic displays related to privacy and data protection. The speakers and panellists are invited by the organisers or selected on the basis of an open call. Selected papers are published afterwards. This has already resulted into three edited volumes: *Reinventing data protection? (2009), Data Protection in a profiled world (2010)* and *Computers, privacy and data protection: an element of choice (2011).* The present volume represents the sequel of the conference held in Brussels from 25–27 January 2011, just prior to the European Privacy Day (28 January 2011). The central theme was to what extent the current regulatory framework and practices are "in good health", and hence fit to cope with the ever changing information society in a time where the reviews of the existing legal framework both in the various EU member states as well as on the European level have become available and the renovation of the Data Protection Directive is in the works.

This book brings together a carefully selected set of papers that fit within the overall theme. Some of the chapters have first been submitted as abstracts and were peer reviewed before being presented at the "PhD evening event" of CPDP2011. They were subsequently resubmitted as full papers. Further chapters were also submitted by participants to the conference. All full papers have been peer reviewed by at least two anonymous readers, of which the comments were sent to the authors, who were required to take them into account (or reason why not). Versions were then subsequently checked for a final review. We are happy to take this opportunity to sincerely thank the reviewers who assisted us in this process: Pedro Bueso, Jean-François Blanchette, Johann Cas, Cecile De Terwangne, Els Debusser, Simone

[1] For more information about the CPDP-conferences, see http://www.cpdpconferences.org.

Fischer-Huebner, Catherine Flick, Raphael Gellert, Gloria Gonzàlez-Fuster, Marit Hansen, Hans Hedbom, Mireille Hildebrandt, Simone van der Hof, Bert-Jaap Koops, Daniel Le Métayer, Leonardo Martucci, Charles Raab, Joseph Savirimuthu, Marc Van Lieshout and Tal Zarsky.[2]

The various contributions have been grouped into three themes. The book's first part focuses on surveillance, profiling and prediction. The information society thrives on the processing of (personal) data. This appears to be an open door, but what many people do not realise is that many data are processed unbeknownst to those involved. One can readily understand that online shops need certain data to perform contracts, but the amount of data processed prior to contracting far surpasses any need. By means of profiling, the internet users are drawn towards service providers and service offers. These profiles are based on behaviour (e.g., mouse clicks on websites), rather than on conscious data entries by those concerned. What's more, people are being monitored and profiled in public and private spaces. The resulting data is used for direct interventions, such as stopping individuals driving cars with license plate numbers found in a black-list in an ANPR system. Such data are also being used however to construct risk profiles used to predict future behaviour of both you and others. This section both describes practices in the public and private sector.

The second part of the book focuses on regulation, enforcement and security. It addresses governance issues and looks at the effectiveness and characteristics of various enforcement instruments, for example self regulation and data protection authorities. It also carves out the possibilities and difficulties of legal (law) enforcement in complex environments, for instance cloud computing and cross border police cooperation.

The third section then turns to some of the fundamental concepts in the area of privacy and data protection. It looks at trust in the context of cloud computing, highlighting that even if the data protection legal framework is suited for this environment, its opacity and complexity requires that users are able to trust service providers to behave appropriately. It also addresses the concept of personal data in addition to discussing the widely felt need for reliable electronic identities and the legal challenges in this area. Furthermore, the scope of data protection rights is scrutinized with a view of protecting individuals rather than protecting data. The prospect of using technology to enforce data protection obligations and rights (privacy by design, privacy enhancing technologies) is often coined as one way to improve the position of European citizens. As such, it is one of the pillars of the renewal of the Directive (COM (2010) 609 final). However, implementing legal provisions in computer systems is far from trivial. The final chapter is an essay on another crucial aspect in the Directive's overhaul: the right to be forgotten.

Reading the various chapters, it appears that the 'patient' needs to be cured of quite some weak spots, illnesses and malformations. European data protection is at a turning point and the new challenges are not only accentuating the existing flaws

[2] In respect of the diversity of nationalities, disciplines, and perspectives represented in this book, the editors and the publisher have left the choices concerning the use of footnote references and/or a bibliography to the authors of the contributions.

and the anticipated difficulties, but also, more positively, the merits and the need for strong and accurate data protection practices and rules in Europe, and elsewhere. We hope that the present book will be useful and contribute to the work done to revise the European Data Protection Directive.

Serge Gutwirth
Ronald Leenes
Paul De Hert
Yves Poullet

Contents

About the Authors

Michel Arnaud is a professor in information and communication sciences at the University of Paris Ouest Nanterre la Défense, where he leads the research on usages of ICT tools for online learning. Specific domains of Michel's interest cover public access to Internet and standards for e-learning. Besides, he has worked on studies on privacy and personal data protection in several French and European projects.

Franziska Boehm is a research assistant at the University of Luxembourg where she is also preparing her PhD thesis on the information sharing in the Area of Freedom, Security and Justice. After having obtained the Licence en Droit in 2003 (University of Nice, France) and the German state exam in law in 2006, she specialized in European data protection law and obtained a Master in this field in 2007 (University of Gießen, Germany). Her research focuses on the data protection rights of individuals, in particular in a law enforcement context.

Sacha Brostoff is a Research Associate currently working on the PVNets project in the Information Security Research Group at the Department of Computer Science, University College London, UK. An Ergonomist by training, he specialised to HCI research, gaining his Ph.D. in the late 90s in the usability of password mechanisms under the supervision of Prof. Sasse. He continues to work on the usability of authentication mechanisms, and anti-phishing/anti-counterfeiting. Homepage: http://sec.cs.ucl.ac.uk/people/sacha_brostoff/.

Claude Castelluccia is senior researcher (directeur de recherché) at INRIA (French National Research Center in Computer Science) in France where he leads a research group on computer and network security/privacy. He has spent 15 years in academic research in several well-known research institutes (such as INRIA, Stanford University and University of California, Irvine). His specific area of expertise includes networking security and privacy.

Jeff Chester is the executive director of the Center for Digital Democracy (CDD), a Washington, D.C. non-profit. CDD's mission is to foster democratic expression and consumer protection in the digital media era. A former journalist and filmmaker, his work has appeared in many publications and on PBS and NPR. He co-founded and was the executive director of the Center for Media Education, a leading force on issues such as Internet privacy, media ownership, and children's TV. In 1996, Newsweek magazine named him one of the Internet's fifty most influential people. He established CDD in 2001 with the support of a Stern Family Foundation "Public Interest Pioneer" grant. He has been the author of a series of reports exposing threats from online marketing, including practices involving finance, health, and children. He was named the 2011 "Domestic Privacy Champion" by the Electronic Privacy Information Center. Chester is the author of Digital Destiny: New Media and the Future of Democracy (New York: The New Press, 2007).

Els De Busser studied Law at Antwerp University and obtained an additional degree in Criminology and an Advanced Master's degree in European Criminology and Criminal Justice Systems from Ghent University, Belgium. From March 2001 to October 2009, she worked as a researcher and professor's assistant in the field of European Criminal Law at Ghent University, Institute for International Research on Criminal Policy where she defended her PhD entitled 'EU internal and transatlantic cooperation in criminal matters from a personal data perspective. A substantive law approach' in May 2009. In November 2009, she joined the European Criminal Law section of the Max Planck Institute in Freiburg, Germany. Her research and publications focus on international cooperation in criminal matters and data protection.

Paul De Hert is an international human rights expert. The bulk of his work is devoted, but not limited, to criminal law and technology & privacy law. At Brussels, Paul De Hert holds the chair of 'Criminal Law', 'International and European Criminal Law' and 'Historical introduction to eight major constitutional systems'. In the past he has held the chair of 'Human Rights', 'Legal theory' and 'Constitutional criminal law'. He is Director of the VUB-*Research group on Fundamental Rights and Constitutionalism* (FRC), Director of the Department of Interdisciplinary Studies of Law (Metajuridics) and core member of the internationally well-accepted VUB-*Research group Law Science Technology & Society* (LSTS) (see: www.vub.ac.be/LSTS). He also holds a position as an associated-professor in the internationally renowned Tilburg Institute for Law, Technology, and Society (TILT) at the Tilburg University (http://tilt.nl). He is member of the editorial boards of several national and international scientific journals such as the *Inter-American and European Human Rights Journal* (Intersentia), *Criminal Law & Philosophy* (Springer). He is co-editor in chief of the *Supranational Criminal Law Series* (Intersentia) and of the New Journal of European Criminal law (Intersentia).

Daniel Guagnin is a junior researcher at the Technical University, Berlin. He is currently working on the FP7 Project PATS on Privacy Awareness. He received his magister in Sociology from the Albert-Ludwigs-University in Freiburg (Germany), his side subjects were computer science and economics. His Magister thesis determined the question about a connection between the freedom of knowledge and the freedom of software as a formalized mode of knowledge. Before he started working at the TUB he was student researcher at a Fraunhofer computer research institute (FIRST) where he supported different projects about Ambient Assisted Living and

Open Source. Besides the privacy and surveillance topic, his research interests are the free content movement and the social implications and effects of technologies.

Serge Gutwirth is a professor of human rights, legal theory, comparative law and legal research at the Faculty of Law and Criminology of the Vrije Universiteit Brussels (VUB), where he studied law, criminology and also obtained a postgraduate degree in technology and science studies. Gutwirth founded and still chairs the VUB-research group *Law Science Technology & Society* (http://www.vub.ac.be/LSTS). He publishes widely in Dutch, French and English. Amongst his recent co-edited publications are *Safeguards in a world of ambient intelligence* (Springer 2008), *Profiling the European citizen* (Springer 2008), *Reinventing data protection?* (Springer 2009), *Data protection in a profiled world* (Springer 2010) and *Computers, privacy and data protection: an element of choice* (Springer 2011). Currently, Serge Gutwirth is particularly interested both in technical legal issues raised by technology (particularly in the field of data protection and privacy) and in more generic issues related to the articulation of law, sciences, technologies and societies.

Leon Hempel is a senior researcher at Centre for Technology and Society at the Technical University Berlin since 1999. His research areas are sociology of technology and innovation, security studies and evaluation methodology. He studied Political Science and Comparative Literature. Hempel co-ordinated the EU funded project URBANEYE on the use and societal impacts of CCTV within public accessible space. Hempel evaluated the compliance impact of Transport for London's camera enforcement systems with the specific focus on mixed uses of cameras and on street officers and also assessed the use of CCTV at Berlin underground. Currently he is leader of the projects PATS (Privacy Awareness through Security Organisation Branding), SIAM (Security Impact Assessment Measure) and others. Hempel is part of the FESTOS consortium, a joint research project that develops threat scenarios of emerging technologies, and of the interdisciplinary doctoral program 'Helmholtz Research School on Security Technologies'. He is in charge of technology assessment within the Fraunhofer Innovation Cluster 'Secure Identity'. He also has provided the Civil Liberties Committee of the European Parliament with external expertise on 'Exchange of Data Between Law Enforcement Authorities'.

Carla Ilten is a junior researcher at the Technical University Berlin currently working on the EU FP7 PATS project. She graduated in Sociology and Technology Studies with a minor in Computer Science from the Technical University Berlin in 2008. Her studies focused on Science and Technology Studies (STS), innovation studies and Constructive Technology Assessment, as well as information and communication technologies. In her diploma thesis, Ilten developed a variant of a CTA-based approach to socio-technical innovation with a special emphasis on civil society actors. She is now working on a PhD thesis on activism for social change and new media after conducting field research as a Visiting Scholar at Temple University.

Charlene Jennett is a Research Associate in the Information Security Research Group at the Department of Computer Science, University College London (UCL), UK. She has a background in Human-Computer Interaction research, gaining her

PhD at the University College London Interaction Centre (UCLIC) in 2010. She started working on the PVNets (Privacy Value Networks) project in 2010. PVNets is a three-year research project (2008–2011) that aims to help government and business understand the value of personal data in different contexts. Homepage: http://sec.cs.ucl.ac.uk/people/charlene_jennett/.

Julien Le Clainche is an ICT specialist with a Master Degree in *Computer Science and Law*, and a PhD in *Private Law* at the Paris II and Montpellier I universities. His research focuses on adapting privacy laws to electronic communications, where he has more than fifteen peer-reviewed publications on topics covering: spam, consent, free speech, IT security, torts and legal information. He worked at the French *National Institute for Research in Computer Science and Control* (INRIA) for the *Legal Issues in Communication and information Technologies* (LICIT) initiative, under the direction of Daniel Le Métayer to foster interactions between research activities in law and ICT. Julien provides consultancy to both French Parliament chambers on matters dealing with ICT laws and also provides training courses to services of the French First Minister and the French Ministry of Culture. He founded and edits the website www.droit-tic.com that deals with interactions between law and technological progress.

Ronald Leenes is professor in Regulation by Technology at TILT, the Tilburg Institute for Law, Technology, and Society (Tilburg University). His primary research interests are privacy and identity management, regulation of, and by, technology. He is also involved in research in ID fraud, biometrics and Online Dispute Resolution. Ronald was work package leader in the EU FP6 PRIME project for socio-cultural aspects of privacy enhanced identity management. He is currently responsible for TILT's contribution to the FP7 project PrimeLife and leads the work package on social networks and collaborative workspaces. He has contributed to and edited various deliverables for the EU FP6 Network of Excellence 'Future of IDentity in the Information Society' (FIDIS).

Daniel Le Métayer is Research Director for INRIA (the French National Institute for Research in Computer Science and Control) and head of a new initiative called LICIT for "Legal Issues in Communication and Information Technologies". The main goal of LICIT is to foster interactions between research activities in law and ICT. From 2000–2006, Daniel Le Métayer worked for Trusted Logic, a leading company in security and open middleware for embedded systems. Daniel Le Métayer has been involved in various international projects on IT security, software design and analysis, testing, etc. He has also served on programme committees of many IT international conferences and he has been the editor of special issues of computer science journals such as ACM Transactions on Software Engineering and Theoretical Computer Science.

Erik Luysterborg is a partner within Deloitte, Belgium. He leads their Security & Privacy group as well as their European Data Protection and Privacy service line. He has somewhat of a hybrid background as he is both a lawyer as well as a security consultant. He has had extensive experience in dealing with the security and privacy issues related to both traditional (out)sourcing as well as Cloud environments, advising both users and Cloud providers. He is an active member of several data protection steering committees and has extensive experience in assisting international clients regarding the cross border and practical/technical aspects of data protection. He has a specific focus on designing operational and pragmatic security management solutions and controls as well as effective risk based legal/compliance strategies both in public and private sector.

Miguel Malheiros is a Research Student in the Information Security Research Group at the Department of Computer Science, University College London, UK. He has an MSc in Computer Engineering from the Technical University of Lisbon. His PhD research focuses on the value that personal data has for individuals and organisations and how privacy protection behaviours on the part of individuals can affect organisations' data quality. He joined the PVNets project in 2008. Homepage: http://sec.cs.ucl.ac.uk/people/miguel_malheiros/.

Vlad Niculescu-Dinca is a PhD researcher within the DigIDeas project. Vlad studied software engineering in the technical universities of Eindhoven and Bucharest and Philosophy of Science, Technology and Society in University of Twente. During his technical studies and professional experience he became interested in the ethical and social aspects of technology design and use, as well as how technology and society mutually influence and shape one another. His thesis at the University of Twente analyzed the structure of ethical debates around new and emerging science and technology, with a case study focused on the converging technologies debate. Within the DigIDeas project, Vlad examines the ethical and social issues at the intersection between new technological developments towards ubiquitous identification and policing practices, aiming to contribute to a value sensitive management of digital identities.

Norberto Nuno Gomes de Andrade is a Scientific Officer at the Information Society Unit of the Institute for Prospective Technological Studies (IPTS) of the European Commission's Joint Research Centre. He graduated in Law at the Faculty of Law of the University of Lisbon, and he holds a Ph.D. in Law from the European University Institute (EUI, Italy), a Master of Arts in International Relations and European Studies from Central European University (CEU, Hungary), as well as a Master of Research in European, International and Comparative Law from the European University Institute. He has previously worked as a legal expert at the External Relations Department of the Portuguese Regulatory Authority for Communications (ANACOM, Portugal). His research interests are focused on law and technology (including biotechnology, neuroscience, artificial intelligence, genetics and genomics, digital environments, ambient intelligence), data protection and privacy law, intellectual property, philosophy of law and legal theory. In 2009 he co-edited and published "Law and Technology: Looking into the Future—Selected Essays".

Ugo Pagallo is a Full Professor in Philosophy of Law at the University of Torino, Law School, and Faculty at the Center for Transnational Legal Studies (CTLS) in London, U.K. He is editor of the *Digitalica* series published by Giappichelli in

Turin, co-editor of the AICOL series by Springer, and member of both the Scientific Advisory Board of the Law, Governance and Technology Series of Springer and the Programme Committee of ETHICOMP. In addition to numerous essays in scholarly journals like *Journal of Business Ethics*, AI & Society, *Philosophy and Technology*, Hobbes Studies, *Journal of Chinese Philosophy*, Knowledge, Technology & Policy, and so forth, he is the author of eight monographs. His main interests are AI & Law, Network theory, Robotics, and Information Technology Law (specially data protection law and copyright).

Maria Grazia Porcedda is a researcher in law at the European University Institute (Italy). She has worked for the Centre de Recherche Informatique et Droit (CRID)—University of Namur for the organization of the workshop "Law Enforcement in the Clouds: Regulatory Challenges" and has coauthored the Working Paper for the event. Previously, she has worked as an intern at the European Data Protection Supervisor (EDPS), mainly on consultation projects in the area of Freedom, Security and Justice/ External Relations. She holds a B.A. in Political Science (University of Cagliari) and an M.A. in International Relations (University of Bologna). She has been an exchange student at Nottingham Trent University, Johns Hopkins School of Advanced International Studies (SAIS—Bologna Center), University of California at Berkeley and Ecole Normale Supérieure—Lyon. She speaks Italian, English and French fluently and also speaks Spanish.

Wolter Pieters is a postdoc researcher in information security at the University of Twente. He studied computer science and philosophy of science, technology and society at the same university, and wrote his interdisciplinary PhD "La volonté machinale: understanding the electronic voting controversy" at the Radboud University Nijmegen. Afterwards he advised the Dutch Ministry of the Interior on electronic voting and electronic travel documents. Since September 2008 he works in the VISPER project at the University of Twente, concentrating on disappearing boundaries in information security. He was program chair of the 2010 CPDP workshop on Security and Privacy in Cloud Computing, and will co-organise the 2011 Dagstuhl seminar on Secure Architectures in the Cloud. He published on electronic voting, verification of security properties, access control, and philosophy and ethics of information security.

Yves Poullet Ph.D. in Law and graduated in Philosophy, is full professor at the Faculty of Law at the University of Namur (FUNDP) and Liège (Ulg), Belgium. He teaches "Sources and Principles of the Law", "Internet Regulations", "International Commercial Law" and "Human Rights in the Information Society". Yves Poullet heads the CRID, since its creation in 1979. He conducts various researches in the field of new technologies with a special emphasis on privacy issues, individual and public freedom in the Information Society and Internet Governance. He is legal expert with the European Commission, the UNESCO and the Council of Europe. He has been during 12 years (1992–2004) member of the Belgian Commission on Data Protection. In addition, he was, since its origin, member of Legal Advisory Board of European Commission. He has received the Franqui Chair in 2004. He also chaired the Belgian Computer Association ABDI (Association Belge de Droit de l'Informatique). Yves Poullet is an active member of the Editorial Board of various famous law reviews.He is a founder of the European Telecommunication Forum, ECLIP and FIRILITE. Recently (2009), he has been nominated as member of the Royal Belgian Academy and as Rector of the University of Namur.

Christian W. Probst is an Associate Professor in the department for Informatics and Mathematical Modelling at the Technical University of Denmark, where he works in the section for Language-Based Technologies. The motivation behind Christian's research is to realize systems with guaranteed properties. An important aspect of his work is questions related to safety and security properties, most notably insider threats. He is the creator of ExASyM, the extendable, analysable system model, which supports the identification of insider threats in organisations.

Arnold Roosendaal LLM MPhil is a PhD Candidate at the Tilburg Institute for Law, Technology, and Society, Tilburg University, The Netherlands. He is currently doing research on digital representations of individuals and shortcomings in current legislation concerning the protection of data that form these representations. A specific focus is on the impact on privacy and autonomy of individuals. Arnold participated in the EU projects FIDIS and PrimeLife and wrote several articles in international journals.

Angela Sasse is the Professor of Human-Centered Technology and Head of Information Security Research in the Department of Computer Science at University College London, UK. A usability researcher by training, she started investigating the causes and effects of usability issues with security mechanisms in 1996. In addition to studying specific mechanisms such as passwords, biometrics, and access control, her research group has developed human-centred frameworks that explain the role of security, privacy, identity and trust in human interactions with technology. Homepage: http://sec.cs.ucl.ac.uk/people/m_angela_sasse/.

Philip Schütz studied political science, English literature and law at the University of Heidelberg and at the Institut d'Etudes Politiques Lille in France. After being a visiting researcher at the University of Cambridge he graduated with an M.A. in political science in 2009. Since March 2010 he is a junior researcher in the Competence Center Emerging Technologies at the Fraunhofer Institute for Systems and Innovation Research ISI in Karlsruhe. His research focuses on data protection in the context of new emerging technologies and governance of privacy. Being supervised by Prof. Andreas Busch from the Political Science Department of the University of Göttingen, Philip has just started his dissertation project that deals with a comparative analysis of data protection authorities (DPAs).

Iván Székely social informatist, is an internationally known expert in the multidisciplinary fields of data protection and freedom of information. A long-time independent researcher, consultant and university lecturer, as well as former chief counsellor of the Hungarian Parliamentary Commissioner for Data Protection and Freedom of Information, Székely is at present Counsellor of the Open Society Archives at Central

European University and associate professor at the Budapest University of Technology and Economics. His studies and publications, as well as his research interests are focused on information autonomy, openness and secrecy, privacy, identity, and archivistics.

Dimkov Trajce is a PhD researcher at the Distributed and Embedded Security Group in University of Twente, The Netherlands. His research interests include physical penetration testing methodologies, social engineering, and formal methods for alignment of security policies.

Mario Viola de Azevedo Cunha holds an L.L.M in Private Law from Rio de Janeiro State University (Brazil) and a Master of Research in European, International and Comparative Law from the European University Institute (Florence, Italy). He is currently concluding his PhD in Law at the European University Institute. In 2009 he was a trainee at the European Data Protection Supervisor and a visiting researcher at the Law Department of the London School of Economics and Political Science, within the Erasmus mobility programme of the EU Commission. He has published many articles and book chapters dealing with consumer law, tort law, data protection and IT law, in Portuguese, English and Italian.

Tal Zarsky is a Senior Lecturer at the University of Haifa—Faculty of Law. In 2010–2011 He is a Global Hauser Fellow, at NYU Law School. His research focuses on Information Privacy, Internet Policy, Telecommunications Law and Online Commerce, Reputation and Trust. He also taught Contract and Property law. He has written and presented his work on these issues in a variety of forums, both in Israel and worldwide. In addition, he has advised various Israeli regulators and legislators on related issues. Dr. Zarsky is also a Fellow at the Information Society Project, at Yale Law School. He completed his doctorate dissertation, which focused on Data Mining in the Internet Society, at Columbia University—School of Law. Most recently, Dr. Zarsky participated, as an affiliate with the Centre for Law in the Information Society in Leiden University, in the research project "Data Mining without Discrimination", funded by the Dutch Research Council (NWO).

Part I
Surveillance, Profiling and Prediction

Chapter 1
We Are All Connected to Facebook ... by Facebook!

Arnold Roosendaal

1.1 Introduction

Tracking and tracing users over the web is a valuable tool for profiling purposes. Based on revealed interests, web users can be targeted for personalized advertisements. Companies that earn their revenues from targeted advertising have a huge interest in using these techniques. It is therefore not surprising that the way these techniques are exploited becomes more and more sophisticated.

The use of cookies and third-party cookies to recognize and track and trace web users is not a new concept (see Dwyer 2009; Martin et al. 2003). Usually, cookies are placed on the user's web browser without any direct visibility. A cookie is a small text file and the process of placing it on the browser is executed without user interaction. The presence and origin of the cookies is not revealed until a user checks his cookies via his browser options. In order to let third-parties place cookies, they have to be allowed to place content on a website. The content is requested from the web server of the third party and is delivered along with a cookie. When a site is visited again, the cookie is sent along in the request for the content. This allows content providers to 'remember' preferences of web users, such as language settings or purchasing history, and to provide the web content according to these preferences. A web browser is recognized through the cookie, which allows the web activity to be monitored.

In order to gain as much reach over the web as possible, the technologies for tracking have become very sophisticated. Sophistication can, however, also be in the presentation of a tracking tool. For instance, Facebook offers content providers to place a Like button on their website. This button is a tool which allows Facebook members to indicate that they like a certain website or item on a website. By clicking the button, a link to the item is placed on their Facebook profile page. In addition, the number of visitors who 'liked' something is indicated next to the button. For content

A. Roosendaal (✉)
Tilburg Institute for Law, Technology, and Society (TILT),
Tilburg University, Tilburg, The Netherlands
e-mail: a.p.c.roosendaal@tilburguniversity.edu

S. Gutwirth et al. (eds.), *European Data Protection: In Good Health?*,
DOI 10.1007/978-94-007-2903-2_1, © Springer Science+Business Media B.V. 2012

providers, the Like button can thus function as an important business tool, because website visitors can contribute to attracting more visitors to a website. This makes the tool valuable for content providers, which is also reflected by the fast increase in web coverage of the Like button. However, even though presented as a nice feature for content providers, the Like button is also used to send cookies and to track and trace web users, regardless of whether they actually use the button. The browsing behavior of individuals can be connected to their Facebook account. If a user has no Facebook account, a separate set of data concerning individual browsing behavior can be created. When a user creates an account later on, the data can be connected to the newly established profile page.

The practice of massively collecting data concerning individual web behavior is an important phenomenon in the Internet realm. It hugely affects the way companies, people, and privacy mutually relate to each other and, at a fundamental level, it influences the abilities of individuals to construct their own identities. The fact that individuals value their privacy and object to these practices also becomes clear from the class action law suit against Facebook, which was filed in California.[1]

In this chapter, the effects on privacy and identity of individuals resulting from hideous tracking technologies will be described. First, a further introduction to the Facebook Like button and its value will be given in Sect. 1.2. In Sect. 1.3, the technical process of placing and replacing cookies with the help of the button will be described, as well as how this facilitates profiling. Subsequently, the way this practice affects the privacy of individuals will be discussed in Sect. 1.4, and finally a conclusion will be drawn in Sect. 1.5.

1.2 The Facebook Like Button

The Facebook Like button is an image displaying a thumbs-up symbol accompanied by the word 'Like.' According to Facebook, "[t]he Like button lets a user share your content with friends on Facebook. When the user clicks the Like button on your site, a story appears in the user's friends' News Feed with a link back to your website."[2] Anyone can display the button on his website by simply implementing the code which is available for free. The button can thus be used by content providers to have web users promote content and create links on their Facebook profile pages. When clicking the Like button, a login field opens in a pop-up window to log on to Facebook. Logging on results in the creation of the link on the Facebook profile page. When a user is already logged on to Facebook, the creation takes place immediately.

In April 2010, at their f8 conference, Facebook announced Instant Personalizer and Social Plugins, two services that allowed partners to leverage the social graph—the information about a user's relationships on the site that the user makes available

---

[1] Ung vs. Facebook, Class action complaint, Superior Court of the State of California, County of Santa Clara, Case No. 111CV200467, filed 05/09/2011.

[2] "Like Button—Facebook Developers," accessed 22 March 2011, http://developers.facebook.com/docs/reference/plugins/like.

to the system—and provide a channel for sharing information between Facebook and third parties. For example, websites could implement a Like button on their own pages that enables users to share content from that site with their connections on Facebook (boyd and Hargittai 2010). The value of displaying the Like button on a website becomes clear from the statistics. Sites that have added such social plugins from Facebook reported increases in traffic in excess of 200%. Besides, the time spent and the number of articles read on websites with Like buttons also increased by over 80%.[3] The button represents 12.9% of the distribution of third-party widgets.[4] It also appears that, within months, the use of social plugins had reached millions of sites.[5] The penetration rate of the Like button in the top 10,000 websites reached over 4% in the first six months after its introduction,[6] and it is likely that it will continue to grow.

While the Like button can help content providers to generate traffic to their websites, it is also a tool for Facebook members to add information about their interests to their personal profile page. Thus, it fits perfectly in the ongoing trend of social networking sites like Facebook encouraging members to share personal information.[7] Obviously, for sharing items from the web, the Like button is a very useful tool, because it allows direct linking without having to copy and paste complete URLs and the content is made up in a readable manner automatically.

1.3 Cookies, Recognition, and Identification

As indicated, there are numerous third parties which deliver content to websites and place cookies. Usually, the function of these third parties is to provide website providers with content such as advertisements or specific functionalities like maps or videos. A piece of content is delivered from the servers of the third party and can be sent together with the cookie. The cookies can be used to generate information on the number of visitors and which items on a website attracted the most attention. In this way, third parties can provide a service to the website provider. A web user is usually not aware of this. He just types in the URL of the website he wants to visit and the

[3] "The Value of a Liker—Facebook," accessed 22 March 2011, http://www.facebook.com/notes/facebook-media/value-of-a-liker/150630338305797.

[4] "Facebook Like Box Usage Statistics," accessed 22 March 2011, http://trends.builtwith.com/widgets/Facebook-Like-Box.

[5] "Facebook Stats Likers," accessed 29 Sept. 2010, http://www.insidefacebook.com/2010/09/29/facebook-stats-likers/.

[6] "Facebook Like Usage Statistics," accessed 22 March 2011, http://trends.builtwith.com/widgets/Facebook-Like.

[7] There are, however, more privacy friendly initiatives which focus on audience segregation and controlled disclosure of personal information. For instance, Clique allows users to have several 'faces' in one account. See http://clique.primelife.eu/. This social networking site is one of the results of the EU FP7 PrimeLife project.

page is loaded. That the loading of the page involves numerous HTTP requests[8] for content from the servers of the visited websites and often several third-party servers is a process which takes place behind the scenes. Or, in more popular terms: that is where the magic happens!

A cookie is placed on the web user's computer via his browser. Each cookie is connected to a web server, so only the server from which the cookie was sent has access to the cookie. The provider of a website does not have access to other cookies placed by third parties via his website. Once a cookie is available on the user's computer, this cookie will be sent together with the HTTP request in each later request for content from the server which installed the cookie. The HTTP request also includes data on the referrer, which is the website on which the content will be displayed. Since the referrer data is always included, third parties can follow exactly which sites a user visits. When data concerning web visits are combined based on the unique cookie, the browsing history of a web user can be mapped. The content is needed to load a page so, for tracking purposes, it is irrelevant whether a user actually clicks a piece of content or not, or whether the content is clickable at all.

1.3.1 Scenarios

The Facebook Like button is also a piece of third-party content. The website provider does not directly place an image of this button on his website. In fact, the button is a piece of HTML code which includes the request to the Facebook server to provide the image when the website is loaded. This implies that the button can be used to set third-party cookies or to recognize them as well. A few different scenarios can be distinguished: (1) a web user has a Facebook account, (2) a web user does not have an account, (3) a web user becomes a member of Facebook, and (4) a member deletes his account. These scenarios have been tested in a practical experiment using Techcrunch.com, CNN.com, and Gizmodo.com.

1.3.1.1 The Web User Has a Facebook Account

The first option is a scenario in which the web user has a Facebook account. When the account is created, Facebook issues a cookie, containing a unique user ID, to the computer of the user. This cookie facilitates the display of a username in the login field at returning visits. When accessing Facebook from another device, a temporary cookie is issued, which is replaced by a cookie with the same ID after logging on to the account. In this way, different devices can be linked to one account and thus one user. Every time the user visits the Facebook website, the cookie is sent together

---

[8] HTTP stands for Hyper Text Transfer Protocol, the programming language used for internet traffic. An HTTP request is a request for a specific piece of content sent from the user's computer to a web server. The web server replies by sending the requested content. If the content is not available, the reply includes an error code.

```
GET
/plugins/like.php?href=http%3A%2F%2Fwww.facebook.com%2FGizmodo&layout=button_co
unt&show_faces=false&width=200&action=like&colorscheme=light&height=21 HTTP/1.1
Host: www.facebook.com
User-Agent: Mozilla/5.0 (Windows; U; Windows NT 5.1; en-GB; rv:1.9.2.10) Gecko/20100914
Firefox/3.6.10
Accept: text/html,application/xhtml+xml,application/xml;q=0.9,*/*;q=0.8
Accept-Language: en-gb,en;q=0.5
Accept-Encoding: gzip,deflate
Accept-Charset: ISO-8859-1,utf-8;q=0.7,*;q=0.7
Keep-Alive: 115
Connection: keep-alive
Referer: http://gizmodo.com/
Cookie: datr=yjPATCXPQuDBLU_J5ZfRsJpd; lu=TgbyaYN2Obo-F4fEBiQTGtwQ;
locale=en_GB; x-
referer=http%3A%2F%2Fwww.facebook.com%2Fhome.php%23%2Fhome.php;
cur_max_lag=20; c_user=100001XXXXXXXXX; sct=1287731574; sid=0;
xs=55dcbdfe4719c2693d477d0c0dd83ab6
Cache-Control: max-age=0
```

Fig. 1.1 The HTTP GET request for the Like button on Gizmodo.com, including the cookie with user ID (anonymized by the author)

with the HTTP request for the site. As a result, Facebook already knows who wants to log in before the actual login has taken place.

However, the cookie is not only sent to the Facebook servers when a member logs on, but also on every occasion when content such as the Like button has to be provided from the Facebook servers (Fig. 1.1). Thus, every single time a website containing the Like button is visited; Facebook receives information concerning the user, including his unique ID, via the cookie. If the user actually clicks the button, he has to provide his Facebook login details, and a message about the 'Like' is posted on his profile page.

Users are often not aware of the fact that data about the user are sent to Facebook regardless of whether the Like button is actually clicked. The cookie contains the unique user ID and thus allows information on browsing behavior to be connected to the account. Even though the user is not involved, Facebook can collect far more individual data than the data made available on the profile page only.

Below is an example of a request for the Like button where the cookie including a unique user ID is sent along.

In this scenario, there is a link between the Internet user and Facebook because there is an account. Now, consider a scenario where there is no membership link.

1.3.1.2 The Web User Does Not Have a Facebook Account[9]

If a user does not have a Facebook account, there is no cookie and no user ID available. A visit to, for example, Techcrunch.com includes an HTTP GET request

[9] This scenario does not apply anymore since Facebook changed its systems after the publication of my initial research findings (Roosendaal 2010). In a communication to the Hamburg Data Protection Authority (Germany) Facebook stated that the tracking of nonusers was the result of a 'bug' in their software development kit.

1. Set-Cookie: datr=ckviTDm3989eNbvw6xMhAWle; expires=Thu, 15-Nov-2012 09:14:26 GMT; path=/; domain=.facebook.com
2. Set-Cookie: datr=ckviTC8tNJ-1ZKqCu_Srlga7; expires=Thu, 15-Nov-2012 09:14:26 GMT; path=/; domain=.facebook.com

Fig. 1.2 A cookie issued via Facebook extern login status (1) and one via Facebook Connect (2) on Gizmodo.com

for the Like button. However, in this scenario, when the button is provided, no cookie is issued. Thus, it seems that the Like button itself is not used to issue cookies. However, when a site is visited which includes Facebook Connect (for instance Gizmodo.com), this application does issue a cookie (Fig. 1.2). From that moment on, visits to other websites which display the Like button result in a request for the Like button to the Facebook server including the cookie. An important part of the process depends on visiting a site which has implemented Facebook Connect. The chance of visiting such a site is considerable. Within a year from its launch in December 2008, Facebook Connect was used on almost 1 million websites and in March 2009 over 40 million unique visitors of Facebook Connect implementations were registered (Burbary 2009). The number of implementations increases exponentially, so the likelihood of accessing such a website is increasing at a fast pace as well.

As indicated, after visiting a website on which Facebook Connect has been implemented, the request for the Like button includes a cookie. This cookie has an expiration date two years from the moment it was issued. However, by browsing across websites, additional cookies can be placed on the user's computer and these can be added later on in new requests. Not all cookies are used in this way. For instance, a cookie issued via the extern login status plugin is not included in later requests.

Based on the cookie, the entire web behavior of an individual user can be followed. Every site that includes some kind of Facebook content will initiate an interaction with the Facebook servers, disclosing information about the visited website together with the cookie.

1.3.1.3 A User Becomes a Facebook Member

It is possible that a web user already has a personal set of data collected by Facebook, based on the mechanism described above. The question is what happens if this user creates a Facebook account. In that case, he first has to go to the Facebook homepage (login page). The cookie on the user's computer is sent to Facebook in the request for the web page to be loaded. The server responds and issues a few new cookies. These new cookies are temporary ones, or session cookies. When the account is actually created, a unique ID number is issued and sent in a cookie. The connection between this ID cookie and the old cookie is made behind the scenes by Facebook's servers. This means that the entire historical information of the user can be connected to the newly created Facebook account. From this moment on, all subsequent requests for Facebook content are accompanied with the unique user ID cookie.

If a user deletes all his cookies, the process starts from the beginning with Facebook Connect placing a new cookie when a site containing Facebook Connect is visited. From the moment on that the user accesses his Facebook account, or connects to this account by clicking the Like button and providing username and password, this cookie is replaced by a cookie containing the unique user ID that belongs to the account.

1.3.1.4 A User Deletes His Facebook Account

A last possibility is that an existing Facebook member decides to exit the network. In this case, the user can delete his account. Facebook offers an easy process to deactivate an account. Deactivation, however, is not similar to deletion. In fact, when an account is deactivated, the account and all its contents are only made unavailable to the network community. The entire account is kept by Facebook just in case the user decides to rejoin the network. In that case, the complete account, including all connections and contents can be reactivated. Clearly, during the inactivity of an account, Facebook is still able to connect data to the account in a way similar to when the account was active.

There is also an option to really delete an account. The deletion process includes a specific request to Facebook that takes two weeks to process. If the account is accessed in this period, the deletion process is stopped. After 14 days, accessing the account is no longer possible and the contents can no longer be retrieved. Whether Facebook keeps any information or even the entire account, probably disconnected from the login credentials, is unclear. However, even if the account is really deleted, the web user can still be tracked and the browsing data can still be connected to an individual data set. This means that, after deleting the account, all services which were connected to Facebook, for instance, by using the Facebook account to sign up, have to be disconnected as well and cookies have to be deleted. Once everything is cleared and disconnected, the web user can be considered to be someone who does not have a Facebook account and the scenario earlier described applies.

1.3.2 Recognition and Identification

Facebook uses cookies for recognition. Web users can be recognized whenever they visit a site with a piece of Facebook content. Facebook members are identified as individual account holders, because the cookie includes their unique user identification number. When different devices are used to access Facebook, such as a home computer, a laptop, or a smart phone, these devices are recognized as belonging all to the same individual, so all web interaction from these different devices is connected as well. Individuals who do not have a Facebook account are recognized as well. Their browsing behavior, however, is not connected to a Facebook account; besides, recognition is machine based and separated for every single device. Since there is no unique user ID in the cookie resulting from a log-on to Facebook, the different devices cannot be connected solely on the basis of the cookies. Single devices can be quite reliable, however, even though they can be used by different persons. More and more devices, such as laptops and smart phones, become personal and are usually used by one single individual. This implies that information collected based on the cookies and browsing behavior results in a very personal profile. Obviously, Facebook can use this to provide their members with targeted advertisements. The information collected about the browsing behavior of nonmembers probably provides a larger sample for profiling and targeting purposes.

The Facebook Like button is not the only button which frequently appears on websites to facilitate sharing or promoting content. Other examples are Twitter's Tweet button, the Digg button, and Google's Buzz, but there are differences. As described above, Facebook Connect is the system that actually issues a cookie the first time. From that moment on, the cookie is sent together with all HTTP requests for content, so also when the Like button is uploaded onto a page. Thus, an additional system is used to initiate the cookie exchange. Twitter, for instance, does not have such a system. The Tweet button does not always send a cookie when the button is requested from the Twitter servers. Only if someone visits the Twitter homepage is a cookie issued which is used in future interactions with the servers, similarly as with the Like button. Logging on or even having a Twitter account is not necessary. A small but important difference with the Like button is that there is at least supposed to be some link to Twitter, because the web user has visited this website. For Facebook, this is not necessary at all, which implies that individuals who consciously choose not to participate in Facebook are still tracked and traced by Facebook. Even if someone does not connect to Facebook himself, Facebook makes the connection.

Another important difference is that Facebook can trace the browsing behavior to member accounts. These accounts are, usually, quite rich concerning disclosed information, but the Like button as exploited by Facebook allows far more information to be collected about individual members than the information disclosed on the personal profile page. Thus, people who have an account, but do not want to disclose much information are still profiled more extensively. Their browsing behavior discloses much information concerning personal interests, and this information can also be collected by Facebook and connected to the individual account. In the end, consciousness in disclosing information, either by not participating on Facebook or by very limited disclosure of personal information, is not sufficient to escape Facebook's tentacles.

An additional point of attention lies in the function Facebook is exploiting as an identity provider. An increasing amount of websites offers the possibility to register or log on with Facebook credentials.[10] The username and password are consequently used at places other than on Facebook only. Obviously, the services that provide this possibility are linked to Facebook as well. However, a more pressing issue is the fact that, for some web services, logging on is *only* possible with a Facebook account. This means that, without a Facebook account, accessing or using the services is simply

[10] For instance: www.slideshare.net.

impossible. If the amount of web services requiring a Facebook account increases, web users will become more dependent on Facebook as an identity provider so users can indirectly be forced to create an account.

1.4 Privacy Implications

The way the Like button is exploited and used to monitor web behavior of individual Internet users raises privacy concerns. In this section, it will be explained how privacy is affected and why this is troublesome. An important starting point in this respect is the link between privacy and identity. The construction of an own identity is only possible when an individual has some privacy. Keeping things private, or at least for some people, enables an individual to present himself in a way he wants and to set long term goals to achieve. Thus, privacy is instrumental to individual identity construction. Because privacy also enables the free and unrestricted determination of goals to achieve by the individual, it is also directly instrumental to individual autonomy. In this chapter, however, the focus will be on privacy and identity.

1.4.1 Privacy and Identity

Making choices and defining wishes and desires is closely related to identity. Identity is who you are as an individual and how you want to be seen by others, so it has an internal and an external element. The internal element can be described as how human beings acquire a sense of self (Hekman 2004, 22). The external element relates to social interaction with others. This interaction, however, is not always similar. When an individual wants to express himself and wants to present himself differently in different roles or contexts, control over data concerning him is a necessary condition. This is where privacy comes in. Agre defines privacy as freedom from unreasonable constraints on constructing identity and control over aspects of identity projected onto the world (Agre and Rotenberg 1997, 7). The world can be the world in general, but usually the world is divided into different contexts which are seen as separate audiences to which a certain (partial) identity or aspect of identity is projected. As Hekman puts it: "I am social in interaction with *specific* others, and understanding identity must attend to both the general (social) and the specific (individual). In other words, we are all embedded but we are all embedded differently at different locations" (Hekman 2004, 23). When approaching identity from a privacy perspective, the external element is the main focus. This is also reflected in Agre's definition where he speaks of projecting onto the world.

In the light of the foregoing, the two main aspects of privacy are informational self-determination and contextual integrity. Before delving into these particular aspects of privacy two open terms in Agre's definition will be briefly discussed. These terms are 'identity construction' and 'unreasonable constraints.'

1.4.1.1 Identity Construction

An important aspect of identity construction is role-playing (Goffman 1959); an individual plays a certain role in social interaction and the role and interaction give clues about the expected behavior of the individual. Depending on how the individual wants to be seen by others, he can decide to behave in accordance with expected behavior or to behave more or less idiosyncratically. This form of self-expression can help to change the way an individual is perceived by others.

A related aspect is audience segregation (Goffman 1959). Individuals tailor their behavior depending on their audience. For instance, the way a person behaves towards family members differs from his behavior in a working context. There are different partial identities for different contexts. An individual is thus always known by his audience as the identity that is shown in the specific context.

When data originating from different roles or contexts are collected and combined by one party, like Facebook, the individual is no longer able to keep roles and contexts separated. As a result, the individual is restricted in his ability to construct an own individual identity or partial identity. This will be further discussed in Sect. 4.2 below.

Given these main aspects of identity construction, it is now time to discuss what constraints on this construction may be unreasonable.

1.4.1.2 Reasonable and Unreasonable Constraints

The fact that the Agre/Rotenberg definition of privacy contains an element called unreasonable constraints implicitly indicates that there are reasonable constraints as well. In practice, I believe there is a sliding scale and that some constraints are definitely reasonable, some are definitely unreasonable, and the major part of constraints is somewhere in between. How reasonable or unreasonable a constraint actually is may depend on the specific circumstances in which there is a constraint. Because the infringement on privacy is taking place without the individual being informed, the reasonableness should be tested from the perspective of the individual user who is affected in his privacy and autonomy by the use of tracking technologies. The individual loses control over his personal data.[11]

Reasonable constraints can be defined as constraints that are defendable or foreseeable for the individual. Being able to predict a constraint or just knowing about it beforehand as well as being able to give grounds for the constraint is an indicator of a reasonable constraint. A clear example can be found in limits that are laid down in law, such as the limitations on fundamental rights. In specific circumstances, for example, involving public order or national security, fundamental rights may be restricted. This means that disclosing personal data to prevent an attack on the national

[11] Another reason to take the individual perspective is that privacy and data protection legislation is based on the privacy interest of individuals. Taking the perspective of a commercial company would come to a weighing of interests (conform Article 7(f) of the Data Protection Directive (Directive 95/46/EC)) and, thus, legally imply an assumption that the commercial business interest is a valid interest. At least, this assumption cannot be made in general.

government is a constraint, because the individual himself does not really have a say in this, but the constraint is reasonable given specific circumstances in which other interests should prevail. Another example directly in the field of data protection is the grounds for legitimate processing of personal data as laid down in the EU Data Protection Directive. Except for the ground of consent, these are constraints related to specific situations or interests where something else prevails over the privacy interest of the data subject. Clearly, the constraints are dictated by the need to maneuver within the rules of the given context.

Unlike reasonable constraints, unreasonable constraints are either not foreseeable or not communicated beforehand, or not defendable, or both. Obviously, even unexpected constraints may be reasonable in the light of specific circumstances. The necessary condition then is that the constraint has to be defendable. For a constraint to be defendable an objective perspective should be adopted, rather than the subjective perspective of the concerned individual.

Taking the example of Facebook, the requirement of using a real name to create a personal profile page may be reasonable. The aim of the social network site (SNS) is to create and maintain networks and find people with whom there is some relationship. Obviously, a name is very helpful in this context. However, taking the perspective of the SNS as a medium to connect people who share a particular interest, the name may be less relevant, but the details of these interests are the most important. In this respect, requiring the use of a real name may be considered to be an unreasonable constraint, because it disables the option to create a partial identity which is not deemed to be known to an individual's friends or family. For instance, when you are highly interested in Disney movies, but do not want your family to know this, looking for other people with the same interest would only require the characteristic of "liking Disney movies" to be known. The real name of the people with whom the interest is shared is of no concern. This constraint is therefore neither completely reasonable nor completely unreasonable.

Another example is when Facebook would require the disclosure of a telephone number. This is not in any way necessary for the function of Facebook and therefore irrelevant for the context to make it a default. As a result, the required disclosure is not objectively defendable as a constraint. In general, default sharing of unnecessary data as well as default disclosure to other contexts can be said to be unreasonable. It hinders identity construction in context and thus limits the individual in creating an identity free from unreasonable constraints.

Having described how conscious, sometimes forced, disclosure of data can be an unreasonable constraint on a person's construction of his identity, it is only a small step towards arguing that invisible data collection, such as is the case with the Like button, can be an unreasonable constraint. In fact, rather than the individual himself, Facebook is building an identity. If the data concern an individual who has a Facebook account, the data complement the data posted on the profile by the individual himself. The fact that the data are combined, however, remains invisible, in contrast to, for instance, wall posts by other Facebook members. The individual has no insight in the data collection, which makes it impossible to construct a separate or different identity.

1.4.2 *Privacy Aspects*

Privacy can be distinguished into different dimensions. Common distinctions are between spatial, relational, communicational, and informational privacy.[12] Informational privacy relates to the protection of personal data and has two main components. The first, which is at the core of the right to privacy, is being free from attention of others and not being watched. The second element comes into play once a third party has information and the individual wants to control the use and dissemination of this information (Lloyd 2008, 7). This element concerns the context to which information belongs. A focus on informational privacy can easily be defended. Obviously, many aspects of an individual's life are captured in data, which implies that information from the other dimensions becomes part of informational privacy as well. Information concerning home environment (smart metering), relationships (social networking sites), and body (medical files) is made compatible with the informational dimension. In the context of informational privacy then, data protection can be seen as an intermediate value, since data protection facilitates privacy. When talking about data, the two abovementioned elements of informational privacy have to be discussed in more detail. I will call these elements informational self-determination and contextual integrity, respectively.

1.4.2.1 Informational Self-determination

Informational self-determination is related to the control of a person over his personal data. In this approach, the individual controls his own personal data and information. However, Rouvroy and Poullet state that informational self-determination means "that an individual's control over the data and information produced about him is a (necessary but insufficient) precondition for him to live an existence that may be said [to be] 'self-determined'" (Rouvroy and Poullet 2009, 51). This approach focuses on the identity aspect and in fact underscores the determination aspect of the 'informational self' rather than the self-determination of information concerning the individual. From that perspective, restricting individual self-determination to control data and deciding what can be done with personal data is far too narrow. Schwarz calls this the 'autonomy trap' and indicates that the "organization of information privacy through individual control of personal data rests on a view of autonomy as a given, pre-existing quality" (Schwartz 1999). However, the problem is that, in the information age, individual self-determination itself is shaped by the processing of personal data. How personal data are used determines the terms under which an individual participates in social and political life. For instance, "the act of clicking through a 'consent' screen on a website may be considered by some observers to be

[12] There have been several efforts define the concept of privacy clearly and concisely. The definition will not be discussed here. For those interested in the discussion and efforts, see, for instance, the valuable work done by Parent (1983), who approaches the concept from different views and disciplines, and the extensive work by Solove (2002, 2006, 2008).

an exercise of self-reliant choice. Yet, this screen can contain boilerplate language that permits all further processing and transmission of one's personal data" (Schwartz 1999). In the end, the autonomy trap refers to a specific form of individual choice being "locked-in." Even though it seems that the individual himself has control over the disclosure of his data simply because he performs a certain action like clicking a button, the control is actually with another party, namely the party who requires the button to be clicked before a certain performance takes place and who decides what conditions are linked to the button being clicked.

The freedom to disclose what you want and to whom you want relates to autonomy and is an active component of privacy. It stresses the action of disclosure initiated by the individual. A passive component lies in the freedom from being monitored or analyzed and can be related to privacy in the sense of being left alone.[13] Next to the active and passive components, there are control mechanisms. These controls can be *ex post*, like access to data and the option to change or to delete them, or *ex ante*, in the mechanism of informed consent. This informed consent can also relate to keeping things to yourself and the mere consideration of whether or not to disclose data.

All components that are of importance for informational self-determination are bound to contexts. The importance of context will be described in the light of contextual integrity.

1.4.2.2 Contextual Integrity

The concept of contextual integrity in informational privacy originates from Nissenbaum (2004), who defines it as "compatibility with presiding norms of information appropriateness and distribution." She specifies the concept by articulating variables which can help determine whether a particular action is a violation of privacy, such as "the nature of the situation, or context; the nature of the information in relation to that context; the roles of agents receiving information; their relationships to information subjects; on what terms the information is shared by the subject; and the terms of further dissemination." Thus, contextual integrity means that information has to be kept within a context and that the way the data are treated has to be in compliance with the general rules and specific agreements concerning that context.

In contextual integrity the emphasis is on the freedom to decide to *whom* an individual discloses information. Evidently, data are usually disclosed within a specific context and to the people that belong in this context. In this respect it is important to understand that disclosing information in a way which makes it accessible to everyone, for instance, by posting something in a public space on the Internet, does not always mean that it is intended to be disclosed to and be made available for use by everyone. A distinction has to be made between the intended audience and the actual audience. The intended audience is the people that belong to the context in

[13] This distinction between active and passive components is inspired by Isaiah Berlin's theory on positive and negative freedoms (Berlin 1958).

which the information is disclosed. The actual audience is the people who in fact have access to the disclosed information, regardless of whether they belong to the specific context in which the information is disclosed.

As can be derived from the variables given by Nissenbaum, purpose binding is an important component. It means that the disclosure of information and its further processing is bound to a specific purpose. This purpose has to be defined before the processing of data takes place. Further dissemination of data, probably to another context, has to be in accordance with the indicated purpose for which the data were disclosed. However, in principle dissemination out of the initial context is not allowed when contextual integrity is the norm. A new context means a new purpose and a new audience.

Someone may browse the web in various contexts: it may be for professional purposes, such as searching for work-related information, or for private purposes, for example, searching for information about a disease a person is suffering from.[14] Obviously, when information about web behavior related to all the different purposes is sent to Facebook, the contexts and purposes change. This implies that the norms that belong to the initial context no longer apply either, resulting in a conflict with contextual integrity.

The fact that the context in which information is processed changes is one thing, but another important issue is at stake here. The collected information is combined with other information by Facebook. As a result, all information is connected and mixed up, so that contexts are collapsed. The distinction between contexts and the consciously or intuitively created boundaries between different contexts are lifted. In the end, the individual can no longer create his own personal identity and cannot even keep partial identities separated. Facebook's Like button interferes with privacy aspects of informational self-determination and contextual integrity and, ultimately, limits individuals in their construction of a personal identity.

1.5 Conclusion and Outlook

This chapter described the purpose and use of the Facebook Like button and how the technical process behind the button works. Four scenarios gave insight into how Facebook is able to monitor individual web behavior, even of nonmembers of the SNS. The scenarios showed that there is no escape from Facebook. The roll-out of the Like button and the potential advantages for web content providers has led to a high implementation rate of the feature. Facebook has a potential connection with everyone, given the fact that actual use of the button is not necessary for it to send information about the web user to Facebook.

Privacy protection is instrumental to the protection of personal identities. Individuals have to be able to construct their own personal identities, free from unreasonable

[14] In this respect, the public outcry on the Like button being available at the website of the National Health Service (NHS) in the UK is a case in point (see, Kirk 2010).

constraints. Forced disclosure of data which, on top of that, may be irrelevant for the purpose for which data sharing takes place may be an unreasonable constraint. This is closely related to informational self-determination and contextual integrity. An individual has to be able to control the disclosure and use of his personal data. If data are disclosed, this has to be done according to the norms and rules that belong to the context in which disclosure takes place and the data may not be disclosed outside the given context. If these conditions are not fulfilled, identity construction is no longer free from unreasonable constraints.

Applying the above to the Facebook Like button shows that there are constraints on the construction of identity. Facebook collects data concerning individual web users and can use these data to construct profiles. In particular, if a web user has a Facebook account, the data can be linked to this account based on the unique identifier that is sent along with the cookie in the HTTP request for the Like button. This implies that Facebook collects far more data than the data disclosed by its members. Moreover, Facebook collects huge amounts of data concerning individuals who do not have a Facebook account at all. As a result, individuals cannot create their own personal identities and cannot keep partial identities within a context.

Even though data collection concerning browsing behavior of web users via third-party cookies is nothing new, the Facebook Like button brings up some slightly different issues. These issues are strongly related to Facebook as a platform and to the presentation of the Like button. As indicated, the presentation of the button as a tool for Facebook members to share the web pages and items they like suggests that actual use of the button is necessary to set up communication with Facebook. Besides, nonmembers will think that they are not involved with Facebook in any case. This is obviously not true.

The other issue, related to the platform Facebook, is also very important. Facebook is a SNS which allows members to share their personal data with other members. These data can contain biographical data, but also pictures, videos, interests, and so on. Even though there is an initial choice on whether to participate in the network or not, there is also some social pressure to create an account.[15] Once you have an account, Facebook tries to let you share more and more data. The introduction of social media plugins, of which the Like button is one, formed a new development in this context. The plugins try to encourage individuals to connect all their web activity, subscriptions, and accounts to their Facebook account. Thus, on the one hand, Facebook possesses extensive information sets of all its members and can supplement these with additional information collected via third-party cookies, even if members do not attach things they like to their account pages. On the other hand, Facebook is trying to become the real digital identity of individuals for all contexts and interactions.

---

[15] Compare the famous quote by Skyler, 18: "If you're not on MySpace, you don't exist!" (Quote posted by her mother Kathy Sierra at http://headrush.typepad.com/creating passionate users/2006/03/ultrafast relea.html, no longer available (cf. boyd 2008).

Because Facebook has thoroughly embedded itself into the personal lives and identities of individuals, its impact reaches much further than the impact of 'traditional' third-party cookies, which are often used for targeted advertisements only. If an individual is not connected to Facebook, Facebook will make the connection instead.

The Facebook Like button is a case making very clear how changes in society and technology change the character of privacy concerns. The Internet has become central to daily life and social media are focused on sharing of personal information. However, the providers of social media are commercial companies which generate profits from the use of personal data. At the same time, these companies succeed in broadening their impact by connecting to other web services and increasing their coverage over the web. It is simply not possible for a web user to escape from being monitored once his browser connects to the Internet, whether the user has a formally established relationship with some services or not.

A related development is the technological trend towards the 'Internet of things' in which the connection to the web is ubiquitous and data are collected everywhere. This development implies that the notion of consent as we know it now becomes unworkable as a central concept in personal data processing. This trend calls for policy changes concerning privacy and data processing, while the need for a workable web environment remains essential. To find proper ways of regulating privacy in an era of ubiquitous information collection and in a society where connectivity is the standard is very challenging. Nevertheless, in light of personal identities, individual autonomy, and privacy, it is of the utmost importance to consider these issues at short notice. For the moment, commercial companies are leading the way with their own interest as a top priority.

References

Agre, Philip E., and Marc Rotenberg. 1997. *Technology and privacy: The new landscape*. Cambridge: MIT Press.

Berlin, Isaiah. 1958. *Two concepts of liberty*. Oxford: Clarendon Press.

boyd, danah. 2008. Why youth heart social network sites: The role of networked publics in teenage social life. In *Youth, identity, and digital media*, ed. David Buckingham, 119–142. Cambridge: MIT Press.

boyd, danah, and Eszter Hargittai. 2010. Facebook privacy settings: Who cares? *First Monday* 15: 8.

Burbary, Ken. 2009. Five reasons companies should be integrating social media with Facebook connect. http://www.kenburbary.com/2009/08/five-reasons-companies-should-be-integrating-social-media-with-facebook-connect/. Accessed 20 Aug 2009.

Dwyer, Catherine A. 2009. Behavioral targeting: A case study of consumer tracking on Levis.Com. Paper presented at the 15th American Conference on Information Systems, San Francisco, California.

Goffman, Erving. 1959. *The presentation of self in everyday life*. Garden City: Doubleday & Company.

Hekman, Susan J. 2004. *Private selves, public identities: Reconsidering identity politics*. University Park: The Pennsylvania State Univ. Press.

Kirk, J. 2010. NHS link to Facebook raises privacy concerns. http://www.pcworld.com/businesscenter/article/211711/nhs_link_to_facebook_raises_privacy_concerns.html. Accessed 7 Dec 2011.

Lloyd, Ian J. 2008. *Information technology law*. Oxford: Oxford Univ. Press.

Martin, David, Hailin Wu, and Adil Alsaid. 2003. Hidden surveillance by web sites: Web bugs in contemporary use. *Communications of the ACM* 46 (12): 258–264.

Nissenbaum, Helen. 2004. Privacy as contextual integrity. *Washington Law Review* 79: 119–158.

Parent, William A. 1983. Privacy, morality, and the law. *Philosophy and Public Affairs* 12 (4): 269–288.

Roosendaal, Arnold. 2010. Facebook tracks and traces everyone: Like this! Tilburg Law School Research Paper No. 03/2011. http://ssrn.com/abstract=1717563. Accessed 30 Nov 2010.

Rouvroy, Antoinette, and Yves Poullet. 2009. The right to informational self-determination and the value of self-development: Reassessing the importance of privacy for democracy. In *Reinventing Data Protection*, ed. Serge Gutwirth et al. 45–76. Berlin: Springer.

Schwartz, Paul M. 1999. Privacy and democracy in cyberspace. *Vanderbilt Law Review* 52: 1609–1701.

Solove, Daniel J. 2002. Conceptualizing privacy. *California Law Review* 90 (4): 1087–1156.

Solove, Daniel J. 2006. A taxonomy of privacy. *University of Pennsylvania Law Review* 154 (3): 477–560.

Solove, Daniel J. 2008. *Understanding privacy*. Cambridge/London: Harvard Univ. Press.

Chapter 2
Behavioural Tracking on the Internet: A Technical Perspective

Claude Castelluccia

2.1 Behavioural Tracking

2.1.1 The Context: Behavioural Profiling

The concept of *Behavioural Profiling* (also known as "targeting") consists of collecting and analysing several events, each attributable to a single originating entity, in order to gain information relating to the originating entity. It consists of, in other words, transforming data into knowledge (Hildebrandt 2006). Behavioural profiling involves collecting data (recording, storing and tracking) and searching it for identifying patterns (with the help of data mining algorithms). The data collection phase is often referred to as *Behavioural Tracking*.

An example of behavioural targeting scenario is provided in Dwyer (2009). A consumer shops online for an airline ticket to New York City. He searches for flights, but does not make any purchase. He subsequently visits the web site of the local newspapers that displays adds offering tickets to New York. While no Personally Identifiable Information (PII) might have been collected, his interest in airline tickets has been noted.

2.1.2 Motivations: Why are We Being Tracked and Profiled?

Profiles are very valuable for many companies in customising their services to suit their customers, in order to increase revenues. The clear intent of behavioural targeting is to track users over time and build profiles of their interests, characteristics (such as gender, age and ethnicity) and shopping activities. For example, advertising or publishing companies use behavioural targeting to display advertisements that

C. Castelluccia (✉)
INRIA Rhone-Alpes, Grenoble, France
e-mail: claude.castelluccia@inria.fr

S. Gutwirth et al. (eds.), *European Data Protection: In Good Health?*,
DOI 10.1007/978-94-007-2903-2_2, © Springer Science+Business Media B.V. 2012

closely reflect the users interests. Online advertising systems are typically composed of three main entities: the *advertiser*, the *publisher* and the *ad network*. The advertiser is the entity, for example a car manufacturer or a hotel, which wishes to advertise a product or service. The publisher is the entity, such as an online newspaper company, which owns one or several web sites and is willing to display advertisements and be paid for it. Finally, the ad network is the entity that collects advertisements from the advertisers and places them on publisher sites. If the user clicks on an advertisement, the ad network collects payment from the corresponding advertiser. There is, therefore, a strong incentive for the ad network to generate very accurate and complete profiles in order to maximise profit. E-commerce sites also use behavioural tracking to recommend products that are likely to be of interest to users. For example, Amazon recommends products to online users based on the individuals past behaviour (personalised recommendation), on the past behaviour of similar users (social recommendation) and, of course, on the searched items (item recommendation; Macmanus 2009).

2.1.3 Tracking and Privacy

It can be argued that the customisations resulting from profiling are also beneficial to the users that only receive information relevant to their interest. However, it creates serious privacy concerns since it allows some companies or institutions to gather and concentrate a huge amount of information about their customers, and about Internet users in general.

The danger is to move into a surveillance society or Internet, where all our online or physical activities are recorded and correlated. Some companies offer various services that gather different types of information from users. The combination and concentration of all this information provides a powerful tool to accurately profile users. For example, Google is one of the main third-party aggregators and tracks users across most web sites (Krishnamurthy and Willis 2009b). In addition, it also runs the most popular search engine and, as such, stores web histories of most users (i.e. their search requests), their map searches (i.e. their requests to the Google map service), their images, etc. (Castelluccia et al. 2010). Web searches have been shown to often be sensitive (Conti and Sobiesk 2007). It has actually been demonstrated that it is quite trivial to derive the identity of a user from his web history (Barbaro and Zeller 2006). Map requests also leak a lot of information, such as the user's home address or his favourite places. Finally, Google runs one of the most popular email systems, *gmail*, and has, therefore, access to emails of millions of users. By combining these different types of information coming from different sources, Google is able to build very accurate profiles of their users. As argued in Hildebrandt (2006), "profiling shifts the balance of power between those that can afford profiling (mostly large organisations) and those that are being profiled (mostly individual citizens), because the profilers have a certain type of knowledge to which those profiled have no effective access."

The advent of *ubiquitous advertising*, which can be seen as the application of *computational advertising*[1] to smart phones, will provide even more sources of profiling information (Krumm 2010). With ubiquitous advertising, advertisements will not only be personalised to users' online profiles, but also to their physical profiles. Advertisements will be customised to users' locations, physical or intellectual activities, interactions and possibly moods. Since, as opposed to a regular computer, a mobile device is usually owned by a single person, more detailed and accurate profiles can be derived from his uses. It is also foreseen that, in the future, sensors on phones will be able to infer users' food habits and preferences (Krumm 2010). These new developments create serious privacy issues that need be studied more carefully (Cleff 2007).

The rest of this chapter considers three of the most popular Internet services, namely the web, location-based services (LBS) and online social networks (OSN). It presents for each of them existing tracking mechanisms. Note that we do not cover the profiling part, which consists on transforming collected data into knowledge. Furthermore, it focusses on technological issues, and eludes legal or policy aspects.

2.2 Web Tracking

One of the main sources of information used for profiling comes from web tracking, i.e., tracking users across different visits or across different sites. Data collected includes the sequence of visited sites and viewed pages, and the time spent on each page. Web tracking is mainly performed by monitoring IP addresses, and using techniques such as *cookies*, *Javascripts* or *supercookies* (McKinley 2008).

Cookies A cookie is a piece of text stored by a user's web browser and associated to a HTTP request. A cookie consists of one or more name-value pairs containing bits of information and is set by a web server. There are two types of cookies: *session* and *persistent* cookies. Session cookies are temporary cookies that are often used to store user preferences. They are set by a service when a user logs in, and are erased when the user logs out. Persistent cookies are often used as authentication tokens to keep an authenticated session with a server. These files stay in the user's browser until they are explicitly deleted or they expire. They are sent back unchanged by the browser each time it accesses that web site and can, therefore, be used by web sites to track users across visits. Persistent cookies raise serious privacy concerns. In the rest of the document, the term cookie refers to persistent cookie, unless explicitly stated.

Cookies are sent only to the web sites that set them or to servers in the same Internet domain. However, a Web page may contain images, links, web bugs (1×1 pixel GIF images), HTML IFrame, Javascript or other components stored on servers

[1] Computational advertising is a new scientific sub-discipline whose main challenge is to find the best ad to present to a user engaged in a given context (Broder and Josifovski 2010).

in other domains. Cookies that are set during retrieval of these components are called *third-party cookies,*[2] in contrast to *first-party cookies*. Some sites, such as advertising companies, use third-party cookies to track users across multiple sites. In particular, an advertising company can track a user across all pages where it has placed advertising images or web bugs. Knowledge of the pages visited by a user allows the advertising company to target advertisements to user's presumed preferences. Third-party tracking raises serious privacy concerns, which are not hypothetical but real. The increasing presence and tracking of third-party sites used for advertising and analytics has been demonstrated in a study (Krishnamurthy and Willis 2009b, 2009c). This study showed that the penetration of the top 10 third-parties growing from 40% in 2005 to 70% in 2008, and to over 70% in September 2009. Another study shows that not only these third-parties are increasing their tracking of users, but also they can now link these traces with identifiers and personal information via OSN (Krishnamurthy and Willis 2009a). In Dwyer (2009), a behavioural targeting study was performed on the *levis.com* site, the e-commerce site for the clothing line. The results show that the web site contains a total of nine tracking tags that link to eight third-party companies.[3]

Javascripts Many web sites contain executable Javascript files that are down-loaded by visiting users. These files, in addition to their computations, sometimes update first-party cookies and send information back to the servers. Javascripts have limited access to user data. However, they can access information stored in the browser including cached objects and the history of visited links. Along with cookies and results of JavaScript execution, the tracking sites have all the regular information available in a typical HTTP request: sender's IP address, user-agent software information, current and previous URL (via Referer header), email address (from header), language preference (Accept-Language header), etc.

Supercookies and Evercookies Use of tracking cookies is fairly ubiquitous and there are known techniques to avoid them (Dixon 2011). Therefore, this is a big impetus in the Internet tracking industry to discover and deploy more robust tracking mechanisms, often referred to as *Supercookies* (McKinley 2008). One of the most prominent supercookies is the so-called "Flash cookie", a type of cookie maintained by the Adobe Flash plug-in on behalf of Flash applications embedded in web pages (Schoen 2009). Since these cookie files are stored outside of the browser's control, web browsers do not directly allow users to control them. In particular, users are not notified when such cookies are set, and these cookies never expire. Flash cookies can track users in all the ways traditionally HTTP cookies do, and they can be stored or retrieved whenever a user accesses a page containing a Flash application. Flash cookies are extensively used by popular sites. They are often used to circumvent

---

[2] Some sites included JavaScript code and third-party cookies from more than ten different tracking domains (Eckersley 2009).

[3] The largest third-party Ad-network companies include *Advertising.com*, *Tacoda*, *DoubleClick* and *Omniture*. Most of these networks are owned by *Google*, *Yahoo*, *AOL* or *Microsoft*. Since Ad-networks are typically partnered with many publishers, they can track users across several publishers and build these users' browsing profiles.

user's HTTP cookie policies and privacy preferences. For example, it was found that some sites use HTTP and Flash cookies that contain redundant information (Ashkan et al. 2009). Since flash cookies do not expire, sites might automatically re-spawn HTTP cookies from Flash ones if they are deleted. The persistence of Supercookies can be further improved as illustrated recent *evercookies* (Kamkar 2010). This new type of cookie identifies a client even when standard cookies, Flash cookies, and others, have been removed. This is accomplished by storing the cookie material in several types of storage mechanisms that are available on the local browser.

Browser fingerprinting A recent study showed that browsers can be identified to a high degree of accuracy without cookies or other tracking technologies (Eckersley 2010). Every Web browser provides enough unique information (User Agent, fonts, screen resolution, ...) to tell one from another. The study shows that a browser fingerprint is unique enough that it can, on the average, identify a browser among a set of 286.777 other browsers. Browser fingerprinting is a powerful tool for tracking users. It should be considered alongside with IP addresses, cookies and supercookies as far as user traceability is concerned.

2.3 Location Tracking

2.3.1 Location Privacy

More and more systems and applications record user's locations and movements in public places. These systems provide very useful and appreciated services, and have come to be regarded as almost essential and inevitable. For example, RFID cards allow users to open doors or pay their transportation ticket; GPS systems help users to navigate and find their ways. Some services tell users where their friends are, or provide personalised services (such as indicating the closest restaurant or hotel). Some wireless parking meters send users a text message when their time is running out (Blumberg and Eckersley 2009). While the benefits provided by these systems are indisputable, they unfortunately pose a considerable threat to location privacy, as illustrated by the recent iPhone and Android controversies (Raphael 2011).

Location privacy is often defined as *the ability of an individual to move in public space with the expectation that their location will not be systematically and secretly recorded for later use*. Location tracking is not a new phenomenon, but new technologies (wireless networks, digital cameras, etc.) make it cheaper and easier to perform. It is this transformation to a world where location is collected pervasively, silently and cheaply that is worrisome (Blumberg and Eckersley 2009).

2.3.2 Location-based Services

Already today, worldwide, hundreds of millions of people permanently hold at least one mobile phone. It is predicted that smartphones will surpass PC sales within two years (Boulton 2010). These mobile phones have increasing computational capacities and are equipped with multiple sensors like microphones, cameras, GPS, accelerometers, etc. As geolocated systems, they already enable individuals and communities to collect and share various kinds of data. Urban sensing is a new sensing paradigm leveraging users as part of a sensing infrastructure (Campbell et al. 2006). In the near future, several urban sensing applications are likely to appear, which will provide extra information about users (Miluzzo et al. 2008). Most users are unaware of the extra information that is collected about them beyond requested data, especially in case of participatory sensing. For example, a picture taken by a user may reveal additional contextual information inferred from the background or the style of any associated text. A recent study showed that most people are unaware of the fact that the photos and videos taken with their smart phones or cameras contain geolocation information (Friedland and Sommer 2010). This information can be used to localise them while they are travelling, or even reveal their home address. This may be considered as a potential source of information leakage and may lead to a privacy breach if used for location tracking or in conjunction with data retrieved from OSN. The risk becomes higher as the border between OSN and LBS becomes fuzzier. For instance, OSN such as *FourSquare*[4] and *Gowalla*[5] are designed to encourage users to share their geolocated data. Information posted on social applications such as *Twitter*[6] can be used to infer whether or not an individual is at home.[7] Other applications, such as *GoogleLatitude,*[8] allow users to track the movements of their friends' cellphones and display their position on a map. In addition to social applications, there are other public sources of information that can be exploited by potential adversaries, such as the free geographic data provided by *Google Maps,*[9] Yahoo! Maps[10] and *Google Earth.*[11]

The W3C geolocation API, which is supported in the Firefox, Opera and Chrome browsers and in Internet Explorer via a plug-in, allows web sites to request geographical information for the clients device. With the approval of the user, the browser sends information like the clients IP address, MAC addresses of connected wireless access points and the cell ids of GSM/CDMA networks within range. With the help of a network location provider, such as Google Location Services, this information can be used to obtain an estimate of the client devices location. While the browser only sends this information to a web site with the users explicit approval, few users realise the accuracy with which these services can often locate a device. For instance, Google Location Services rely on the MAC addresses of wireless access points detected during the Google Street View data collection to locate client devices within

---

[4] http://foursquare.com/.

[5] http://gowalla.com/.

[6] http://twitter.com/.

[7] http://pleaserobme.com/.

[8] http://www.google.com/latitude/.

[9] http://maps.google.com/.

[10] http://maps.yahoo.com/.

[11] http://earth.google.com/.

the range of an 801.11 wireless-base station (i.e. tens of meters). Furthermore, a growing number of sites now provide public APIs to their geolocalised content. For example, *Flickr*, *YouTube* and *Twitter* allow queries for results originating at a certain location. PicFog, for example, uses one of these APIs to provide real-time location-based search of images posted on Twitter. As shown in Friedland and Sommer (2010), these APIs can also be used to identify the current location of a user while he or she is away from home.

The emergence of *Reality Mining* raises even more privacy concerns (Greene 2008). As Greene (2008) explained, reality mining infers human relationship and behaviour from information collected by cellphones. This information include data collected by cellphone sensors, such as location or physical activity, and data recorded by phones themselves, such as call duration and numbers dialled. Reality mining could help users identify things to do or new people to meet. It could also help to monitor health. For example, monitoring a phone's motion might reveal changes in gait, which could be an early indicator of ailments or depression. The idea of *autonomous search* is a first step toward reality mining. With *autonomous search*, the search engine will conduct searches for users without them having to manually type anything (Boulton 2010). For example, a user could be walking down a street and receive personalised information about the places in the vicinity on his or her mobile phone, without having to click any buttons. While the promise of reality mining is great, the idea of collecting so much personal information naturally raises many questions about privacy.

2.4 Social Network Tracking

2.4.1 Online Social Networks

OSN have gained an immense popularity in recent years. Social-based services such as *Facebook,*[12] *Twitter*, *MySpace*[13] and *Orkut,*[14] just to name a few, allow millions of individuals to share some of their personal information with a multitude of other entities, such as their friends, companies or even the public at large. The common characteristic of these OSN is that users can make contacts and share easily personal information on a large scale. More specifically, people can meet old as well as new friends (*Facebook*, *MySpace*), find new jobs (*LinkedIn*[15]), or receive and provide recommendations (*Tribe*[16]). In a near future, many more complex services are likely to appear, which will tap on the power of the social connection and personal information provided by OSN.

---

[12] http://facebook.com/.

[13] http://www.myspace.com/.

[14] http://www.orkut.com/.

[15] http://www.linkedin.com/.

[16] http://www.tribe.net/.

As the primary objective of most of these services is to make individuals or groups visible, people need to share personal information to ensure some form of identifiability. Hence, most OSN encourage users to publish personal information, which may enable anyone accessing this information to infer further private information, thus causing a privacy breach. On top of that, the majority of users are not only willing but also pleased to disclose their personal information to as many users as possible and some OSN make this information public by default. Moreover, compared to traditional off-line, real-life, social networks, OSN are usually larger and contain more ties. For instance, people easily classify thousands of users as "friends", or as "friends of friends", when they probably would not qualify some of these users as friends in their real life. These facts inherently entail the question of trust and privacy in OSN.

Generally, average users do not have a clear idea of who accesses their private information, or what portion of it really needs to be accessed by applications. For instance, in Facebook, the terms of use of some applications clearly state that these applications can access any personal information put by the user, even though it may not be required. Although most sites provide coarse-grained privacy controls, the majority of users do not use this feature because they find it too complex (Gross et al. 2005). Moreover, these sites are permissive and allow anyone to access user's profile data, which means that, by default, it is accessible by any other user in the network. In addition, it is difficult for an average user to know and control users or groups of users who can access his information and to limit this access without losing the benefits of the various features of OSN.

Another problem stems from the fact that while a user's profile may be set to be inaccessible for other users, the friendship links and group affiliations often remain public. This public social information can leak further information about the private attributes of a profile. For instance, Zheleva and Ghetoor (2009) have shown that the structure of the social network and group information leak a surprisingly large amount of personal information. Moreover, even if a user makes some parts of his profile private, the person's membership in a particular group remains publicly accessible from the group profile. Another study lead by MIT students, called the Gaydar project, has shown that it is possible to predict with a fairly high accuracy the sexual preferences of an individual. This is possible even if his profile is private, just by looking at the amount of gay friends it includes, compared with a person sampled randomly from the population (Johnson 2009).

Furthermore, much like traditional web sites, third-party aggregators track user activity pervasively on OSN (Krishnamurthy and Willis 2008). Third-party domains are then not only able to track the web sites that a user visits, but also the OSN sites that he connects to. In a follow-up work (Krishnamurthy and Willis 2009a), the same authors demonstrate that PII belonging to any user, such as name, gender or OSN unique ID, is also being directly leaked to these third-party servers via the OSN. This leakage happens via a combination of HTTP header information and cookies being sent to third-party aggregators. This result implies that third parties are not only able to view the surfing habit of some users, but are also able to associate the habits with a specific habit and potentially gather much more personal information. This ability to link information across web sites and OSN raises important privacy concerns.

2.4.2 Mobile Online Social Networks

Mobile Online Social Networks (MOSN) have recently grown in popularity. Mobile devices provide ubiquitous access to the web and naturally to social networks. There are typically two classes of mobile OSN: (1) traditional OSN (such as *Facebook*, *Twitter*) that have created content and access mechanisms tailored to mobile devices, and (2) new MOSN, such as *Foursquare* and *Loopts*,[17] created to deal with the new mobile context. These new MOSN tend to customised their content to the location and the user's community (friends). For example, using the phone's self-location features, as well as information about the prior activities of the user's friends, some MOSN propose new places to explore or activities to try. Other MOSN allow a user to locate his friends that are currently in his or her vicinity. The predominant concepts of new MOSN are *presence* and *location* (Krishnamurthy and Willis 2010). Presence allows a user to know the current status of his or her friends. The indication of presence allows the expectation of a quick response. Location allows a user to locate his friends and obtain LBS, such as the closest restaurants or hotels. A recent study showed that most MOSN leak some kind of private information to users within the same MOSN, to users within other OSN via the interconnect features and, and more importantly, to third-party tracking sites. In many cases, data given out contained user's precise location, his gender or name, and even subject's unique social networking identifier, which could allow third-party sites to connect the records they keep of users' browsing behaviour with their profiles on the social networking sites.

The combination of location information, unique identifiers of devices, and traditional leakage of other personally identifiable information now give third-party aggregation sites the capacity to build a comprehensive and dynamic portrait of MOSN users.

2.5 Discussion

As illustrated in this report, users are being constantly tracked and profiled when using the Internet. This profiling will increase with the development of ubiquitous advertising and personalised services.

Unfortunately, there is no easy way to use modern, cookie- and JavaScript-dependent web sites and social networking sites and avoid tracking at the same time (Eckersley 2009). However, although not perfect (Aggrawal et al. 2010), private browsing mode of major browsers, that disable cookies, should be used when possible. Also, the popular Firefox *NoScript* extension should be considered. *NoScript* (2010) is a Firefox add-on which allows executable content such as JavaScript to run only if it is being hosted on a trusted domain. Finally, anonymisation networks, such as *TOR* (Dingledine et al. 2004), and network/web proxies that allow users to surf the Internet anonymously, mitigate some of the highlighted privacy issues.

[17] http://www.loopts.com/.

As suggested in Schoen (2009), privacy-invasive marketing practices need greater scrutiny. More research is needed to reveal how the other kinds of cookies described in McKinley (2008) are also being used to track users. There is a lot of work to be done to bring these next-generation cookies even to the same level of visibility and control that users experience with regular HTTP cookies. Application and Browser developers should do more to let users control how they are being tracked. However, this is not an easy task since, as shown previously, some of these tracking cookies, such as the Flash ones, are stored outside of the browser. The *BetterPrivacy* Firefox plug-in tries to address this problem by finding Flash cookies on the hard drive and regularly deleting them.

In this context, it is going to be challenging to protect users' privacy. Some people argue that abstinence or withdrawal from the online world is the only method guaranteed to work (Conti 2009), or that users should lower their privacy expectation. According to Eric Schmidt, executive chairman of Google, it is possible to identify a person from 14 of his photos and then search the Web for more content about this user. Furthermore, he argues that, in the future, not only we will be able to identify a person but also predict, from his messaging and location, where that person is going to go (Kirkpatrick 2010).

Users should be given the ability to control access and distribution of their personal data. Once data is used without the knowledge or consent of the user, privacy is clearly compromised. Solving these privacy issues will be beneficial not only to users but also to service providers. In fact, as argued in Cleff (2007), users might react to this privacy fear by restricting the information they provide or by providing false information. This would have for effect to limit business, and to affect the validity of customer databases and profiles.

Users must also be able to choose what data is collected about them. They must keep the right to access, modify and delete them. Users should be explicitly informed about how they are being tracked, how their data is being sent/leaked out of their social network sites, by advertisers or others, and the corresponding destination. For example, users should need to acknowledge usage of their location on a per-application basis, or even, for some applications, each time location information is used. A simple, yet promising, approach is the *Do Not Track* (DNT) initiative. DNT gives users a way to opt out of behavioural tracking universally. In its simplest form, DNT is implemented as a HTTP header. This header contains a "Do-Not- Track" flag that indicates to web sites the user's wish to opt out of tracking. This extension is simple to implement in the web browser. As a matter of fact, there is already a Firefox add-on that implements such a header. However, this solution will only be effective if advertisers will respect the user's preference of not being tracked. As discussed in Narayanan (2010), there are several possibilities to enforce it, ranging from self-regulation via the Network Advertising Initiative, to supervised self-regulation or direct regulation.

Furthermore, more tools to help users making informed decisions about the publication of their data or their online activities should be developed. These tools should, for example, inform users whether the information to be published can potentially be combined with other data on the Internet to infer sensitive information

(Chew et al. 2008). *ReclaimPrivacy*[18] is an example of such tools. *ReclaimPrivacy* is an open tool for scanning Facebook privacy settings and warn users about settings that might be unexpectedly public.

Finally, services and networks should follow the "privacy by design" concept (Le Métayer 2010). Privacy should be seen as a main design requirement, not as an add-on. For example, data collection should be minimal and only performed when necessary. Services should potentially be distributed and open-source to minimise data monitoring and collecting.[19] They should request and use users' identities only when strictly necessary. For example, most LBS request users to provide their identity before offering their services. This is required for accounting and billing purposes. However, the only thing that service operators actually need is an anonymous proof that the user is a registered subscriber (Blumberg and Eckersley 2009). This can be achieved, without revealing the user's identity, by using existing cryptographic primitives (Zhong et al. 2007).

In summary, networks and services should be designed to limit unnecessary data collection and give individuals control over their data (Castelluccia and Kaafar 2009; Schneier 2009). Indeed as argued by Bruce Schneier (2009), Privacy is not something that appear naturally online, it must be deliberately architected. Privacy issues in behavioural profiling are complex and cannot be treated exclusively by technological means. There is a need for a true research approach that considers educational, policy, legal and technological aspects.

Acknowledgement The author would like to thank the members of the INRIA Planete group for discussions and for proofreading this chapter. He would also thank Levente Buttyan, Imad Aad, Aurelien Francillon, Bala Krishnamurthy, Emiliano De Cristofaro and many others for providing comments on this chapter. Finally, the author would like to thank ENISA and more particularly Rodica Tirtea who was at the origin of this work and chapter. This chapter was published as a section of the Privacy, Accountability and Trust Challenges and Opportunities report, published by ENISA (2011).

References

Aggrawal, G., E. Bursztein, C. Jackson, and D. Boneh. 2010. An analysis of private browsing modes in modern browsers. Proceedings of 19th Usenix Security Symposium. Washington D.C., U.S.A.

Ashkan, S., S. Canty, M. Quentin, T. Lauren, and J. Chris. 2009. *Flash cookies and privacy.* Technical report, University of California, Berkeley. http://papers.ssrn.com/sol3/papers.cfm?abstract_id=1446862. Accessed in November 2010.

Barbaro, M., and T. Zeller. 2006. A face is exposed for AOL searcher no. 4417749. *New York Times,* 9. August.

Blumberg, A., and P. Eckersley. 2009. On locational privacy, and how to avoid losing it forever. http://www.eff.org/wp/locational-privacy. Accessed in November 2010.

[18] http://www.reclaimprivacy.org/.

[19] The Disapora project, see http://www.joindiaspora.com/.

Boulton, C. 2010. Google CEO Schmidt Pitches autonomous search, flirts with al. http://www.eweek.com/c/a/Search-Engines/Google-CEO-Schmidt-Pitches-Autonomous-Search-Flirts-with-AI-259984/1/. Accessed in November 2010.
Broder, A., and V. Josifovski. 2010. Introduction to computational advertising. http://www.stanford.edu/class/msande239/. Accessed in November 2010.
Campbell, A. T., S. B. Eisenman, N. D. Lane, E. Miluzzo, and R. A. Peterson. 2006. People-centric urban sensing (invited paper). Proceedings of the Second ACM/IEEE International Conference on Wireless Internet. Boston, MA, U.S.A.
Castelluccia, C., E. De Cristofaro, and D. Perito. 2010. Private information disclosure from web searches. Proceedings of the 2010 Privacy Enhancing Technologies Symposium (PETS). Berlin, Germany.
Castelluccia, C., and D. Kaafar. 2009. Ocn: Owner-centric networking. In Future Internet Security and Trust (FIST) workshop. Seattle, WA, U.S.A.
Chew, M., D. Balfanz, and B. Laurie. 2008. (under) mining privacy in social networks. Web 2.0 Security and Privacy workshop. Oakland, CA, U.S.A.
Cleff, E. B. 2007. Privacy issues in mobile advertising. *International Review of Law, Computers & Technology* 21 (3): 225–236.
Conti, G. 2009. *Googling security: How much does Google know about you?* Boston: Addison-Wesley.
Conti, G., and E. Sobiesk. 2007. An honest man has nothing to fear: User perceptions on web-based information disclosure. Proceedings of the 3rd SOUPS' 07, New York, pp. 112–121.
Dingledine, R., N. Mathewson, and P. Syverson. 2004. Tor: The second-generation onion router. Proceedings of Usenix security symposium. San Diego, CA, U.S.A.
Dixon, P. 2011. Consumer tips: How to opt-out of cookies that track you. http://www.worldprivacyforum.org/cookieoptout.html. Accessed in July 2011.
Dwyer, C. 2009. Behavioral targeting: A case study of consumer tracking on levis.com. Proceedings of Fifteen Americas Conference on Information Systems. San Francisco, CA, U.S.A.
Eckersley, P. 2009. How online tracking companies know most of what you do online. https://www.eff.org/deeplinks/2009/09/online-trackers-and-social-networks. Accessed in November 2010.
Eckersley, P. 2010. How unique is your web browser? Proceedings of the 2010 Privacy Enhancing Technologies Symposium (PETS). Berlin, Germany.
ENISA. 2011. *Privacy, accountability and trust challenges and opportunities*. Technical report, ENISA.
Friedland, G., and R. Sommer. 2010. Cybercasing the joint: On the privacy implication of geotagging. Usenix Workshop on Hot Topics in Security. Washington D.C., U.S.A.
Greene, K. 2008. Reality mining. http://www.technologyreview.com/read_article.aspx?id=20247&ch=specialsections&sc=emerging08&pg=1. Accessed in November 2010.
Gross, R., A. Acquisti, and H. Heinz. 2005. Information revelation and privacy in online social networks. WPES. Alexandria, VA, U.S.A.
Hildebrandt, M. 2006. Profiling: from data to knowledge. *DuD: Datenschutz und Datensicherheit* 30(9).
Johnson, C. 2009. Project Gaydar. http://www.boston.com/bostonglobe/ideas/articles/2009/09/20/project_gaydar_an_mit_ experiment_raises_new_questions_about_online_privacy/. Accessed in November 2010.
Kamkar, S. 2010. Evercookie—never forget. http://samy.pl/ evercookie/. Accessed in November 2010.
Kirkpatrick, M. 2010. Google CEO Schmidt: "people aren't ready for the technology revolution",. http://www.readwriteweb.com/archives/google _ ceo_schmidt_people_arent_ready_for_the_tech.php. Accessed in November 2010.
Krishnamurthy, B., and C. Wills 2008. Characterizing privacy in online social networks. In WOSN' 08: Proceedings of the first workshop on Online social networks. Seattle, WA, U.S.A.

Krishnamurthy, B., and C. Wills 2009a. On the leakage of personally identifiable information via online social networks. In WOSN' 09: the second workshop on Online social networks. Barcelona, Spain.

Krishnamurthy, B., and C. Wills. 2009b. Privacy diffusion on the web: a longitudinal perspective. In WWW' 09: Proceedings of the 18th international conference on World wide web. ACM. Madrid, Spain.

Krishnamurthy, B., and C. Wills. 2009c. Privacy diffusion on the web: A longitudinal perspective (updated graphs). http://www.ftc.gov/os/comments/privacyroundtable/544506–00009.pdf. Accessed in November 2010.

Krishnamurthy, B., and C. Wills. 2010. Privacy leakage in mobile online social networks. In WOSN' 10: Proceedings of the third workshop on Online social networks. Boston, MA, U.S.A.

Krumm, J. 2010. Ubiquitous advertising: The killer application for the 21st century. IEEE Pervasive Computing.

Le Métayer, D. 2010. Privacy by design: A matter of choice. In *Data protection in a profiled world*, ed. S. Gutwirth, Y. Poullet, P. De Hert, 323. Verlag: Springer.

Macmanus, M. 2009. A guide to recommender systems. http://www.readwriteweb.com/archives/recommender_systems.php. Accessed in November 2010.

McKinley, K. 2008. *Cleaning up after cookies*. Technical report, iSEC PARTNERS. https://www.isecpartners.com/files/iSEC_Cleaning_Up_After_Cookies.pdf. Accessed in November 2010.

Miluzzo, E., N. Lane, K. Fodor, R. Peterson, H. Lu, M. Musolesi, S. B. Eis, X. Zheng, S. EisenMan, and A. Campbell 2008. Sensing meets mobile social networks: The design, implementation and evaluation of the cenceme application. Proceedings 6th ACM Conference on Embedded Networked Sensor Systems (SenSys' 08). Raleigh, NC, U.S.A.

Narayanan, A. 2010. Do not track explained. http://33bits.org/2010/09/20/do-not-track-explained/. Accessed in November 2010.

Raphael, J. R. 2011. Apple vs. Android location tracking: Time for some truth. http://blogs.computerworld.com/18190/apple_android_location_tracking. Accessed in July 2011.

Schneier, B. 2009. Architecture of privacy. *IEEE Security and Privacy*.

Schoen, S. 2009. New cookie technologies: Harder to see and remove, widely used to track you. http://www.eff.org/deeplinks/2009/09/new-cookie-technologies-harder-see-and-remove-wide. Accessed in November 2010.

Zheleva, E., and L. Getoor. 2009. To join or not to join: The illusion of privacy in social networks with mixed public and private user profiles. In International World Wide Web Conference (WWW). Madrid, Spain.

Zhong, G., I. Goldberg, and U. Hengartner. 2007. Louis, lester and pierre: Three protocols for location privacy. Proceedings of the 2007 Privacy Enhancing Tsechnologies Symposium (PETS). Ottawa, Canada.

Chapter 3
Privacy for Loan Applicants Versus Predictive Power for Loan Providers: Is It Possible to Bridge the Gap?

Charlene Jennett, Miguel Malheiros, Sacha Brostoff and M. Angela Sasse

3.1 Introduction

Consumers have to trust that financial services will work for, rather than against them. In a recent speech, Mark Hoban (2010) MP, Financial Secretary to the UK Treasury, stated that "*We need a financial sector that works for consumers—one that earns their confidence, competes for their services, and keeps them properly informed.*"

The collection, use, maintenance, and disclosure of consumer information, is an essential part of any financial transaction (MacCarthy and Gellman 2010). However, recent research suggests that more needs to be known about the public's worries about how their personal information is used and protected (Raab 2004)—and that applies to financial services.

This chapter explores consumers' privacy concerns about information requested on loan applications. Currently, loan applicants have low expectations of privacy—they are expected to: (1) answer all questions, without exception; (2) consent to all terms and conditions (which often includes their data being passed onto third parties); and (3) accept that their credit record will be checked. Based on our interviews and surveys, we argue that it is possible to maintain the efficacy of the loan risk assessment process and respect applicants' privacy at the same time.

In Sect. 3.2, we review existing literature on the perspectives of loan providers and loan applicants, and identify a discrepancy between information that loan providers and loan applicants consider relevant. To explore this discrepancy, we conducted

C. Jennett (✉) · M. Malheiros · S. Brostoff · M. A. Sasse
Department of Computer Science, University College London (UCL),
Gower Street, WC1E 6BT, London, UK
e-mail: c.jennett@cs.ucl.ac.uk

M. Malheiros
e-mail: m.malheiros@cs.ucl.ac.uk

S. Brostoff
e-mail: s.brostoff@cs.ucl.ac.uk

M. A. Sasse
e-mail: a.sasse@cs.ucl.ac.uk

S. Gutwirth et al. (eds.), *European Data Protection: In Good Health?*,
DOI 10.1007/978-94-007-2903-2_3,

three studies. Study 1 was a survey study with 283 participants, investigating how comfortable participants felt about fulfilling loan information requests on application forms. Study 2 was an interview study with 10 participants, investigating participants' perceptions of loan information requests and whether they thought it was ever acceptable to lie on an application form. Study 3 was a survey study with 298 participants, investigating whether participants had ever decided not to apply for credit because of the information that was requested on the application form. The aims, methodology and results of these studies are presented in Sects 3.3 and 3.4.

In Sect. 3.5, the results of the studies are discussed within the context of three privacy issues: (1) perceived relevance of the information; (2) expected usage of information; and (3) perceived accuracy and fairness of the application process. In Sect. 3.6, we present the main conclusions of the chapter: (1) to improve applicants' confidence in the lending system, consumers' privacy concerns should be acknowledged; and (2) it is possible for loan providers to do this without reducing the predictive power of credit scoring. It can be achieved by: (1) letting applicants specify how and when they want to be contacted; (2) obtaining informed consent for data sharing with third parties; and (3) allowing applicants some degree of application form customization, e.g., making some data items optional and allowing a wider range of items.

3.2 Literature Review

3.2.1 Loan Providers

In the UK, total consumer credit lending to individuals at the end of June 2011 was £210 billion.[1] Having a loan request accepted can improve someone's circumstances—for instance, because it enables them to buy a car, a house, or start their own business. However, if a borrower is unable to re-pay the loan, credit can lead to negative outcomes, including bankruptcy. To predict whether a new applicant is a "good" or "bad" credit risk, loan providers employ a method known as credit scoring—a statistical model that converts applicants' personal data into an estimate of risk. These scoring systems differ from loan provider to loan provider, and even though the exact scoring methods are not publicized, the statistical modeling draws on three sources of information: (1) the application form; (2) past dealings with the company; and (3) credit reference agency (CRA) files.[2]

For information about how loan providers use applicants' information, we conducted interviews with four experts.[3] Expert 1 was a risk management consultant for a financial services authority. Expert 2 was an executive from a peer-to-peer lending

[1] *Credit Action*, "Debt statistics", December 2010.

[2] *Martin Lewis' MoneySavingExpert.com*, "Credit rating: How it works and how to improve it". http://www.moneysavingexpert.com/loans/credit-rating-credit-score.

[3] Expert interviews were conducted in a previous case study (not yet published).

company. Expert 3 was a board member from a credit union. Expert 4 was a university professor with a background in consumer finance statistics research. Information from particular experts will be indicated using the footnotes.

The experts agreed that data quality is an important issue for loan providers. CRAs are used to cross-reference application data, for example for error and fraud prevention. Discrepancies or flags will trigger examination of the applicant.[4] CRAs are viewed as the most reliable source of information because they are supplied by "disinterested" third parties, whereas applicants might lie on application forms.[5] For example, when applicants are asked about their income, a significant portion state an exaggerated figure to boost their apparent ability to re-pay.[6] To deter applicants from "gaming" the application process in this way, loan providers tend to give little or no explanation on why certain items of information are requested; the industry believes that—if applicants knew how specific information items were used—they would selectively report information to appear a better credit risk than they really are. This would undermine the predictive power of the score card. Loan providers sometimes even include misleading questions in the loan application to reduce applicants' gaming.[7]

3.2.2 Applicants

Past research suggests that people believe that loan officers consider fewer variables in evaluating credit applications than they actually do (Kamleitner and Kirchler 2007). In particular, people do not think that *the number of dependents*, *the time on current job*, and *the time at current address* matter, but these items are commonly used in the risk assessment. Some applicants want to disclose more information than is requested: one lending brokerage service decided to remove a field about *additional sources of income*, but had to re-introduce it after applicants complained about its absence.[8]

Previous research (Adams and Sasse 2001; Culnan 1993) found that a disclosing party's labeling of information as "sensitive" or "private" is vital to how a request for that information is perceived by that individual. *Information sensitivity* depends on how "personally defining" the information is perceived to be, and how the individual thinks others will interpret the information: data that could portray an individual in a negative way is perceived as more sensitive, whereas information considered *relevant* to the interaction is considered less sensitive. When personal information collected for one purpose is subsequently used for another one (e.g., marketing), this may be viewed as an invasion of privacy. The nature of the relationships with the

[4] Expert 3, a representative from a credit union.

[5] Expert 1, a risk management consultant for a financial services authority.

[6] Expert 2, a representative from a peer-to-peer lending company.

[7] Expert 1, a risk management consultant for a financial services authority.

[8] Expert 2, a representative from a peer-to-peer lending company.

information receiver, and in particular the level of trust the individual has in them (based on past experience or reputation) also influences an individual's assessment of the privacy implications. A third factor is *information usage*. Thus, giving *informed consent* for data use requires: (1) *disclosure* about the potential benefits and harms expected from the action; (2) *comprehension* of the potential benefits and harms; (3) *voluntariness*, meaning the individual can resist participation if they wish to; (4) mental, emotional and physical *competence* to give informed consent; (5) *agreement*, meaning clear and physical opportunity to accept or decline the participation; and (6) *minimal distraction*, doing all of the above without diverting the individual's attention from the task at hand (Friedman et al. 2005).

3.3 Research Questions

The literature review suggests that there could be a discrepancy between what information loan providers request to assess risk, and what information applicants consider relevant in this context. Perceived relevance, in turn, determines how sensitive information is deemed to be. But to protect the predictive power of their scoring systems, loan providers only provide minimal information about why specific information is requested.

Our research aims to answer the following questions:

1. Which information do loan applicants feel most and least comfortable with disclosing, and why?
2. Do applicants think it is acceptable to lie on application forms, and if so, why?
3. Have applicants ever not applied for credit because of the information that is requested on the application form?

In particular, we want to uncover whether it is possible for loan providers to request information in a way that does not invade applicants' privacy, and maintain the efficacy of their score cards at the same time.

Three studies were conducted:

1. In Study 1, 283 participants filled in a survey where they rated 59 loan information requests in terms of how comfortable they felt giving this information to loan providers. They were also asked what effect they thought their information disclosure would have on the chances of their loan request being accepted.
2. In Study 2, 10 participants were interviewed, exploring the topics of Study 1 in more depth. This included being asked about their perceptions of loan information requests and whether it was acceptable to lie on an application form.
3. In Study 3, 298 participants filled in a survey about their experiences of being denied credit. This included being asked whether they had ever chosen not to apply for credit because of the information that was requested.

The methodology and results of these studies will now be described.

3.4 Research Studies

3.4.1 *Study 1: Assessing Level of Comfort for Loan Information Requests*

Study 1 aimed to extend previous research by investigating which loan information requests people feel most and least comfortable giving to loan providers, and why. As financial data is sensitive, a key aspect of our elicitation was not to ask participants for their actual financial information itself, but elicit their perception of sensitivity of the information requested.

A survey was created using Limesurvey[9] that took approximately 15 min. for participants to complete. The survey had two main components: (1) annual equivalized income calculation (before housing costs, using the OECD scales), and (2) ratings of loan information requests for comfort. The equivalized income calculation (Department for Work and Pensions 2009) was used so that each participant could be asked to imagine a reasonable loan amount—i.e., an amount they could possibly be offered in real life. The equivalized income was computed based on the participant's weekly net household income, and how many adults and children live with them. This figure was then used to set the loan amount that participants were asked to imagine they were applying to £500, £2,000, or £5,000. Based on the calculation, 46 participants were quoted £500, 148 were quoted £2,000 and 89 were quoted £5,000.

In the second part of the survey, participants were shown a list of 59 *loan information requests*—information items requested on real credit card and loan application forms—examples include *title, employer name* and *monthly income*. For each loan information request, participants were asked to rate on a 5-point scale to what extent they were comfortable with giving the loan provider this information, where −2 = "very uncomfortable," −1 = "uncomfortable," 0 = "neither comfortable or uncomfortable," 1 = "comfortable," and 2 = "very comfortable." After rating all 59 loan information requests, participants were asked to write a brief summary regarding which items they were most comfortable and least comfortable giving the loan provider, and why.

Participants were recruited according to a nationally representative sampling frame via the market research company e-Rewards[10] and were rewarded by e-Rewards for their participation. Three hundred and seventy-five survey responses were collected; however 92 were excluded due to incomplete/nonsense responses for the open text questions. Therefore the analysis is based on 283 participants (107 male, 176 female, age range 18–60 + years, mode age "40–59"). Seventy-five percent had experience of applying for credit.

Mean comfort ratings were computed for all 59 loan information requests. Note that if a person chose not to give a rating, by selecting "not applicable," they were

[9] "Limesurvey." http://www.limesurvey.org/.

[10] The "e-Rewards" business, subsequently re-branded as "Research Now." http://www.researchnow.com/.

Table 3.1 Means and SDs for the 5 items rated highest for comfort, in descending order, where −2 = "very uncomfortable" and +2 = "very comfortable"

Loan item	N	Mean	SD
Title (Mr, Ms, etc.)	283	+1.32	1.06
Are you currently living in UK (yes/no)	281	+1.25	1.04
First name	283	+1.25	1.11
Surname	283	+1.23	1.11
Gender	283	+1.22	1.07

Table 3.2 Means and SDs for the 5 items rated lowest for comfort, in ascending order, where −2 = "very uncomfortable" and +2 = "very comfortable"

Loan item	N	Mean	SD
Work phone number	228	−1.50	1.20
Value of other assets	283	−1.64	1.17
Total balance of investments	277	−1.69	1.20
Total savings balance	280	−1.75	1.24
Mobile phone number	270	−1.99	1.17

excluded from the data for that particular loan information request. As can be seen in Table 3.1, the 5 information requests that participants were most comfortable giving loan providers, in descending order, were: (1) *title*; (2) *currently living in the UK*; (3) *first name*; (4) *surname*; (5) *gender*. One can suggest that these items are primarily related to public identity.

As can be seen in Table 3.2, the 5 information requests that participants were least comfortable giving loan providers, in ascending order, were: (1) *work phone number*; (2) *value of other assets*; (3) *total balance of investments*; (4) *total savings balance*; (5) *mobile phone number*. One can summarize these information requests as phone numbers (excluding the house phone) and information about applicants' additional finances (savings, assets, investments).

Participants' written responses revealed that a common worry was that phone numbers might be used to contact them at awkward times. For example P210: *"Happy giving general information about my finances, do not like to give work details as I work in an open plan office and everyone would be able to hear my personal details on a telephone call."*

Another concern was that phone numbers provided might be passed onto third parties and used for sales calls. For example P166: *"I am fairly comfortable with giving most information, they need it to do their job and work out if you are a risk. The thing I hate the most is if then afterwards my details are passed on and I get unsolicited emails/phone calls."*

Information about savings, investments and assets were also rated as uncomfortable. Some participants felt that they might be denied credit because of the amount they had saved. For example P219: *"not so comfortable with them knowing how much I have saved in case they decide not to give me a loan."* Other participants felt that this information was irrelevant. For example P144: "*Least comfortable with questions about other assets/savings which aren't immediately relevant in my view.*" Similarly, P109: "*Least comfortable with savings & investment—none of their business!!!*"

In a third part of the survey, participants were shown the list of 59 loan information requests again but this time asked to rate them in terms of what effect they thought their information disclosure would have on the chances of their loan request being accepted. Information requests were rated on a 5-point scale, where −2 = "my answer would show me in a very negative light to a loan provider" and + 2 = "my answer would show me in a very positive light to a loan provider." Pearson correlations comparing the comfort ratings and the effect ratings detected statistically significant correlations for 56 of the 59 loan information requests (the items for which no statistically significant effects were found being *surname, first name* and *middle name*). This suggests that there is relationship between comfort and effect—if a person thinks that the loan information request will show them in a positive light then they feel more comfortable giving that information to the loan provider.

Finally, participants thought that not answering all questions on the application form would be viewed negatively by the loan provider. For example P217: *"... the information I did not wish to give could be construed in a bad light."*

3.4.2 Study 2: Perceptions of Loan Information Requests and Lying on Loan Applications

To explore the findings of Study 1 in more depth, we conducted interviews with 10 participants in Study 2. The interviews were conducted either face-to-face or over the phone. With the participant's permission, the interview was audio recorded and later transcribed.[11] Each interview lasted approximately 30–60 min. and was semi-structured, covering several issues around personal finance. Transcripts were analyzed using a qualitative methodology known as thematic analysis (Braun and Clarke 2006). In this chapter, we will report responses regarding: (1) perceptions of loan information requests; and (2) lying on application forms.

Participants were recruited via the UCL Psychology Subject Pool[12] and received a £10 Amazon gift voucher for their participation. There were 10 participants interviewed (2 male, 8 female, age range 19–59 years, mode age "25–39"). Regarding employment status, 4 were students, 2 were in part-time employment, 2 were unemployed, 1 was in full-time employment, and 1 was a homemaker.

7 out of the 10 participants had experience of applying for credit, including bank loans, overdraft extensions and mortgages. These 7 participants were asked to reflect on what they could remember about the application process. The 3 participants that had no experience of applying for credit were asked to consider how they would feel if they had filled in a loan application form (all 3 claimed to have seen loan application forms before).

All participants said that they felt that the majority of information requested on credit application forms was relevant. Reasons given included *"the bank needs to*

[11] "Fingertips Typing Services." http://www.fingertipstyping.co.uk/.

[12] "UCL Psychology Subject Pool." http://uclpsychology.sona-systems.com/.

know whether they can trust you" and *"if I don't pay the bank needs my information to catch me."* The loan amount appeared to be an influential factor—the bigger the loan, the more reasonable it was perceived to give loan providers the information.

Participants questioned the relevance of questions; including *next of kin* and *marital status*. For example P2: *"Did not think next of kin was important, not sure why they needed to know that."* P7: *"I don't know, if they ask about your relationship status, but I guess they might do and I think that might kind of be irrelevant [. . .] Well, unless you're married, I guess you're kind of just applying for a loan for yourself anyway, you wouldn't, you know, if someone else needed a loan, they could apply for themselves."*

P1 questioned the level of detail needed: "*I guess it was just, um. . . probably the level of detail that they wanted, it was quite, um. . . quite, like I said, you'd have to go and look things up, and settle them [. . .] and that was probably, um, made the form more difficult to fill out. . .* "

Five participants said that they would have liked to provide more information to improve their chances of getting a loan, such as information about future employment, or the likelihood of a well-paid job after finishing their studies. For example P1: *"Hypothetically, if I was applying for a loan now, I'd have to say unemployed but I would want to tell them I have employment lined up for when I finish my studying [. . .] I think it would be a positive—in a more positive light, because I would have . . . kind of the security of employment for the future so I'd be more able to pay back my debt, I think that'd be positive, rather than saying that I'm a student, I don't know if I'd be able to find a job when I finish, don't know how long it'd take."*

Another student in the sample wanted to show loan providers that they have no outstanding debts or bills. 2 participants (not students) said that they wanted to provide more information about their personal circumstances, and how they planned to re-pay the loan. For example P3: *"I think that I was looking for work at the time, and I did get a job quite quickly, so I probably would have been alright paying it back. They don't know stuff like that, do they? Whereas I suppose other people get mortgages on huge houses and then they can't pay it back. It's not like I was asking for loads of money. So I think, yeah, in that respect it should maybe have been different. . .* "

All of the participants said that they would always tell the truth on a credit application form. When asked about their views on other people not telling the truth, 7 participants said that lying was always unacceptable. They mentioned the consequences of defaulting and the possibility of getting a criminal record as reasons why lying was wrong. For example P2: "*It is constantly wrecking society when people make fraudulent applications, because if they default then we all have to pay more.*" Similarly, P5: *". . . they are like fraudsters [. . .] it's always wrong because it's against the law."*

Two participants said that they could understand why people might "*tweak*" information about themselves. Such tweaks were referred to as "*white lies,*" as the person is confident that they can re-pay the loan and just wants to improve their chances of being approved. For example P3: "*I suppose if you definitely know, maybe it's okay to do it, I wouldn't say the right thing to do it, if you know you can pay it back, so it's kind of like a white lie. I suppose it's wrong if you definitely can't pay it back, then I don't think you should, because it has a knock-on effect.*" Similarly, P6: "*Sometimes they do it for good reason, and I can understand them telling little white lies [...] If you've got a basic wage, but perhaps you do overtime once every two weeks, you might exaggerate a little and say that every week, you're getting X amount overtime, so you've got more income than strictly you have. What else might apply? You might forget to admit that you're paying X amount of your credit card every month, so one of your outgoings isn't included in your budget...*"

P10 felt that honesty does not pay off: "*I think today, right, the bank seems to want you to actually be completely honest with them and when they are actually completely honest with them they don't actually get what they want, so some people just know how to manipulate and get an advantage so I don't feel sorry for the bank really in that sense because they give it out to the wrong people [...] When they have no intention of actually paying the money back at all, then that's absolutely... I do believe in paying things back if you are going to use it.*"

P1 said that people might feel less of a need to lie if they were able to provide more information to loan providers about their personal circumstances and how they planned to pay back the loan: "*...obviously, I don't think lying is the right way to go about it [...] but I mean, where if you're able to provide additional information to explain your circumstance, for instance, like, there might have been a case where you hadn't been able to pay your loan back, so where that would just count against you, if you maybe stick with your circumstances why and maybe how that wouldn't happen again, and so on—then that would make people less willing, less, um, less likely to lie.*"

3.4.3 Study 3: Choosing Not To Apply for Credit Because of Loan Information Requests

Study 3 was a survey exploring experiences of being denied credit. We focus here on a subset of the results associated with applicants choosing not to apply for credit. Participants were asked whether they had ever not applied for credit because of the information that was requested on the application form. If they answered "yes," they were presented with an open text box and asked to write about the experience—what kind of information did the financial service ask for? Why did they not want to disclose this information?

Like Study 1, the survey for Study 3 was created using Limesurvey[13] and took approximately 15 min. to complete. Participants were recruited via the market research company e-Rewards[14] and were rewarded by e-Rewards for their participation. Three hundred and twenty survey responses were collected according to a nationally

[13] "Limesurvey." http://www.limesurvey.org/.

[14] The "e-Rewards" business, subsequently re-branded as Research Now. http://www.researchnow.com/.

representative sampling frame; however 22 were excluded due to incomplete/nonsense responses for the open text questions. Therefore the analysis is based on 298 participants (96 male, 202 female, age range 18–60 + years, mode age 25–39 years). 158 of the sample were in full-time employment, 52 were part-time employed, 17 were self-employed, and 4 were temporary employed. Regarding the other employment categories, 30 were homemakers, 14 were permanently sick/disabled, 12 were students, and 9 were retired (note that participants could select more than one category). All 298 participants had experience of being denied credit, this being a pre-requisite for taking part. Regarding their current financial circumstances, 168 described themselves as being in "manageable debt," 60 as being in "problem debt," 52 as "debt free," 13 were on an Individual Voluntary Agreement, and 5 were bankrupt.

Thirty six (12%) reported that they did not proceed with an application due to the information requested. 28/36 provided clear descriptions of what happened:

- 12 participants "*knew*" that they would be rejected due to their previous debts and did not want their credit record to deteriorate further by having a refusal of credit added to their record. For example P37: *"I no longer apply for any credit as I do not want to make my credit rating worse by being refused."*
- 7 participants did not want to disclose information that they thought would put them in a negative light. For example P160: *"Overdraft extension. Did not wish to disclose that I was unemployed."* P93: *"It was a personal loan. When I still had a CCJ [county court judgment] on my record I hated to have to tell anybody because it did not reflect my current attitude to borrowing, or ability to re-pay."* P182: *"I am a recipient of income support and when applying for a loan from the social fund I set the application aside because I did not wish to explain my debts."*
- 4 participants described a *"fear of rejection."* Having been denied credit in the past, they did not want to go through the embarrassment again. For example P55: *"Every time I go into a store and they offer me a store card I refuse because I am scared of being rejected."*
- 3 participants did not apply for credit because they felt that the information being asked for was not relevant, particularly when it was information about their partner or spouse. For example P38: *"Completing surveys, buying a car, home insurance. None of their business how old I am or what my profession is or my wife's."* P202: *"Credit card companies always want to know about your spouse's income/debts etc, which I don't feel should be relevant if you are applying for a card yourself and you have income."* P50: *"Store credit agreement. Wanted information about my husband whilst I was separating from him. I explained this but they would not continue the sale."*
- 2 participants did not apply for credit because of the level of detail the loan provider wanted. For example P194: *"[Store] credit card, they wanted 3 months of bank statements so I didn't progress with the application."* P176: *"An unsecured loan with a guarantor and they wanted too much information regarding my guarantor's mortgage details."*
- 1 participant did not apply for credit because she did not want to give her phone number.

Table 3.3 Study findings and privacy issues

Finding		Study	Privacy issue
1	Applicants do not consider all of the information in application forms relevant	1, 2, 3	Perceived relevance of information requested
2	Applicants are discouraged by the level of detail that is needed for some loan information requests	2, 3	Perceived relevance of information requested
3	Applicants are uncomfortable giving information that could portray them in a negative light to a loan provider	1, 3	Expected usage of information provided
4	Applicants have concerns that phone numbers will be used to contact them at awkward times, or passed onto third parties	1, 3	Expected usage of information provided
5	Applicants feel like their personal circumstances are not fully considered in application forms	2	Perceived accuracy and fairness of the application process

3.5 Discussion

Five key insights can be drawn from the research studies regarding applicants' perceptions of loan application items, see Table 3.3. These findings relate to three privacy issues:

1. Perceived relevance of information requested;
2. Expected usage of information provided;
3. Perceived accuracy and fairness of the application process.

For the purposes of this chapter, one privacy issue has been assigned to each finding. It is important to note however that there is likely to be some degree of overlap between all three privacy issues. For example, for Findings 1 and 2, perceived relevance indirectly relates to whether participants thought the items were going to be used to assess them fairly (e.g., expected usage and perceived accuracy). Similarly, for Finding 3, some participants thought that the usage of their information was unfair (perceived accuracy).

The privacy issues, and the study findings related to these issues, will now be discussed.

3.5.1 Perceived Relevance of Information Provided

The perceived relevance of information requested on application forms was an important factor for how comfortable applicants felt with disclosing information. The following information requests were viewed as irrelevant: *value of other assets*, *total balance of investments*, *total savings balance* (Study 1); *next of kin*, *marital status*

(Study 2); information about partner's finances (Study 3). In some cases—e.g., *bank statements*, *mortgage details*—participants viewed the level of detail requested as unnecessary (Study 2, Study 3); providing this information represented a lot of effort for the participants, and at the same time, it was not clear to respondents why it was needed.

From the loan providers' perspective, all of these items are relevant, necessary and fair (see Sect. 2.1)—thus applicants' perceptions of how loan providers use information are inaccurate. It is the lack of transparency surrounding *why* these items are needed that creates inaccurate perceptions and leads to applicants second-guessing. As stated in the introduction, loan providers currently do not explain the purpose of these questions because they fear applicants might "game" their answers. However our findings suggest that applicants being asked for information that they do not think is relevant leads to a negative perception of the motivation of loan providers, and—as shown in Study 3—it can put applicants off applying altogether. While it is a good idea not to encourage applicants that have no chance of being approved (because being refused a loan can lower credit rating further)—putting off eligible applicants is a poor outcome for both loan providers and applicants. Greater transparency of the loan application procedure would help, if this can be done without enabling applicants to "tweak" their answers.

3.5.2 Expected Usage of Information

The expected usage of the information was another important factor for how comfortable participants felt with disclosure. Participants were uncomfortable with disclosing information that they thought would show them—unfairly in their view—in a negative light (Study 1); and this was a reason why 7 participants chose not to apply for credit (Study 3). An example of these diverging interpretations is that loan providers see savings as a positive indicator of ability to re-pay, whereas applicants think they will be denied on the grounds of "no need" (Study 1). Some participants thought that if they left any of the information requests blank this would also be viewed negatively by the loan provider (Study 1). Again these findings suggest that greater transparency is needed in terms of why information is requested, because currently: (1) applicants are second guessing the fairness of the assessment; and (2) applicants are unsure whether they need to fulfill all information requests (Kerr et al. 2006).

A further issue was the use of phone numbers. The loan information requests *work phone number* and *mobile phone number* were rated negatively for comfort (Study 1). One participant chose not to apply for credit because she did not want to disclose her phone number (Study 3). These findings suggest that loan providers should explain how phone numbers are going to be used and allow applicants more control over when they will be contacted: if applicants think the information being disclosed might be used to contact them at awkward times, or subsequently passed onto third parties for another purpose (e.g., marketing), then they will view it as a potential privacy risk.

3.5.3 Perceived Accuracy and Fairness of the Application Process

In Study 2, five participants said that they would like to give additional information in support of their loan application. This included: (1) providing details of future employment or the likelihood of a well-paid job after completing studies; and (2) providing more information about their personal circumstances and how they plan to re-pay the loan. The ability to volunteer relevant details about their personal circumstances might reduce the perceived need to "tweak" the responses to questions that are seen to be too narrow.

Our findings suggest that loan providers must make more effort to make applicants feel *understood*: (1) currently some applicants do not feel that they are able to accurately express their ability to re-pay in credit application forms; and (2) this inability to fully express themselves is part of the reason why a small proportion of applicants decide to "tweak" their responses, which can have a negative impact on data quality.

3.6 Conclusions

Overall, our use of both quantitative and qualitative methods, and comparing results across studies, has allowed us to gain a more detailed understanding of applicants' perceptions of information requested on loan application forms. Perceived relevance of information requested, expected usage of information provided, and perceived accuracy and fairness of the application process, are all factors that influence how *sensitive* an applicant perceives the application form to be. These privacy issues are similar to those discussed in other privacy contexts such as multimedia communication,[15] providing support for Adams and Sasse's privacy model.

Our research findings suggest that, when privacy issues are not addressed, this leads to second guessing, wrong perceptions of how information is used, and a negative perception of loan providers. Some viable applicants are put off the application process altogether, which represents lost business for loan providers.

In the remaining part of this chapter, we propose three recommendations of how loan providers could improve the perceptions and satisfaction of potential customers without necessarily reducing the efficacy of the risk management process (see Table 3.4).

[15] See Chap. 2.2, note 13. (1) Adams and Sasse, "Privacy in multimedia communications: Protecting users, not just data", 49–64. (2) Culnan, "How did they get my name? An exploratory investigation of consumer attitudes towards secondary information use", 341, 363.

Table 3.4 Privacy issues and recommendations

Privacy issue	Recommendation
Expected usage of information provided	1 Let applicants specify how and when they want to be contacted 2 Informed consent for data sharing with third parties
Perceived relevance of information requested Perceived honesty and fairness of the application process	3 Application form customization, e.g., making some data items optional and allowing a wider range of items

3.6.1 *Let Applicants Specify How and When They Want To Be Contacted*

Loan providers should allow applicants to state how and when they want to be contacted in credit application forms, i.e., indicating preferred phone number and preferred time of contact. This would improve customer satisfaction and potentially lead to lower costs as less effort is wasted on unsuccessful calls.

3.6.2 *Informed Consent for Data Sharing With Third Parties*

The UK Data Protection Act of 1998 states that "*Personal data shall be obtained only for one or more specified lawful purposes, and shall not be further processed in any manner incompatible with that purpose or those purposes.*"[16] This means that if a loan provider collects data for risk assessment purposes, it cannot be passed onto a marketing company for telemarketing purposes. However many applicants might be giving away their consent without realizing it, by accepting the terms and conditions of the loan without reading the small print. Also applicants might view third party data use as something one has to accept in order to get the loan; again the application form being viewed as a type of "*all-or-nothing*" transaction (Kerr et al. 2006). Therefore, even if the information sharing is legal, it is evident that individuals are not really exercising informed consent.

In order for applicants to give informed consent,[17] loan providers must: (1) give an explanation of how the information will be used; and (2) if there is other usage, such as marketing, it should be made clear that refusal will have no implication for the loan application. Recently it has been suggested that technological systems can increase transparency in data sharing. For example, the EnCoRe technical architecture

[16] *UK Data Protection Act 1998*, c. 29. http://www.legislation.gov.uk/ukpga/1998/29/section/29.

[17] See Chap. 2.2, note 14. Friedman, Lin and Miller, "Informed consent by design", 495–521.

(Mont et al. 2010) enables clients to view and edit how their information is shared; allowing clients to express their consents and revocations about the usage, processing and disclosure of their personal data. A similar system could possibly be used in the context of lending.

If loan providers did allow applicants to enact true informed consent, there could potentially be commercial resistance. For example, when Bankcorp was caught selling its data to telemarketers in 1999, Gellman (2002) writes that "*it is hard to believe that many U.S. Bancorp customers would have agreed to the data sharing of credit card and Social Security numbers.*" There is evidence, however, that people are willing to give information away when they can see the benefits from sharing (e.g., receiving special promotions) or are given small incentives (e.g., prize draws; Acquisti 2009). Previous research also suggests that satisfaction with a loan application process plays a central role to customers' loyalty to a financial service provider, and its profitability (Johnson et al. 1996).

3.6.3 Application Form Customization

The industry has moved from relationship banking to transactional banking due to the sheer volume of customers—instead of seeing a bank manager in a High Street branch, most transactions are conducted online or via call centers. Our research suggests that customers' mental models have not caught up with this development—they want a relationship with their financial service provider, and want to feel that they are being considered and treated as individuals. Gaining applicants' confidence should therefore be a key consideration for loan providers.

Currently applicants are viewing the application form as a type of "*all-or-nothing*" transaction.[18] Past research, however, suggests missing data is not necessarily a reliable indicator of how risky a customer is: from the point of view of the score card, there are ways to cope with missing data.[19] Therefore, allowing applicants not to answer certain questions might resolve the transparency-gaming dilemma. As applicants might try to game the system—choosing to not answer items that might show them in a negative light—it would be important to keep key items compulsory. Score card developers would need to invest time researching how many items, and which items, could be made optional without reducing the score card's efficacy.

Past research also suggests that sometimes loan providers allow applicants to give extra information, even though the loan provider does not actually intend to use it, because it makes the applicant feel more comfortable.[20] However this does not solve the problem—people do not want to just "talk," they want to actually be listened to. Therefore the challenge for loan providers is to give applicants better quality

[18] See Chap. 6.1, note 26. Kerr, Barrigar, Burkell and Black, "Soft surveillance, hard consent", 1–14.

[19] Expert 4, a university professor with a background in consumer finance statistics research.

[20] See Chap. 2.2, note 12. Expert 2, a representative from a peer-to-peer lending company.

of care and a more personal experience, but in a way that is manageable and can be conducted on a large-scale. One route is to investigate making available a wide range of optional items, allowing applicants some degree of customization of the application form. Our group is currently investigating the acceptability to applicants of a wide range of hypothetical items (Jennett et al. 2010). Any items found to be acceptable enough would then need to be further investigated for their predictive power.

Overall, our recommendations demonstrate that there are potential ways to maintain the efficacy of the loan risk assessment process, and at the same time respect applicants' privacy and choice on what they disclose. Future research is needed to check how such measures could be implemented effectively—however, when considering privacy for applicants versus loan providers' needs for predictive power, one can conclude that it does seem possible to "bridge the gap."

References

Acquisti, Alessandro. 2009. Nudging privacy: The behavioural economics of personal information. *IEEE Security and Privacy* 7:82–86. doi:10.1109/MSP.2009.163.

Adams, Anne, and Angela Sasse. 2001. Privacy in multimedia communications: Protecting users, not just data. In *People and computers XV—interaction without frontiers: Joint Proceedings of HCI 2001 and IHM 2001*, ed. Ann Blandford, Jean Vanderdonckt and Phil Gray, 49–64. London: Springer. ISBN-13: 978–1852335151.

Braun, Virginia, and Victoria Clarke. 2006. Using thematic analysis in psychology. *Qualitative Research in Psychology* 3 :77–101. doi:10.1191/1478088706qp063oa.

Culnan, Mary J. 1993. How did they get my name? An exploratory investigation of consumer attitudes toward secondary information use. *MIS Quarterly* 17:341–363. http://www.jstor.org/stable/249775. Accessed 10 Dec 2011.

Department for Work and Pensions. 2009. Households below average income (HBAI), 1994/95–2007/08", ed. Nick Adams, George Johnson, Peter Matejic, Rupesh Vekaria and Julia Whatley, 190. Crown Copyright, 2009. http://research.dwp.gov.uk/asd/hbai/hbai2008/index.php?page=contents. Accessed 10 Dec 2011.

Friedman, Batya, Peyina Lin, and Jessica K. Miller. 2005. Informed consent by design. In *Security and usability: Designing secure systems that people can use*, ed. Lorrie Cranor and Simson Garfinkel, 495–521. Sebastopol: O'Reilly. ISBN-13: 978–0596008277.

Gellman, Robert. 2002. Privacy, consumers, and costs: How the lack of privacy costs consumers and why business studies of privacy costs are biased and incomplete. *Digital Media Forum*, Ford Foundation. http://epic.org/reports/dmfprivacy.html. Accessed 10 Dec 2011.

HM Treasury. 2010. Speech by the Financial Secretary to the Treasury, Mark Hoban MP, at Reform. http://www.hm-treasury.gov.uk/speech_fst_011210.htm. Accessed 10 Dec 2011.

Jennett, Charlene, Sacha Brostoff, Miguel Malheiros, and M. Angela Sasse. 2010. Investigating loan applicants' perceptions of alternative data items and the effect of incentives on disclosure." *Privacy and Usability Methods (PUMP) Workshop*, British HCI Conference. http://scone.cs.st-andrews.ac.uk/pump2010/papers/jennett.pdf. Accessed 10 Dec 2011.

Johnson, Michael D., Georg Nader, and Claes Fornell. 1996. Expectation of perceived performance, and customer satisfaction for a complex service: The case of bank loans. *Journal of Economic Psychology* 17 :163–182. doi:10.1016/0167–4870(96)00002–5.

Kamleitner, Bernadette, and Erich Kirchler. 2007. Consumer credit use: A process model and literature review. *Revue Européenne de Psychologie Appliqué* 57:267–283. doi:10.1016/j.erap.2006.09.003.

Kerr, Ian, Jennifer Barrigar, Jacquelyn Burkell, and Katie Black. 2006. Soft surveillance, hard consent. *Personally Yours* 6:1–14. http://papers.ssrn.com/sol3/papers.cfm?abstract_id=915407. Accessed 10 Dec 2011.

MacCarthy, Mark, and Robert Gellman. 2010. The consumer financial protection bureau needs a privacy office. *Privacy & Security Law Report* 2010, Bureau of National Affairs Inc. http://explore.georgetown.edu/publications/51958/. Accessed 10 Dec 2011.

Mont, Marco Cassasa, Yun Shen, Gina Kounga, and Siani Pearson. 2010. Technical architecture for the first realized case study. *EnCoRe Project Deliverable* D2.1, version 1.0. http://www.encore-project.info/deliverables_material/D2.1%20EnCoRe%20Architecture%20V1.0.pdf. Accessed 10 Dec 2011.

Raab, Charles. 2004. The future of privacy protection. *Cyber Trust & Crime Prevention Project*. http://www.mendeley.com/research/future-privacy-protection/#. Accessed 10 Dec 2011.

Chapter 4
Cookie Wars: How New Data Profiling and Targeting Techniques Threaten Citizens and Consumers in the "Big Data" Era

Jeff Chester

4.1 Introduction

Digital marketers have unleashed a powerful and far-reaching data collection, profiling, and targeting apparatus. Interactive advertising techniques incorporate some of the latest developments in such fields as semantics, artificial intelligence, auction theory, social network analysis, data mining, and neuroscience. Consumers and many policymakers are largely unaware of how online advertising operates, let alone its impact. Tracking technologies regularly monitor our travels on the Internet, generating information that forms digital targeting profiles. Unlike more traditional ads, digital marketing *watches us*, relying on such techniques as data optimization, "self-tuning" algorithms, "intent" data, and "immersive" multimedia (Econsultancy 2011, 41). This data collection and targeting apparatus has already been purposely migrated into the core business models shaping social media, mobile devices, gaming platforms, virtual worlds, and online video. Some digital data marketers refer to this as a new kind of "Wild West" and the era of "Big Data," as both conglomerates and start-ups vie to develop even more methods to "monetize" consumer information online (Hutchinson 2011).[1]

Since the emergence of the commercial Internet in the early 1990's, I have followed closely the role online advertising has played in shaping the foundations of our digital culture. While back in the 1990's there was tremendous international enthusiasm for the Internet's democratic potential, there was far less concern over the ultimate impact of the interactive advertising business model at the core of the new medium. As a campaigner on media issues who has worked for decades trying to promote "public interest" policies for US television, and who knew the lessons of

[1] See also eXelate Networks, "Targeting Exchange, Digiday 2009" http://www.slideshare.net/DM2EVENTS/exelate-networks-1556802 (viewed 23 Mar. 2011).

J. Chester (✉)
Center for Digital Democracy, 1220 L Street, NW,
Washington, DC 20005, US
e-mail: jeff@democraticmedia.org

S. Gutwirth et al. (eds.), *European Data Protection: In Good Health?*,
DOI 10.1007/978-94-007-2903-2_4,

twentieth-century American communications history, I recognized that many of the same forces that shaped broadcasting and multichannel communications would be at work in the new environment. The combined imperatives of marketing, advertising, and entertainment—fueled by the technological capabilities that would deliver personalized interactive content—and coupled with a growing consolidation of digital media ownership and control by leading companies, would likely unleash a highly commercialized digital environment. Since the mid 1990s, I have played a leading role trying to temper the impact of an unfettered online data collection system by promoting privacy regulation, as well as exposing how the goals of much of Internet marketing pose threats to consumers and citizens (Barnouw 1968; Federal Trade Commission 2007; McChesney 1995; Singer 2010; Starr 2005).

Online advertising companies, such as Google, Facebook, Yahoo, and Microsoft, routinely offer the public and government officials a glossy version of digital reality that purposely evades how their tactics and techniques threaten privacy and have other problematic consequences. They claim that there are only benefits derived from access to the abundance of information readily available online. In their worldview, the ad-supported Internet has now freed consumers and citizens to make more informed choices, immune even from the persuasive lure of marketing messages that may have influenced their behavior in the past. This essay attempts to challenge such storybook claims, relying on my investigatory work to peer behind the Wizard's curtain and discover what kind of digital "Oz" we may find. For what online marketers say to the public, as we shall discuss, is very different from the discourse they have with each other and their allies. In that conversation, ultimately we believe the more truthful one, the public is at the mercy of advanced technologies designed to move them through a "purchase funnel," whether on their computers, mobiles, game players, or digital TVs.

The leading global companies and brands, including those from the financial, health, consumer product, and entertainment sectors, have enthusiastically embraced online advertising, with nearly $ 71 billion (US) expected to be spent worldwide in 2011 (MagnaGlobal 2010). US Internet marketing companies, including Facebook, Google, Microsoft, and Yahoo, operate a range of digital data-targeting services throughout the world, including in the European Union, the Asia-Pacific region, and Latin America. Advanced techniques for the buying and selling of individuals online for targeted advertising, developed initially in the US, are now found in EU countries as well as new markets such as China. Digital marketing trade associations, the networks of "Interactive Advertising Bureaus," have banded together to fight against privacy legislation and regulation, including in the US and EU. In an attempt to counter critics, the industry has developed a new self-regulatory scheme relying on "icons" to signal to consumers that they are being tracked (Interactive Advertising Bureau 2010a; Lau 2011; Microsoft 2010; Facebook 2011b; Google 2011b; Interactive Advertising Bureau 2011a; EU has Trouble Digesting New Law on Internet Cookies—IAB Europe Offers Solution 2010).[2]

---

[2] Sociomantic Labs, "Social Targeting," http://www.sociomantic.com/social-targeting.

The potential impact of these tiny graphic icons on websites will likely be overwhelmed by the everyday practices of contemporary online marketing. The commercial digital media system is largely designed to promote data collection through "360-degree" online marketing strategies (Advertising Research Foundation 2010, 2011). While the debate on privacy and online marketing has focused on behavioral profiling—so called Online Behavioral Advertising (OBA)—such practices are only a part of the overall data collection apparatus. From social media surveillance tools and "in-game" advertising tracking, to online video measurement and location tracking, a bevy of increasingly inter-connected user data services are deployed to track us throughout the interactive landscape. Our Internet experiences are also shaped, invisibly, by technologies that "optimize" how we interact with Web pages, to help manage our online journeys so we will "convert" to whatever the digital marketer desires us to do (such as buying a product or filling out a form). A growing range of "immersive" and neuromarketing-based applications, designed to convince us to accept the enjoyable pleasures of much of contemporary online marketing-based content, has added new forms of "subliminal persuasion" to the data collection equation (Facebook 2011c; Omniture 2011).[3]

Interactive marketing campaigns rely on all of these techniques and more to encourage users to provide their information, including through a process known as "engagement." A new approach developed for the Internet era, the goal of engagement is to create deep emotional connections between brands, products, and users. Strategies employed to promote engagement include the creation of "immersive" online environments—various forms of interactive multimedia such as video, games, or animation—that spur "data capture" (DoubleClick 2011). Increasingly, digital advertising also relies on forms of neuromarketing. Ads for both online and offline are thoroughly tested using fMRIs and other brain scan diagnostic tools. The goal is to perfect an ad that bypasses the consumer's rational decision-making process and is absorbed into the subconscious (Chester and Montgomery 2007; Montgomery and Chester 2009; Montgomery et al. 2011).[4] An emerging genre of techniques focused on a user's "social graph"—one's connections of friends and networks—increasingly enable marketers to reap a treasure trove of highly specific social media data (Facebook Developers 2011).

---

[3] Samira Lama, "Lennox Invests in a Mobile Strategy and Sees Lower Cost per Conversions," Google Mobile Ads Blog, 1 July 2011, http://googlemobileads.blogspot.com/2011/07/lennox-invests-in-mobile-strategy-and.html;; Neurosense, "Applications," http://www.neurosense.com/apps.html (all viewed 5 July 2011).

[4] See also "Advertising Research Foundation Develops Standards for Neuromarketing Research," 22 Mar 2011, http://www.prnewswire.com/news-releases/advertising-research-foundation-develops-standards-for-neuromarketing-research-118423879.html (viewed 23 Mar. 2011); A. K. Pradeep, *The Buying Brain* (Hoboken, NJ: Wiley, 2010).

4.2 Cookies on Digital Steroids

The fundamental foundation of online marketing is based on the concept of "one-to-one marketing," popularized during the first dot-com boom of the 1990s (Peppers and Rogers 1999). The Internet was seen early on as the perfect vehicle to conduct new forms of "consumer relationship marketing," given all the granular details that could be collected via the Web (Chester 2007). Since that time, the over-arching goal of online marketing has been the collection and use of ever-greater amounts of consumer information, across all platforms and many applications. Major online ad companies have invested significant resources to expand the capacity of digital advertising. Microsoft has established one of its "labs" for data mining and ads in Beijing; Yahoo's Bangalore facility in India works on "computational advertising"; Google has an extensive global ad research apparatus that includes the funding outside scholars (Matias 2011; Google Research 2011).[5]

One of the ironies of the debate about behaviorally targeted (BT) advertising and privacy is that marketing industry representatives primarily tell regulators that such data techniques aren't targeted to individuals. When pressed about the privacy concerns connected to BT, they generally retort that there is a misunderstanding. Such targeting is both "anonymous and innocuous," and is only aimed at providing consumers with ads they will find of greater interest. In the US, what is currently considered "personally identifiable" information (or PII) is undergoing review. But traditionally it has meant one's physical and email address, birth date, and Social Security number. Online marketers cling to a claim that most, if not all, of the information they collect on a user is non-personally identifiable (non-PII). But such arguments don't hold up to serious scrutiny (not to mention the claims marketers make to each other and to prospective clients). Dozens of online ad firms openly speak of their use of "Unique personal data" in their targeting equations. Beyond all the data they compile, digital advertisers also recognize that in today's Internet-connected environment, it isn't necessary to know someone's real name in order to target them. Through cookies, Web beacons, and IP addresses, marketers know the key aspects of our digital identities: the content we favor or ignore; the amount we are willing to spend in shopping carts; the search terms we use; and favored sites and the like. Increasingly, we also willingly provide additional details that can be used in our profiles, through social media, mobile devices, and by obligingly filling out online forms, questionnaires, and entering online sweepstakes.

[5] In its joint academic grant program operated with ad giant WPP, Google supports scholars who can help make interactive ads more precise. For example, in a recent research round, Google listed as "Topics of interest" that included: What do we know and what more do we need to know about on-line audiences? How can advertisers be welcome in social networks? How do teens interact with digital media and what are the implications? How can pharmaceutical brands engage more effectively online? What are the unique marketing and targeting opportunities for other verticals: financial services, insurance?.

The actual role BT plays as a form of commercial surveillance can be confirmed from the "glossary" provided to members of the US Interactive Advertising Bureau (IAB). It explains that BT "uses information collected on an individual's web browsing behavior such as the pages they have visited or the searches they have made to select which advertisements to be displayed to that individual" (Interactive Advertising Bureau 2001). In its 2010 "Networks and Exchanges" guidelines, the IAB further defines BT as a "user-initiated action which may include, but not limited to: searches, content views, clicks, purchases, form-based information and other interactions." Stored in a "user profile," it explains, are data that can "consist of demographic information (e.g., age, gender, geographical location), segment or cluster information (e.g., auto enthusiast), and retargeting information (e.g., visited Site X two days ago)" (Interactive Advertising Bureau 2011b). Both the IAB UK and Microsoft Advertising describe BT as a "form of online marketing that uses advertising technology to target web users based on their previous behaviour. Advertising creative and content can be tailored to be of more relevance to a particular user by capturing their previous decision making behaviour (e.g., filling out preferences or visiting certain areas of a site frequently) and looking for patterns" (IAB UK 2011; Microsoft Advertising 2011).

Yahoo, like other leading digital marketing companies, claims its BT approach is anonymous. But a more candid admission by Yahoo can be found in a 2007 presentation to advertisers from the United Kingdom. Yahoo boastfully described its behavioral targeting efforts as a form of "intelligent user profiling," explaining that it captures user "DNA" from "registration and behaviours" (including such online activities as page views, ads clicked, search queries, and clicks) (Behavioural Targeting 2009).[6] More recently BT ads have been transformed into so-called "Smart Ads," as Yahoo calls them. Data collected from a user helps transform the creative copy into a more precise interactive pitch in real-time. Yahoo explains that by "using Yahoo! Data" for these Smart Ads it can push "valuable prospects deeper into the purchase funnel."[7] Google has also entered into the "smart ads" business through its 2009 acquisition of Teracent. The Google subsidiary enables advertisers to deploy "an unlimited number of ad creative combinations. . . through a single ad unit. Then, sophisticated machine learning algorithms instantly select the optimal creative elements for each ad impression—based upon a real-time analysis of which items will convert from impressions into sales."[8]

Not only are online ads compiling data about us, the Web pages and sites we access are often stealthily designed to ensure we leave our digital fingerprints behind. The design of a site includes analyzing how best to place various banners, buttons, and videos, and other navigation tools, in order to structure what's known as the "user's

---

[6] See also Yahoo, "Behavioural Targeting," http://advertisingcentral.yahoo.com/en_GB/products/behaviouraltargeting (viewed 23 Mar. 2011).

[7] Yahoo, "Yahoo! Smart Ads," http://advertisingcentral.yahoo.com/en_GB/products/retargeting_smartads (viewed 23 Mar. 2011).

[8] Teracent, "Advertiser Solutions," http://www.teracent.com/advertiser-solutions/ (viewed 23 Mar. 2011).

journey." Many online services use a variety of tools to structure the composition of what's known as "landing pages" in order to facilitate the "on-site behavioral targeting" of a user. Various data "optimization" techniques are routinely used, including evaluating how our eyes move across a page—"eye-tracking"—in order to make sure we favorably interact with the site. The goal of such practices, as marketing firm Web Trends recently explained, is to "maximize conversions." These conversions aren't about religion or politics—it's about what online advertisers call moving a consumer through the "pathways to purchase" (Enquiro 2011; Garcia 2010).[9]

4.3 The Right Ad, Right Time, Right Price, and Right Audience

Contemporary online data collection practices have more recently crossed a digital Rubicon. Advertisers are now able to track, buy, and sell an individual in real-time, through what's known as digital ad exchanges. In just milliseconds, a user is subject to an invisible auction process, where advertisers—armed with copious amounts of information on that person—compete in a bidding process for the ability to serve them an ad. Real-time bidding is available for targeting consumers whether they are visiting a website, watching an online video, or using their mobile phone. As one industry executive explained, we now find ourselves unwitting participants in the "Cookie Wars." James Lancelot of Invite Media (now owned by Google), observed that these battles are taking place because "a major shift is happening currently in the industry away from buying 'inventory' and moving towards buying 'audiences.' From a technical perspective, buying audiences means bidding on users, and how you bid on a user is based off of the cookies that have been dropped on that user's computer" (Lancelot 2009). Competition for the "best cookies," in effect (i.e., the prime prospects for any particular good or service), has become fierce, leading to what Lancelot expects will be an even larger consolidation within the digital ad industry—and more exchange and aggregation of personal data as the control over user cookies falls into fewer corporate hands.

Online ad industry consolidation has *already* helped transform the industry, as leading companies positioned themselves in the lucrative consumer online data collection market. For example, Google now operates DoubleClick and Admob; Yahoo acquired Blue Lithium and Right Media; AOL owns Tacoda; WPP took over 24/7 Real Media; Adobe bought Omniture and Demdex; Apple purchased Quattro; IBM acquired Coremetrics and Unica; Akamai owns Acerno; and Microsoft bought aQuantive—to name only a few. There has also been a boom in venture capital investment for both existing and start-up digital advertising companies. Owning a piece of the digital "data ecosystem" is seen as a necessity if companies are to thrive in the interactive advertising era (Ebbert 2011; M&A & Venture Capital 2011; Terence 2010).

[9] For an example of eye-tracking, see Enquiro (2011).

It's not just technology companies or digital marketers like Google that are enhancing their data-targeted assets. Leading global advertising agencies are also buying and selling consumer data for online targeting. For example, WPP's "Zeus Advertising Platform" (ZAP) enables its clients to use advanced data-mining techniques "to track the effectiveness of each individual digital marketing element in the purchase funnel; to identify precisely which factors affect their audience at what times, and if/how they ultimately lead to conversion. ZAP provides a holistic view of *site analytics and campaign data for a comprehensive understanding of every individual consum*er. . . . within many live campaigns that reach hundreds of millions of unique users per month, and the solution is expanding in both data volumes and capabilities" (Netezza 2009).[10] Through the "Zeus data warehouse, advertisers can action consumer and advertising data as well as integrate and action external data. . . . Third party data is layered on top of aggregated user level data. . . to form a record for each user, marrying audience data with performance metrics" (Google Extends VivaKi Partnership 2010; Adnetik 2011; Cadreon 2011; Econsultancy 2011, 76–77).[11]

As a January 2011 report on audience buying platforms explains, "Data has become one of the most valuable commodities in the real-time bidding system. There is a fundamental shift in media buying from buying placements to buying audiences" (Econsultancy 2011, 3). Complex array of data are used for consumer profiling, tracking, and targeting on these "exchange" and "demand-side" platforms. Data collected on an individual, including via behavioral tracking, "intent" data warehouses, and outside databases, are used to determine the value of an individual targeting "impression." In the words of computational advertising company Rocket Fuel, companies can buy "individual impressions of the users that matter most—the ones . . . determined [to] fit [a] customized data-driven audience profile" (Econsultancy 2011, 92).

Among the leaders of this new marketplace are such companies as Admeld, Data XU, the Rubicon Project, Mediamind, and Turn. They are part of the "audience buying" system predicted to benefit from an increase in spending in real-time ad buying from $ 2.2 billion in 2010 to $ 5.2 billion in 2014. But the growing reliance on superfast computers that can constantly track us wherever we are, compile and analyze sets of online and offline data, and then offer us up for sale to the highest bid for ads underscores the urgent need to protect privacy.[12]

Consumers should not be expected to understand the privacy dimensions of a "custom targeting" system that uses wide-ranging data sets to determine "the absolute value of each impression" for an advertiser. How and why should any user

---

[10] See also Media Innovation Group, http://www.themig.com/mobile/zap.php (both viewed 15 Feb. 2011).

[11] Other ad giants operating their own data platforms or "audience buying" services include Havas (Adnetick), IPG (Cadreon), and Publicis (VivaKi). VivaKi, http://www.vivaki.com/ (all viewed 23 Mar. 2011). In November 2010, Google extended its digital ad targeting partnership with agency giant Publicis and its "VivaKi Nerve Center Trading Desk." They are buying video and mobile ads via Google Doubleclick's ad exchange for data targeting.

[12] Ebbert, "eXelate CEO Zagorski Discusses New DataLinX Platform and Company Strategy."

have to know how a data-targeting “demand-side platform” operates and will affect their privacy and consumer decision-making? (Econsultancy 2011, 10).[13] Even technology-savvy users may be hard-pressed to address the consequences to privacy of automated decision systems able to cull data in a flash of an eye, but online publishers, marketers, and data brokers understand the benefits in better targeting users. They can now “precisely identify and target desired audiences and behavior, without using content as a proxy”; use “Impression-level-bidding [to] make cookie retargeting more scalable and powerful; Execute cross-sell, up-sell and retention campaigns by leveraging customer relationship management databases and third-party data” (Econsultancy 2011, 23).[14]

4.4 BYOBD: Bring Your Own Behavioral Data

New advanced approaches for targeting consumers, such as “demand-side platforms,” have not displaced behavioral targeting. The global growth of real-time digital ad exchanges depends on their ability to seamlessly access both online and offline consumer information. To better serve the twenty-first-century digital marketing industry, behavioral targeting warehouses and “co-ops” have been formed. Such services are a kind of data-mining “one-stop-shopping” for online targeting. For example, BlueKai claims to operate “the largest data exchange focused on identifying consumer intent in the advertising world as well as bringing to market the most advanced data management platform available to marketers” (BlueKai 2011d). US-based BlueKai assures prospective clients that they will be able to “access actionable audience data on more than 200 million users” (BlueKai 2011b). BlueKai offers

[13] As Econsultancy describes it, a demand-side platform includes: Connects to multiple inventory sources (e.g., ad exchanges, optimizers), creating a significant pool of impressions; Calculates the value of an impression relative to its characteristics in real-time; Makes decisions on what impressions to bid for and what price to bid for each in real-time; Enables data integration with third-party data providers, agencies, analytics companies and clients; Integrates data, targeting, optimization, analytics, impression attribution, and reporting; Makes the media and data buying process more transparent and efficient; Enables media buyers to manage and optimize their campaigns in real-time though a single interface; Provides better insight into users’ behavior and allows retargeting across numerous platforms.

[14] In deciding which advanced online targeting technology company to use, marketers are told they should ask themselves a range of data-related questions, including “Who are their data partners? Is the company able to integrate and manage first-party data as well as third-party data sources? Can you use data from any third-party provider or are you limited to certain providers only? What types of data can the platform integrate, e.g., intent data, unique personal data? Does the platform have predictive targeting… capabilities? Are cross-platform buying capabilities (e.g., Facebook, Google Adwords) offered?” Questions that should be asked on “targeting and optimization” include “Is the optimization approach rules-based or algorithmic-based? Are the algorithms static or dynamic? Does the DSP offer real-time assessment, page-level optimization and automated optimization? …What targeting approaches does the DSP offer (e.g., demographic, contextual, behavioral, geo-targeting, retargeting, multivariate targeting)?” Econsultancy, “Demand-Side Platforms Buyer’s Guide,” (p. 33).

marketers the ability to track and target a consumer's financial interests through the sale of their data related to credit cards, mortgages and refinancing, retirement, and other financial service products (BlueKai 2011c).[15]

eXelate, similarly, enables "data buyers [to] build an instant behavioral targeting function and optimize their campaign delivery, while data sellers gain direct control over their audience data distribution...." Its "eXchange includes over 50 top ad network, agency and demand-side platform buyers, and dozens of leading publishers, who deliver targeting data on nearly 200 million US unique users in verticals including Business-to-Business, Auto, Travel, Finance, Shopping, and registration-based Demographics" (eXelate 2011a; b).[16]

Across the world, both established companies and new entrants are now part of a consumer data outsourcing supply chain. So-called "third parties" collect and sell information that can be used by ad networks, audience buying platforms, and other data buyers (Econsultancy 2011). For example, Experian—long known for its credit reports—now operates several corporate divisions focused on online commerce. Experian Interactive collects information on consumers who are interested in a loan or in buying a product through its Lowermybills.com, Pricegrabber.com, and other subsidiaries (Experian Interactive 2011). It also offers marketers real-time data for targeting through its "Audience IQ" product. Consumers are likely unaware that websites using Experian will have data that can "predict the location of the consumer at a resolution that varies from five-digit ZIP Code to household," and that will help generate a "score" based on what is known about their "lifestyle" and "credit" (Experian 2011). Experian is part of a "Data Partner" online marketing system available to advertisers, where data from different sources can be mixed and matched. Little known database companies such as Bizo, Rapleaf, AlmondNet, TARGUSinfo, eBureau, Datalogix, and Acxiom, as well as Bluekai, and eXelate, can be tapped in an instant to help better profile a user.

The combination of all these data used for real-time targeting should be a central focus for the privacy policy debate. Given the consolidation within the online marketing industry, advances in advertising technologies, the growth of new online ad markets (such as Asia-Pacific), and the dizzying data-chain of partnerships and alliances, it is vital for regulators to develop appropriate rules that reflect today's challenges to privacy; however, with the online ad industry largely united in claiming that its profiling practices are based on anonymous data, it is useful to examine how one company compiles targeting information.

Turn is "one of the largest marketing platforms on the Internet." It operates what it says is a "data-driven" ad-targeting platform that "crunches 2000+ behavioral,

---

[15] http://www.bluekai.com/intentdata_bluekaiinside.php (both viewed 9 Feb. 2011).

[16] How does eXelate collect all these data? As it explains, "All of eXelate's online-based activity data is directly sourced from online publisher partners via tags located on web pages in which consumers interact with relevant content or queries. Via this tag, eXelate is able to drop a 'targeting cookie' which collects relevant activity...." The company uses a consumer's data for targeting that "may be limited to a specific deep action (such as a shopping search, or lead generating auto interaction), while in others, such as age or gender, multiple registration-based data points may be accumulated on the user in the segment".

contextual, inventory, and ad selection variables within 25 milliseconds... all to determine the right ad, right time, right price, and right audience."[17] A recent research paper by Turn discusses how its "data mining solution enables marketers to cost-effectively identify interactions and variables of thousands of data points. It also allows them to look at the entire user profile at the time of impression receipt and do a thorough analysis of the impact of all the variables on a campaign (including latent variables which go beyond the audience segmentation and are often times overlooked)."[18] Turn explains that its "secret sauce" is a "scalable infrastructure [that] enables us to read an individual user's data profile from among hundreds of millions of profiles within a very small time frame, generally 2 or 3 milliseconds. And, we do this over 100,000 times a second (8 + billion times a day)."[19]

In its privacy statement, Turn notes that it "does not collect PII," while saying it collects the following non-personal information: "... the IP address used to access the Internet, the type of browser used, which and how many Business Partner web pages have been viewed, search terms entered on Business Partner websites, referring/exit pages, and the date and time a Turn Ad was viewed."[20] In its discussion of the use of cookies and Web beacons, the company claims that such tracking and analysis isn't personally identifiable. But Turn's claim that it's targeting is all based on non-PII data needs to be evaluated by what its "data partners" can provide (as well as its own pronouncements concerning its ability to track and target an "entire user profile"). Turn uses Bizo, IXI, TARGUSinfo, Polk, Datalogix, Almondnet, Bluekai, and eXelate for its data operations.[21] The data provided by a single partner of Turn, let alone the impact of its combination, should raise questions about whether regulators—and the public—should accept the claims that all this information is "anonymous and innocuous."

Bizo, for example, provides "business demographics of a person that may include, but is not limited to job function, seniority, company size, industry, geography, etc" (Bizo 2011). IXI's digital ad data enables online marketers to "target only the consumers that have the right financial profile for each offer and brand.... [with] real-time user classification capabilities.... [that] ranks online consumers based on their expected ability to pay their financial obligations... [and] provides a powerful, complete and accurate estimate of your prospects' and customers' total household income... [along with an] estimate of a household's spending after accounting for the fixed expenses of life (housing, utilities, public transportation, personal insurance and pensions)" (IXI Corporation 2011a, b, c). TARGUSinfo's

[17] Turn, "Turn Media Platform Overview," http://www.turn.com/?p=3055; Turn, "The Ingredients of Our Secret Sauce: Part 1," http://www.turn.com/?p=5973 (both viewed 15 Feb. 2011).

[18] Turn, "Mining Data for Digital Advertising," http://www.turn.com/?p=4014 (viewed 15 Feb. 2011).

[19] Turn, "The Ingredients of Our Secret Sauce: Part 1."

[20] Turn, "Site Privacy Policy," http://www.turn.com/?page_id=534 (viewed 15 Feb. 2011).

[21] Turn, "General Info," http://www.turn.com/?page_id=532; Turn, "Info Collection & Use," http://www.turn.com/?page_id=536; Turn, "Site Privacy Policy"; Turn, "Data Partners," http://www.turn.com/?p=1392 (all viewed 15 Feb. 2011).

data include "names, addresses, landline phone numbers, mobile phone numbers, email addresses, IP addresses and predictive attributes" (continually updated "10 times daily").[22] TARGUSinfo also facilitates the collection of "audience targeting data high-quality, offline attributes—including demographics, shopping behaviors, lifestyles, preferences and brand affinities—that are verified. . . to accurately identify Internet users and link them to attributes—such as demographics, buying behaviors and attitudes—in a real-time. . . manner. . . . enabling you to target the most relevant ad to every user regardless of location or media buying methodology."[23] "AdAdvisor services use cookies that give you a window to rich, predictive data on **over 50 million unique US users**."[24] Polk provides "consumer detail (e.g., age, household income, gender), phone numbers, email addresses," along with "comprehensive customer profiles with unique automotive variables. . . . The number of registered vehicles in a household, When a household will likely be in the market for their next vehicle purchase, How much will likely be spent on the next vehicle purchase," and "reliable and extensive ethnic data including those with the highest levels of purchasing power—Hispanics and Asians" (R. L. Polk & Co. 2011a, b).[25] Datalogix, "a source for real-world data for online targeting," uses "tens of millions of . . . Affiniti Cookies to support online targeting" (Datalogix 2011b, d). "DataLogix' audience platform is powered by a database with over $ 1 trillion dollars in consumer spending behavior" (Datalogix 2011a). "Available data spans hundreds of product categories and a host of recency, frequency and monetary value data elements" (Datalogix 2011c). AlmondNet "partner(s) with Data-Owners & Media-Owners to facilitate the delivery of relevant, targeted (based on recently-conducted searches for products/services) ads to consumers wherever they go. . . ," ". . . based on their observed online behavior wherever they may be found" (AlmondNet 1998, 2010). We've already discussed the data collected and sold by both BlueKai and eXelate, which can be configured for Turn's targeting mix. Only a semantic revisionist specializing in Orwellian "Doublespeak" could claim this aggregation of information on a single user is somehow striking naively in the dark!

4.5 Subprime Data

The debate about privacy online has primarily focused on how data are collected from a user without their informed consent—and less about how all that information is ultimately used. Some in the online ad industry claim that their data collection

[22] TARGUSinfo, "About Us: Our Data," http://www.targusinfo.com/about/data/ (viewed 15 Feb. 2011).

[23] TARGUSinfo, "Solutions: On-Demand Scoring: Display Advertising Optimization," http://www.targusinfo.com/solutions/scoring/optimization/default.aspx (viewed 15 Feb. 2011).

[24] TARGUSinfo, "Solutions: On-Demand Scoring: Advertisers," emphasis in the original, http://www.targusinfo.com/solutions/scoring/optimization/advertisers.aspx (viewed 15 Feb. 2011).

[25] "Targeted Marketing Lists," http://usa.polk.com/Industries/Media/Communicate/TargetMkt/ (viewed 10 Feb. 2011).

practices are relatively harmless, and merely about providing a consumer with more relevant advertisements. Such arguments fail to acknowledge the real-world implications to our personal lives of digital data collection, as we increasingly rely on the Internet and mobile devices to engage in key transactions. Already, major business models for digitally delivering a wide range of financial products have been developed, including for loans, credit cards, and mortgages. During the height of the housing boom in the US, from 2005 to 2007, online mortgage services companies Countrywide Mortgage and LowRateSource increased their online spending (from $ 18.3 million to $ 35.5 million and $ 17.9 million to $ 51.7 million, respectively). Four mortgage or financial services companies were in the top five on online ad spending in August 2007. Consumers were unaware of the role played by behavioral marketing and online lead generation (where a lead or prospect is identified as a target for a financial product) in the marketing of subprime mortgages. To date, the part played by online marketers in what became the global economic crisis has not received the proper scrutiny (Chester 2009).

Consumers, who were victimized during the subprime mortgage era, or those who were sold unaffordable loans for education, should not have to remain vulnerable to new forms of database marketing that combine offline and online data. For example, in a new partnership involving eBureau and BlueKai, they explained,

> Through this partnership, marketers are no longer forced to make a tradeoff between precision and scale. Because eBureau Custom Audiences are built using tens of thousands of predictive factors to identify ideal customers and new prospects, the addressable audience is dramatically larger than a simple demographic approach. To build a Custom Audience, a marketer defines their customer profile, using input from their historical performance data, customer lists or demographic and/or psychographic criteria. eBureau's predictive analytics platform amasses the client data with eBureau's extensive amount of offline data to define the marketer's unique target market. The results are distilled into a single, custom targeting attribute representing the Custom Audience and made available only to those clients through the BlueKai Platform. (eBureau and BlueKai Partnership Provides New Level of Targeting Precision at Scale for Digital Advertisers 2010)

Drug companies increasingly take advantage of online data collection for the promotion of branded pharmaceuticals for serious illnesses, which also illustrates privacy and consumer protection concerns. Digital marketers tell drug companies they can help manage the "online patient journey," as well as influence prescribing by medical professionals, to spur the demand for branded pharmaceuticals and medical treatments. New forms of "'condition' or 'disease' targeting" are now part of the behavioral advertising's arsenal (Center for Digital Democracy and U. S. PIRG 2009b; Center for Digital Democracy, U.S. PIRG 2010, Consumer Watchdog, and World Privacy Forum 2010).

4.6 Targeting Hits the Road: Mobile Behavioral Profiling

Many of the same consumer data collection techniques that have raised privacy concerns on the Internet have also been brought into the mobile marketplace. Mobile devices, which know our location, are being turned into portable behavioral tracking and tracking tools (Center for Digital Democracy and U.S. PIRG 2009a; Velti 2011).[26] By combining behavioral tracking with our location, digital marketers can promise advertisers that they can influence a consumer regardless of where they happen to be at any given moment. Campaigns are increasingly designed to shape what's called "The New Shopper Journey," including tracking how digital marketing can help move a consumer to a store or make a purchase, document what has been bought (such as through the encouragement of using mobile barcodes to gather price or product information), and then use a range of digital applications to convince them to repeat the process. Mobile marketers are also able to take advantage of social media, offering discounts, coupons, and incentives when users encourage their friends to visit commercial establishments they have "friended" or otherwise endorsed. Left unsaid to mobile consumers, however, is that the data collected about their location and habits can be added to their data profile and sold in real-time to the highest advertising bidder (Constine 2011; Heine 2010; PSFK 2011; Microsoft 2011; Mobclicx 2011).

Mobile marketers have already deployed a dizzying array of targeted marketing applications, involving so-called rich media, mobile video, and games. They have developed an array of standardized techniques designed to foster a user to "opt in" for data-driven advertising and other services, through such formats as "Click-to-video: click initiates an advertiser's video commercial for a product or service; Click-to-SMS: click initiates an SMS for a user to send a keyword to a shortcode to request more Information; and Click-to-storyboard: click transitions to a second interstitial ad (which itself may provide additional actions)" (Mobile Marketing Association 2011).

Mobile advertisers are working together to develop innovative multi-media services that can lure a user into consenting for their information to be used. For example, the recently created Open Rich Media Mobile Advertising (ORMMA) initiative is setting a new standard for the creation and delivery of interactive mobile marketing applications that have a direct impact on privacy (Google 2011c).[27] Helping fuel

[26] As mobile marketer Velti noted in a filing for the U.S. Securities and Exchange Commission, according to ABI Research, mobile marketing and advertising spending is expected to increase from $ 1.64 billion in 2007 to nearly $ 29 billion in 2014. Unlike other media platforms, mobile devices cover a very large installed base and enable access to consumers virtually anytime and anywhere, allowing real-time interaction and engagement. By using a mobile device, campaigns can be further targeted to consumers based on interest, demographic profile, and behavioral characteristics, thereby enabling brands, advertising agencies, mobile operators, and media companies to effectively engage consumers in interactive, measurable advertising, and marketing campaigns. Measure the consumer engagement. Unlike other media platforms, the mobile device is used by the consumer more frequently and over longer periods, providing greater opportunities to generate data on where, when, and how a consumer responds to a marketing or advertising message. Brands, advertising agencies, mobile operators and media companies can leverage this data to motivate a specific consumer action (e.g., a product purchase) at critical moments (e.g., when choosing between products) or at a distinct location (e.g., a nearby retailer).

[27] As the new collaborative project explains, "Mobile Rich Media ad units are mobile compatible ad units with various measurable, interactive options which drive greater brand engagement and messaging across to end-users compared to basic banner ads.... Optionally, the ad unit can capture

the growth of mobile marketing is considerable investment from companies such as Google and Apple, which recently acquired leading mobile ad networks (Takahashi 2010; Wojcicki 2010).

4.7 Surveillance on Social Media

Social media marketing has developed as an extensive but too little scrutinized digital data collection apparatus. Companies such as Facebook suggest that somehow consumers of what they call the "social web" operate with a different set of expectations for privacy. As Facebook recently explained to the Obama administration's Internet Policy Task Force, "certain aspects of the social web... exist precisely because people want to share rather than limiting the sharing of their information to others.... [I]mposing burdensome privacy restrictions could limit Facebook's ability to innovate, making it harder for Facebook to compete in a constantly evolving industry" (Facebook 2011a).

Facebook has been continually pressed to improve privacy practices—especially given its strategy of pushing the limits of using member data. Since the introduction of its now well-known "Beacon" and Facebook Advertising programs in 2007, the social networking leader has been engaged in extensive data mining of its users for advertising purposes (Facebook 2007). Facebook reportedly made $ 1.86 billion from advertising in 2010 (and that excludes revenues from sales of virtual currency) and delivered 1 billion ads on its platform (O'Dell 2011). That's one reason why Facebook is currently expanding its already five-football-field-large data center in the US, and is expanding its Dublin-based EU operations (Letzing 2011, Facebook 2011b).

There are now a host of techniques for "social media marketing" designed to elicit data from users of Facebook and similar sites. New sophisticated "enterprise intelligence" applications have already transformed the nature of marketing and data collection, enabling companies to develop and fine tune far-reaching social media campaigns in real-time. Dozens of companies, with names like Buddy Media, Radian 6, and Rapleaf, vie to provide leading global brands the ability to identify and target the hundreds of millions of users. Techniques have been perfected to identify what are called key "influencers"—individuals whose comments have an impact on the point of view—and buying habits—of many consumers. Algorithms are generated that help target the "influential nodes on networks" and provide the basis for predictive modeling to further implement social media campaigns. New forms of "social contagion" that promote the interests of advertisers are part of this still largely stealth work of the digital marketing industry. Determining the economic value of a Facebook user—especially one considered a "fan" of a brand—is now the subject of research by Nielsen and others. Companies such as Pepsi have established

---

information from the end-user to continue engagements at other times or via other media.... [and] can be dynamically composed so the ad content is targeted to the end-user".

social media "command centers" that operate around the clock monitoring consumer "buzz" about their products, with the ability to respond in real-time to either positive or negative sentiment (Sinan and Walker 2010; Gibs and Bruich 2010; Leskovec 2011; Ostrow 2010).[28]

While Facebook regularly touts its interest in protecting user privacy, its continual changes to its user interface are designed to harvest greater amounts of member data. For example, the Facebook Marketing Bible recently explained how advertisers can take advantage of the data available via Facebook's recent "user profile" redesign:

The December user profile redesign leads users to provide more personal information which can be targeted through Facebook ads.

Previously, personal info was only shown in the secondary Info tab, meaning users and their friends rarely saw it during typical browsing. Users would often go months or years without updating their information to reflect changes in location or employer. Others who only entered the required name, gender, email, and date of birth when signing up for Facebook had little to encourage or remind them to list additional information.

Accurate and plentiful personal information allows advertisers to target users with more relevant ads. Here are the ways in which the new redesign coaxes additional information out of users:

- The Profile Info Summary makes personal info more visible to a user and their friends;
- Users see prompts to add missing information on their own Profile Info Summary;
- The Featured Friends panel prominently displays a user's significant other and family members;
- The enhanced Work and Education section encourages users to add their employers and schools;
- The Likes and Interests section now shows images for each Like;
- The new "Sports You Play" Likes category could become a targeting parameter in the future.

Users can now list additional information about their work, such as projects they've undertaken and friends who helped them, and about their education, such as classes and classmates. This information can be a strong indicator of socioeconomic class (Facebook Marketing Bible 2011). Few users—or regulators—however, are informed about how such changes permit Facebook and its partners to take greater advantage of the wealth of data for marketing purposes (Constine 2010; Kontagent 2011; Refresh Partners 2011).[29]

---

[28] Webtrends, "Analytics," http://www.webtrends.com/Products/Analytics/Facebook.

[29] Social media users are also tracked based on data collected on them that measures their "Viralocity," (viral coefficient), whether they are "social influencers," "daily active users," and other social engagement metrics. One technique used by Facebook and its advertisers to elicit data is "incentivizing social action with rewards." Such techniques can use third parties to install "tracking pixels" on a Facebook page, which "automatically contact and rewards users when pixels are triggered or activity is observed".

Increasingly, marketers are merging insights gathered via behavioral targeting and analysis of users' social media actions. For example, Adobe's Omniture SiteCatalyst, which offers BT tracking, now incorporates a range of "social media analytics" for Facebook marketers, so they can "gain deeper insights into user behavior" and "understand how apps 'go viral' amongst... users" (Smith 2010).

4.8 The Limits of Self-regulation and Voluntary Codes

The threat to privacy of consumers and citizens throughout the digitally connected world grows daily. In the US and the EU, digital marketers have banded together to offer various self-regulatory plans designed to blunt new regulatory safeguards (Dixon 2007).[30] The IAB on both sides of the Atlantic have offered a new self-regulatory system using graphical "icons" to inform online users that data are being collected. The real goals of such a program is to offer a set of self-regulatory privacy principles and an "opt-out" scheme that will blunt the growing support for serious reform designed to protect Internet privacy (EU has Trouble Digesting New Law on Internet Cookies—IAB Europe Offers Solution 2010).

Online advertisers have engaged in self-regulation for more than a decade, with little success. Over the last few years, there have been growing calls by Congress, the FTC, and the public at large for new legislation or rules to regulate commercial online data collection. Sensing that they were losing the battle for digital "hearts and minds," and that the growing concern over privacy threatened their economic self-interest, online advertisers came up with yet another self-regulatory approach. The "Self-Regulatory Principles for Online Behavioral Advertising," offered in 2009 and developed in the US by the IAB and others, has breathed new life into the industry's efforts to oppose new regulation (Interactive Advertising Bureau 2009). Among its more prominent flaws is a failure to protect sensitive information, including data related to finances, health, and families. The woefully inadequate "Sensitive Data Principle" reflects the narrowest range of sensitive information, requiring consent "for the collection of financial account numbers, Social Security numbers, pharmaceutical prescriptions, or medical records about a specific individual for online behavioral advertising purposes" (Interactive Advertising Bureau 2009, 4). The principles likely embraced such a limited definition of sensitive information in order to ensure that consumer data can continue to be collected without consent for the online marketing of financial and health products (as well as from adolescents, racial/ethnic groups, and others who rightly should have their information classified as sensitive). Online marketers in the US spent some $ 1 billion targeting online users seeking medical condition and health-related information last year, and more than

---

[30] Research conducted by the World Privacy Forum on the Network Advertising Initiative (NAI), the US self-regulatory group created in 1999 (and whose members include Google, Microsoft, and other leading players), has documented the limitations of its approach. See also Network Advertising Initiative, "About the NAI," http://www.networkadvertising.org/about/ (viewed 24 Mar. 2011).

$ 2 billion for financial digital advertising during the first half of 2010 alone (Interactive Advertising Bureau 2010b).

Evidon (formerly "Better Advertising"), which implements the new self-regulatory program for the "Digital Advertising Alliance" (the group of marketing trade associations backing the icon plan), says it has created the equivalent of a nutrition food label for online privacy. But, in reality, it is as if that soup label failed to inform a consumer about the salt, fat, and additive content used to make the product. The new self-regulatory approach relies primarily on a triangulated graphical icon that appears on display ads and is called "Ad Choices." The icon generates no information on the actual techniques used to collect data, leaving a user to wonder what it might actually mean to their privacy. The system also fails to address how it can ensure that a consumer will even notice the icon, while they are likely purposefully distracted with various interactive design techniques (such as rich media, online video, and the like). As the Evidon site illustrates, when users click on a Bank of America ad, they first read the following: "This ad has been matched to your interests. It was selected for you based on your browsing activity." A further click of the overlay generates the following headline: "how data powers your experience" (Evidon 2011b).[31]

If a user seeks to learn from Evidon "How Interest-based Advertising Works," one sees a presentation that does not comport many of the techniques used for behavioral and digital marketing nor does the section candidly discuss the privacy and consumer protection concerns. Instead, Evidon uses sanitized statements such as "Some companies collect data and sell it to other companies; being familiar with company privacy policies helps people protect their privacy.... Companies usually provide their own opt-out mechanisms through their web sites. A good place to start is a company's privacy policy" (Evidon 2011c).

If one links to Evidon's section on participating data company BlueKai, a consumer initially sees a description lifted out of obtuse privacy policies: "BlueKai operates an auction based, online data exchange... connecting advertisers to ad networks and data aggregators (online and off)... [which] collects data from online publishers and provides data to advertisers directly or via exchange...." (Evidon 2011a). For those determined to proceed to declare an ad preference, one has to click to learn what profiling categories one was placed in, in order to decide whether to edit them. But missing from this new self-regulatory system are any of the details a company such as BlueKai actually tells its customers—which is a description a consumer deserves to be told. A consumer would learn directly that by using BlueKai, "For the first time in history, advertisers can target individual consumers independent of their media choices...." (BlueKai 2011a). BlueKai provides "... the single, largest source of Intent data qualified by in- market actions and keyword searches in the world. It is real- time data from top tier websites with unique access to purchase, shopping comparison, and product research behavior from their users...." (BlueKai

[31] See also The Self-Regulatory Program for Online Behavioral Advertising, "Welcome to the Online Home of the Self-Regulatory Program for Online Behavioral Advertising," http://www.aboutads.info/ (viewed 24 Mar. 2011).

2011c). With access to "[m]ore than 30,000 data attributes," moreover, "... a marketer defines their customer profile, using input from their historical performance data, customer lists or demographic and/or psychographic criteria" (BlueKai 2010).

A similar set of principles and self-regulatory practices have been deployed as well in the European Union, but like their American cousin, marketers in the EU offer a purposely sanitized fairy-tale version of their "online behavioral advertising" practices. Although many leading companies, including Google, Microsoft, and the aforementioned Blue Kai, signed the 2011 "Transparency and Control for Consumers" document, the same purposefully disingenuous claims offered by US digital marketers are echoed. Online behavioral advertising, the IAB EU claims, is really only about providing the consumer "advertisements on the websites you visit and making them more relevant to your interests." Little or nothing is said about the actual data practices, including information gathered via social media or through neuromarketing tactics, that would actually encourage a consumer to opt-out of targeted marketing. Despite the admirable framework established by the EU protecting data privacy, marketers such as Google continue to tout their ability to track and target EU consumers across the Internet. New forms of self-regulation have not damped the growth of data exchanges in the EU selling users in real-time to advertisers (Cole 2011; Durrani 2011; IAB Europe 2011).[32]

Research on the new self-regulatory system already indicates that few consumers ever proceed with opting out, illustrating its ineffectiveness (Marshall 2010; RESEARCH: Consumers Feel Better about Brands that Give Them Transparency and Control Over Ads 2010). "The pilot test data shows that consumers want to learn more about behavioral advertising but that only a small percentage, once informed, will change their preferences," explained Fran Maier, president of the self-regulatory privacy group TRUSTe. "This low rate of preference change indicates that an effective ad notice may actually increase trust without any negative impact on advertising revenues" (Consumers Find Behavioral Advertising Choices Compelling with TRUSTe TRUSTed Ads Privacy Platform 2010).[33]

But US online marketing companies are worried about the potential impact of the EU's growing consumer privacy framework, including requirements from the new E-Privacy Directive. Google, Microsoft, and Facebook, among others, have proposed that the US engage in negotiations with the EU on consumer privacy that will lead to a revamped "safe-harbor" regime (NTIA 2011). What US online marketers hope to achieve is a new treaty that creates a "separate, but equal" privacy regime, enabling them to conduct business in the EU as unfettered as possible by rules on data collection. This approach argues that if the US enacts a federal privacy law—even a

---

[32] "Your Online Choices," http://www.youronlinechoices.eu/; For a regular report on online ad exchanges in the EU, see ExchangeWire, http://www.exchangewire.com/ (viewed 5 July 2011).

[33] "Consumers Find Behavioral Advertising Choices Compelling With TRUSTe TRUSTed Ads Privacy Platform," Marketwire, 16 Nov. 2010, http://www.marketwire.com/press-release/Consumers-Find-Behavioral-Advertising-Choices-Compelling-With-TRUSTe-TRUSTed-Ads-Privacy-1354242.htm (viewed 16 Feb. 2011).

weak one relying on self-regulation and those twinkling icons—it should be treated as the equivalent of the civil liberties-based EU system (Kennard 2010).[34]

US online companies are especially concerned about the data privacy framework to be chosen for the vital Asia-Pacific region. Throughout Asia, there is a growing population of youthful online users who have enthusiastically embraced mobile phones, social networks, and videogames. Marketers hope that any cross-border data protection agreement made by the Asia-Pacific Economic Cooperation (APEC) economic forum will rely more on the US than on the EU approach to privacy (Asia-Pacific Economic Cooperation 2010; Schweizer 2010).

For many US privacy advocates, the Obama administration has a crucial responsibility to ensure that it respects and supports the EU data framework; that it leads the development of privacy safeguards for the US that match or exceed what has been articulated by the EU; and that it plays a leadership role supporting a privacy policy regime for the Asia-Pacific market that reflects the highest possible standards for consumer privacy protection (Center for Digital Democracy and US PIRG 2011).

But just as during the 1990s, when the online marketing industry initially opposed consumer privacy rules at the FTC, digital advertising companies claim that enacting appropriate privacy safeguards will (as Google puts it), "thwart the ability of companies to develop new services and tools, and in turn make US Internet companies less competitive globally and make the Internet a less robust medium. . . . [A]n anti-innovation framework would counterproductively choke off the development of new tools and services to protect personal privacy" (Google 2011a). The facts—as Google undoubtedly knows—show this not to be the case. First, online marketers did not build serious privacy and consumer protection safeguards into their online marketing products. All the innovation has been, and continues to be, focused on expanding the data collection, profiling, and targeting of users across multiple platforms and applications. Google, Yahoo, Microsoft, ad agencies, and digital marketing companies have significantly invested in creating new forms of digital data collection and new ways to measure it.

Can the digital marketing "ecosystem," as online advertisers have called it, be transformed so it balances the interests and rights of consumers and citizens while it also expands its data collection capabilities? Right now, there are few regulatory or practical impediments to a system that requires people to share greater details about themselves. The lures of technological innovation, entertainment, and convenience—as well as the economic and political clout of the global marketing business—will make truly protecting our privacy a formidable endeavor. But much is at stake, including preserving individual autonomy and assuring freedom of expression, in the outcome of this debate. A new level of candor is required from digital marketers, where they readily identify all the data collection techniques now hidden from the public. Until that time, the citizens and consumers who now rely on the Internet as an essential part of their daily lives will be the unwilling victims of the "Cookie Wars" (Bartz 2011; Interactive Advertising Bureau 2007).

---

[34] See, for example Kennard (2010).

References

Aral, Sinan, and Dylan Walker. 2010. Creating social contagion through viral product design: A randomized trial of peer influence in networks (30 Nov 2010). http://papers.ssrn.com/sol3/papers.cfm?abstract_id=1564856. Accessed 29 Mar 2011.

Asia-Pacific Economic Cooperation. 2010. APEC cross-border privacy enforcement arrangement (CPEA). http://www.apec.org/en/Groups/Committee-on-Trade-and-Investment/Electronic-Commerce-Steering-Group/Cross-border-Privacy-Enforcement-Arrangement.aspx. Accessed 24 Mar 2011.

Barnouw, Erik. 1968. *The golden web: A history of broadcasting in the United States*, vol. 2, 1933–1953. New York: Oxford University Press.

Bartz, Diane. 2011. Google hires 12 lobby firms in wake of FTC probe (Reuters 1 July 2011). http://www.reuters.com/article/2011/07/01/us-google-lobbyists-idUSTRE76056T20110701. Accessed 5 Jul 2011.

BlueKai. 2010. eBureau and bluekai partnership provides new level of targeting precision at scale for digital advertisers (8 Dec 2010). http://www.bluekai.com/newsandmedia_pressreleases_20101208.php. Accessed 24 Mar 2011.

Center for Digital Democracy and U.S. PIRG. 2009a. Complaint and request for inquiry and injunctive relief concerning unfair and deceptive mobile marketing practices. Federal Trade Commission Filing (13 Jan 2009). http://www.democraticmedia.org/current_projects/privacy/analysis/mobile_marketing. Accessed 18 Oct 2010.

Center for Digital Democracy and U.S. PIRG. 2009b. Cookie wars, real-time targeting, and proprietary self learning algorithms: Why the FTC must act swiftly to protect consumer privacy. Comments of the Center for Digital Democracy and U.S. PIRG to the Federal Trade Commission Privacy Roundtables, Project no. P095416 (4 Nov 2009). http://www.ftc.gov/os/comments/privacyroundtable/544506-00013.pdf. Accessed 24 Mar 2011.

Center for Digital Democracy and U.S. PIRG. 2011. CDD and U.S. PIRG urge commerce department to protect consumers online (28 Jan 2011). http://www.democraticmedia.org/information-privacy-and-innovation-in-the-nternet-economy. Accessed 24 Mar 2011.

Center for Digital Democracy, U.S. PIRG, Consumer Watchdog, and World Privacy Forum. 2010. In the matter of online health and pharmaceutical marketing that threatens consumer privacy and engages in unfair and deceptive practices. Complaint, request for investigation, public disclosure, injunction, and other relief: Google, Microsoft, QualityHealth, WebMD, Yahoo, AOL, HealthCentral, Healthline, Everyday Health, and Others Named Below. Federal Trade Commission Filing (23 Nov 2010). http://www.democraticmedia.org/files/u1//2010-11-19-FTC-Pharma-Filing.pdf. Accessed 24 Mar 2011.

Chester, Jeff. 2007. *Digital destiny: New media and the future of democracy*. New York: The New Press.

Chester, Jeff. 2009. Digital dollars: Why the marketing and ad industry are afraid of new regulatory watchdogs (AlterNet, 8 Dec 2009). http://www.alternet.org/economy/144416/digital_dollars%3A_why_the_marketing_and_ad_industry_are_afraid_of_new_regulatory_watchdogs/. Accessed 24 Mar 2011.

Chester, Jeff, and Kathryn Montgomery. 2007. Interactive food & beverage marketing: Targeting children and youth in the digital age (May 2007). http://www.digitalads.org/documents/digiMarketingFull.pdf. Accessed 24 Mar 2011.

Cole, Sally. 2011. Criteo gets great results retargeting audiences at scale with real-time bidding. DoubleClick advertiser blog (1 June 2011). http://doubleclickadvertisers.blogspot.com/2011/06/criteo-gets-great-results-retargeting.html. Accessed 5 July 2011.

Constine, Josh. 2010. Incentivizing social action with rewards. *Facebook marketing bible* (Dec 2010). http://gold.insidenetwork.com/facebook-marketing-bible/?p=2815. Accessed 12 Jan 2011.

Constine, Josh. 2011. Facebook acquires hyper-local mobile advertising startup rel8tion (Inside Facebook, 25 Jan 2011). http://www.insidefacebook.com/2011/01/25/acquires-mobile-advertising-rel8tion/. Accessed 5 July 2011.

Consumers Find Behavioral Advertising Choices Compelling With TRUSTe TRUSTed Ads Privacy Platform. 2010. Marketwire (16 Nov 2010). http://www.marketwire.com/press-release/Consumers-Find-Behavioral-Advertising-Choices-Compelling-With-TRUSTe-TRUSTed-Ads-Privacy-1354242.htm. Accessed 24 Mar 2011.

Datalogix. 2011a. Datalogix taps consumer packaged goods and retail vet David Sommer as general manager of Datalogix CPG (24 Jan 2011). http://www.datalogix.com/assets/files/press/Datalogix-Sommer-final.pdf. Accessed 24 Mar 2011.

Dixon, Pam. 2007. The network advertising initiative: Failing at consumer protection and at self-regulation (World Privacy Forum, 2 Nov 2007). http://www.worldprivacyforum.org/behavioral_advertising.html. Accessed 24 Mar 2011.

Durrani, Arif. 2011. Behavioural ads in Europe to be flagged by icon (Brand Republic, 14 Apr 2011). http://www.brandrepublic.com/news/1065528/Behavioural-ads-Europe-flagged-icon/. Accessed 5 July 2011.

Ebbert, John. 2011. eXelate CEO Zagorski discusses new Datalinx platform and company strategy (AdExchanger.com, 22 Mar 2011). http://exelate.com/new/2011/03/22/exelate-ceo-zagorski-discusses-new-datalinx-platform-and-company-strategy/. Accessed 24 Mar 2011.

eBureau and BlueKai Partnership Provides New Level of Targeting Precision at Scale for Digital Advertisers. 2010. Business wire (8 Dec 2010). http://www.businesswire.com/news/home/20101208005581/en/eBureau-BlueKai-Partnership-Level-Targeting-Precision-Scale. Accessed 24 Mar 2011.

Econsultancy. 2011. Demand-side platforms buyer's guide (purchase required). http://econsultancy.com/us/reports/dsps-buyers-guide.

Enquiro. 2011. White paper summary: Enquiro eye tracking report I: Google. http://pages.enquiro.com/whitepaper-enquiro-eye-tracking-report-I-google.html. Accessed 24 Mar 2011.

EU has Trouble Digesting New Law on Internet Cookies—IAB Europe Offers Solution. 2010. http://www.iabeurope.eu/news/eu-has-trouble-digesting-new-law-on-internet-cookies.aspx. Accessed 22 Nov 2010.

eXelate. 2011a. eXelate launches premier media partnership (18 Jan 2011). http://www.exelate.com/home/inside-press-releases-28.html. Accessed 24 Mar 2011.

Facebook. 2007. Facebook unveils facebook ads (6 Nov 2007). http://www.facebook.com/press/releases.php?p=9176. Accessed 24 Mar 2011.

Facebook. 2011a. Comments regarding commercial data privacy and innovation in the internet economy: A dynamic policy framework (28 Jan. 2011). http://www.scribd.com/doc/47918734/Facebook-Comments-Commerce-Dept-Dynamic-Privacy-Framework. Accessed 24 Mar 2011.

Facebook Marketing Bible. 2011. http://gold.insidenetwork.com/facebook-marketing-bible/?p=3096. Accessed 3 Jan 2011.

Federal Trade Commission. 2007. Federal Trade Commission closes google/doubleclick investigation: Proposed acquisition 'unlikely to substantially lessen competition.' (20 Dec 2007). http://www.ftc.gov/opa/2007/12/googledc.shtm. Accessed 5 July 2011.

Garcia, Bob. 2010. Improving on-site targeting results—engage EMEA 2010. http://www.slideshare.net/WebTrends/engage-emea-2010-improving-onsite-targeting-results. Accessed 24 Mar 2011.

Gibs, Jon, and Sean Bruich. 2010. Nielsen/facebook report: The value of social media ad impressions. Nielsen wire (20 Apr 2010). http://blog.nielsen.com/nielsenwire/online_mobile/nielsenfacebook-ad-report/. Accessed 5 July 2011.

Google. 2011a. Comments regarding Information privacy and innovation in the internet economy. Department of Commerce Filing (28 Jan 2011). http://www.ntia.doc.gov/comments/101214614-0614-01/attachments/FINALCommentsonDepartmentofCommercePrivacyGreenPaper%20(3).pdf. Accessed 24 Mar 2011.

Google Extends VivaKi Partnership. 2010. Warc (8 Nov 2010). http://www.warc.com/LatestNews/News/ArchiveNews.news?ID=27471. Accessed 24 Mar 2011.

Heine, Christopher. 2010. Olay, gap, pepsi run geo-social campaigns—but not on foursquare, ClickZ (22 June 2010). http://www.clickz.com/clickz/news/1721804/olay-gap-pepsi-run-geo-social-campaigns-not-foursquare. Accessed 5 July 2011.

Hutchinson, James. 2011. Big data to get even bigger in 2011. *InfoWorld* (20 Jan 2011). http://www.infoworld.com/d/data-explosion/big-data-get-even-bigger-in-2011-064. Accessed 24 Mar 2011.

IAB Europe. 2011. Europe's online advertising industry releases self-regulation framework (14 Apr 2011). http://www.iabeurope.eu/public-affairs/top-stories/self-regulation-framework.aspx. Accessed 5 July 2011.

Interactive Advertising Bureau. 2007. Key initial deliverables from industry-wide study on marketing and media ecosystem 2010 confirm digital's prominence (23 Oct 2007). http://www.iab.net/insights_research/iab_news_article/64401. Accessed 5 July 2011.

Interactive Advertising Bureau. 2009. Self-regulatory principles for online behavioral advertising (July 2009). http://www.iab.net/media/file/ven-principles-07-01-09.pdf. Accessed 5 July 2011.

Interactive Advertising Bureau. 2010a. IAB and NAI release technical specifications for enhanced notice to consumers for online behavioral advertising (14 Apr 2010). http://www.iab.net/about_the_iab/recent_press_releases/press_release_archive/press_release/pr-041410. Accessed 24 Mar 2011.

Interactive Advertising Bureau. 2010b. Internet ad revenues break records, climb to more than $ 12 billion for first half of 10 (12 Oct 2010). http://www.iab.net/about_the_iab/recent_press_releases/press_release_archive/press_release/pr-101210. Accessed 24 Mar 2011.

Kawaja, Terence. 2010. The science-ification of media. http://www.slideshare.net/tkawaja/the-scienceification-of-media. Accessed 24 Mar 2011.

Kennard, William E. 2010. Data protection in a transatlantic perspective. Remarks of William E. Kennard, U.S. ambassador to the EU before the Committee on Civil Liberties, Justice, and Home Affairs (25 Oct 2010). http://www.europarl.europa.eu/document/activities/cont/201010/20101027ATT90670/20101027ATT90670EN.pdf. Accessed 24 Mar 2011.

Lancelot, James. 2009. Cookie wars: How audience targeting is creating intense competition for cookies (AdExchanger.com, 1 Oct 2009). http://www.adexchanger.com/data-driven-thinking/cookie-wars/. Accessed 23 Oct 2009.

Lau, Adaline. 2011. A U.S. DSP seeks its fortunes in Asia, ClickZ (10 Mar 2011). http://www.clickz.com/clickz/news/2033322/dsp-seeks-fortunes-asia. Accessed 24 Mar 2011.

Leskovec, Jure. 2011. KDD 2011 tutorial: Social media analytics—tracking, modeling and predicting the flow of information through networks (Stanford University, 21 Aug 2011). http://snap.stanford.edu/proj/socmedia-kdd/index.html. Accessed 29 Mar 2011.

Letzing, John. 2011. Facebook data center is boon for Oregon town. *Wall Street Journal* (21 Jan 2011). http://online.wsj.com/article/SB10001424052748704881304576094222157412808.html. Accessed 24 Mar 2011.

McChesney, Robert. 1995. *Telecommunications, mass media, and democracy: The battle for the control of U.S. broadcasting (1928–1935)*. New York: Oxford University Press.

MagnaGlobal. 2010. Global advertising forecast 2010 (6 Dec 2010). http://www.magnaglobal.com/magnaglobal-news/global-advertising-forecast-december-2010. Accessed 24 Mar 2011.

Marshall, Jack. 2010. Few opt out of behavioral ads, ClickZ (20 Dec 2010). http://www.clickz.com/clickz/news/1933561/opt-behavioral-ads. Accessed 24 Mar 2011.

Matias, Yossi. 2011. Games, auctions and beyond (Google research blog, 16 Mar 2011). http://googleresearch.blogspot.com/2011/03/games-auctions-and-beyond.html. Accessed 24 Mar 2011.

Microsoft. 2010. Microsoft advertising strengthens Asia pacific team (26 Aug 2010). http://advertising.microsoft.com/asia/NewsAndEvents/PressRelease.aspx?pageid=2592&Adv_PressReleaseID=1326. Accessed 24 Mar 2011.

Montgomery, Kathryn, and Jeff Chester. 2009. Interactive food & beverage marketing: Targeting adolescents in the digital age. *Journal of Adolescent Health* 45 (3): S18–S29.

Montgomery, Kathryn, Sonya Grier, Jeff Chester, and Lori Dorfman. 2011. A conceptual framework for food marketing in the digital age (Unpublished manuscript).
Netezza. 2009. Media innovation group case study (2009). http://www.netezza.com/documents/MIG_CaseStudy.pdf. Accessed 24 Mar 2011.
NTIA. 2011. Information privacy and innovation in the internet, docket # 101214614–0614-01: Comments of google inc. (28 Jan 2011). http://www.ntia.doc.gov/comments/101214614-0614-01/comment.cfm?e=10FE3003-691B-4E2E-9685-87D7DB413C1D. Accessed 24 Mar 2011.
O'Dell, Jolie. 2011. Facebook's ad revenue hit $ 1.86b for 2010. (Mashable, 20 Jan 2011). http://mashable.com/2011/01/17/facebooks-ad-revenue-hit-1-86b-for-2010/. Accessed 24 Mar 2011.
Ostrow, Adam. 2010. Inside Gatorade's social media command center. (Mashable, 15 June 2010). http://mashable.com/2010/06/15/gatorade-social-media-mission-control/. Accessed 5 July 2011.
Peppers, Don, and Martha Rogers. 1999. *The one to one future*. New York: Random House.
PSFK. 2011. Future of mobile tagging. http://www.psfk.com/future-of-mobile-tagging. Accessed 5 July 2011.
Research: Consumers Feel Better about Brands that Give Them Transparency and Control Over Ads. 2010 Evidon's corporate blog (3 Nov 2010). http://blog.evidon.com/2010/11/10/research-consumers-feel-better-about-brands-that-give-them-transparency-and-control-over-ads/. Accessed 24 Mar 2011.
Schweizer, Kristen. 2010. Asia-Pacific to pass North America as biggest ad market in 2014. (Bloomberg, 13 Sept 2010). http://www.bloomberg.com/news/2010-09-12/asia-pacific-to-pass-north-america-as-biggest-ad-market-in-2014.html. Accessed 24 Mar 2011.
Singer, Natasha. 2010. Privacy groups fault online health sites for sharing user data with marketers. *New York Times* (23 Nov 2010). http://www.nytimes.com/2010/11/24/business/24drug.html. Accessed 5 July 2011.
Smith, Justin. 2010. Analytic tools for developers. *Facebook marketing bible* (Sept 2010). http://gold.insidenetwork.com/facebook-marketing-bible/?s=Third-Party+Facebook+Platform+Analytics+Providers. Accessed 12 Jan. 2011.
Starr, Paul. 2005. *The creation of the media: Political origins of modern communication*. New York: Basic Books.
Takahashi, Dean. 2010. Apple to buy Quattro wireless mobile ad firm for $ 275 M (VentureBeat, 4 Jan 2010). http://venturebeat.com/2010/01/04/apple-to-buy-quattro-wireless-mobile-firm-for-275m/. Accessed 24 Mar 2011.
Velti. 2011. SEC filing pursuant to rule 424(b)(1), registration no. 333–166793. http://www.sec.gov/Archives/edgar/data/1490412/000104746911000342/a2201716z424b1.htm#ea45601_business. Accessed 24 Mar 2011.
Wojcicki, Susan. 2010. We've officially acquired admob! (official google blog, 27 May 2010). http://googleblog.blogspot.com/2010/05/weve-officially-acquired-admob.html. Accessed 24 Mar 2011.

Websites Consulted

Adnetik. 2011. How it works. http://adnetik.com/how-it-works/. Accessed 24 Mar 2011.
Advertising Research Foundation. 2010. 360 media and marketing council. http://www.thearf.org/assets/360-media-council. Accessed 24 Mar 2011.
Advertising Research Foundation. 2011. The ARF inaugural neurostandards retreat. http://www.thearf.org/assets/neurostandards-meeting. Accessed 24 Mar 2011.
AlmondNet. 1998. http://www.almondnet.com/Home.aspx. Accessed 24 Mar 2011.
AlmondNet. 2010. AlmondNet partners with invite media (25 May 2010). http://findarticles.com/p/articles/mi_m0EIN/is_20100525/ai_n53774289/. Accessed 24 Mar 2011.
Behavioural Targeting. 2009. Advertising.com. http://uk.advertising.com/publishers/behavNetwork.php. Accessed 13 Oct 2008.

Bizo. 2011. Bizo membership agreement. http://www.bizo.com/partner/membership_terms. Accessed 24 Mar 2011.

BlueKai. 2011a. About us. http://www.bluekai.com/aboutus.php. Accessed 24 Mar 2011.

BlueKai. 2011b. The bluekai exchange. http://www.bluekai.com/exchange.php. Accessed 24 Mar 2011.

BlueKai. 2011c. Intent data. http://www.bluekai.com/intentdata.php. Accessed 24 Mar 2011.

BlueKai. 2011d. Jobs: Client service manager. http://www.bluekai.com/aboutus_jobs.php#account_executive_chicago_nyc. Accessed 24 Mar 2011.

Cadreon. 2011. http://www.cadreon.com/. Accessed 24 Mar 2011.

Datalogix. 2011b. http://affiniti.datalogix.com/. Accessed 24 Mar 2011.

Datalogix. 2011c. Data append. http://nextaction.datalogix.com/index.php?id=93. Accessed 24 Mar 2011.

Datalogix. 2011d. DLX platform. http://affiniti.datalogix.com/what-is-dlx-platform. Accessed 24 Mar 2011.

DoubleClick. 2011. Rich media gallery. http://www.google.com/doubleclick/gallery/features/data_capture.html. Accessed 24 Mar 2011.

Evidon. 2011a. About bluekai. http://info.evidon.com/companies/bluekai. Accessed 24 Mar 2011.

Evidon. 2011b. Build trust. Grow your business. http://www.evidon.com/solutions/overview. Accessed 24 Mar 2011.

Evidon. 2011c. How interest-based advertising works. http://info.evidon.com/about_behavioral_advertising/section1?n=103. Accessed 24 Mar 2011.

eXelate. 2011b. Data 101 FAQs. http://www.exelate.com/home/advertiser-data-101-faqs.html. Accessed 24 Mar 2011.

Experian. 2011. Audience IQ for customer and website experience. http://www.experian.com/marketing-services/customer-experience.html?cat1=marketing-services&cat2=digital-advertising. Accessed 24 Mar 2011.

Experian Interactive. 2011. http://www.experianinteractive.com/. Accessed 24 Mar 2011.

Facebook. 2011b Careers: Dublin, Ireland. http://www.facebook.com/careers/department.php?dept=dublin. Accessed 24 Mar 2011.

Facebook. 2011c. Preferred developer consultant program. http://developers.facebook.com/preferreddevelopers/. Accessed 5 July 2011.

Facebook Developers. 2011. Preferred developer consultant program. http://developers.facebook.com/preferreddevelopers/. Accessed 24 Mar 2011.

Google. 2011b. Head of display, google display media team—Mexico City. http://www.google.com.mx/jobs/adsales/head-of-display-google-display-media-team-mexico-city/index.html. Accessed 24 Mar 2011.

Google. 2011c. Ormma: Description of issues and solutions. http://code.google.com/p/ormma/wiki/Description. Accessed 24 Mar 2011.

Google Research. 2011. Google and WPP marketing research awards. http://research.google.com/university/marketingresearchawards/. Accessed 24 Mar 2011.

Interactive Advertising Bureau. 2001. Glossary of interactive advertising terms v. 2.0. http://www.iab.net/media/file/GlossaryofInteractivAdvertisingTerms.pdf. Accessed 24 Mar 2011.

Interactive Advertising Bureau. 2011a. International IABs. http://www.iab.net/about_the_iab/international_iabs. Accessed 24 Mar 2011.

Interactive Advertising Bureau. 2011b. Networks & exchanges quality assurance guidelines. http://www.iab.net/ne_guidelines. Accessed 24 Mar 2011.

IAB UK. 2011. Jargon buster. http://www.iabuk.net/en/1/glossary.html. Accessed 24 Mar 2011.

IXI Corporation. 2011a. AudienceIXInsights. http://www.ixicorp.com/ixi-digital/solutions-for-advertisers-and-agencies/audienceixinsights/. Accessed 24 Mar 2011.

IXI Corporation. 2011b. IXI digital targeting options. http://www.ixicorp.com/ixi-digital/ixi-digital-targeting-options/. Accessed 24 Mar 2011.

IXI Corporation. 2011c. Solutions for advertisers and agencies. http://www.ixicorp.com/ixi-digital/solutions-for-advertisers-and-agencie. Accessed 24 Mar 2011.

Kontagent. 2011. The kontagent fact sheet. http://www.kontagent.com/about/. Accessed 24 Mar 2011.

M&A & Venture Capital. 2011. paidContent.org. http://paidcontent.org/topic/ma-venture-capital/. Accessed 24 Mar 2011.

Microsoft. 2011. The new shopper journey. http://advertising.microsoft.com/how-shoppers-use-media?uuid=d0c69450-6786-4dcb-ba77-56dc46402e6f. Accessed 5 July 2011

Microsoft Advertising. 2011. Glossary of terms. http://advertising.microsoft.com/uk/glossary-of-terms. Accessed 24 Mar 2011.

Mobclicx. 2011. About us. http://www.mobclix.com/company. Accessed 5 July 2011.

Mobile Marketing Association. 2011. Mobile advertising guidelines. http://www.mmaglobal.com/mobileadvertising.pdf. Accessed 24 Mar 2011.

Omniture. 2011. The adobe online marketing suite, powered by omniture. http://www.omniture.com/en/products/online_marketing_suite. Accessed 5 July 2011.

Refresh Partners. 2011. Refresh analytics: Facebook application demographics. http://refreshpartners.com/products. Accessed 24 Mar 2011.

R. L. Polk & Co. 2011a. Data enhancement services. http://usa.polk.com/Industries/Dealers/Communicate/DataEnhance/. Accessed 24 Mar 2011.

R. L. Polk & Co. 2011b. Profiling. http://usa.polk.com/Industries/Finance/Analyze/Profiling/. Accessed 24 Mar 2011.

Chapter 5
The Data Mining Balancing Act

Tal Z. Zarsky

5.1 Introduction: The Lure and Confusion of Governmental Data Mining

Governments are facing new and serious risks when striving to assure the security and safety of their citizens. Perhaps the greatest concern is the fear of terrorist attacks. Various technological tools are being used or considered as means to meet such challenges and curb these risks. Of the tools discussed in the political and legal sphere, data mining applications for the analysis of personal information have probably generated the greatest interest. The discovery of distinct behavior patterns linking several of the 9/11 terrorists to each other and other known operatives has led many to ask: What if data mining had been applied in advance? Could the attacks and their devastating outcomes have been avoided?

Data mining has captured the imagination as a tool which can potentially close the intelligence gap constantly deepening between governments and their new targets—individuals posing a risk to security and the public's well-being (Jonas and Harper 2006; Schneier 2006).[1] Data mining is also generating interest in other governmental contexts, such as law enforcement and policing. In recent years, law enforcement has shifted to "Intelligence Led Policing ("ILP"; Cate 2008). Rather than merely reacting to events and investigating them, law enforcement is trying to preempt crime. It does so by gathering intelligence, which includes personal information, closely analyzing it, and allocating police resources accordingly—all tasks which could be enhanced by data mining technology (IBM 2010).[2] The growing appeal of data mining in all these contexts results from similar reasons—the development of cutting edge technologies, advances in mathematics,

[1] For a countering view, see Jonas and Harper (2006). See also commentary Schneier (2006).

[2] For a paper discussing these initiatives in the Netherlands, see van der Veer et al. (2009).

T. Z. Zarsky (✉)
Faculty of Law, University of Haifa, Mount Carmel, Haifa, Israel
e-mail: tzarsky@law.haifa.ac.il

S. Gutwirth et al. (eds.), *European Data Protection: In Good Health?*,
DOI 10.1007/978-94-007-2903-2_5,

statistics, and computer science, and the sinking costs of the hardware, software, and manpower needed for their implementation (Zarsky 2002–2003).[3] The reports on the success of data mining in predicting human behavior (Ayres 2007; Baker 2008)[4] in the commercial realm have also strengthened these models' appeal.

It thus should come as no surprise, that in the United States, data mining initiatives are popping up everywhere. A recent GAO report indicates current data mining initiatives in a broad array of contexts (U.S. General Accounting Office 2004). DARPA has famously promoted the Total (later changed to "Terrorist") Information Awareness ("TIA") Program—an ambitious project which planned to analyze vast amounts of personal information from governmental and commercial sources. This project was catastrophically handled in terms of public relations. Public concerns and outrage led to Congressional intervention and the project's quick demise (Cate 2008, 441). However, it is broadly understood that similar projects are living on, under different names and acronyms.

The reaction to the data mining of personal information by governmental entities came to life in a flurry of reports, discussions, and academic papers. The general notion in these sources, as well as the one in the public sphere[5] is that of fear and even awe. Information privacy, which many feel is under constant attack in both the commercial and governmental realm, seems to be utterly compromised. Many share the visceral feeling that the outcome of data mining analyses, which enable the government to differentiate among individuals and groups in novel ways, is extremely problematic. The quick demise of the TIA program serves as a case in point.

Understanding what stands behind this strong visceral response is a difficult task. Even though governmental data mining is extensively discussed in recent literature (Cate 2008; Ramasastry 2004; Slobogin 2008; Solove 2008), an overall sense of confusion is ever present. Given the fact that data mining will probably prove necessary (or a "necessary evil" for some) scholars have moved to examine whether the problems it generates could be mitigated and how its risks and benefits should be balanced. While mapping out these matters, scholars as well as policy makers will be required to further establish which paradigms of legal thought are most fitting to address these matters. For that, they will examine constitutional law, privacy law, anti-discrimination law, and other matters. Yet as this discourse unfolds, something is still missing. An important, yet often overlooked, methodological step must be part of the inquiry mentioned above—the adequate consideration of *alternatives*. Scholars and policy makers swiftly point out the troubles of data mining as well as the dangers of ignoring it. Yet they are not equally quick to consider the detriments and shortcomings of alternatives which will surely be applied by governments setting data mining aside. Understanding the importance of this analytical step follows

[3] For a discussion of the building blocks of data mining, see Zarsky (2002–2003).

[4] Such success has been recently detailed in several popular books—see Baker (2008).

[5] This outcome is interesting, as stories related to privacy in general have generated limited interest, less they involve an actual catastrophe—personal data about a judge blocks his nomination, information regarding the address of an actress leads to her murder, and many other examples. Yet the data mining stories here addressed focus on potential harms, which have yet to materialize. This outcome tells an interesting story about the risks of data mining.

from acknowledging that the challenges bringing data mining to the forefront of our discussion are not going away. Governments must address new security and law enforcement challenges and pressure to take action. They must also face the challenges of optimally utilizing the vast volumes of personal information at their disposal. Considering alternatives is also helpful in sharpening our understanding of the benefits, determinants, traits and qualities of data mining itself.

This chapter strives to bring the methodology of examining alternatives to the attention of academics and policy makers. It provides basic tools for engaging in this important analytic exercise. To do so, the chapter proceeds as follows: In this section, it briefly demonstrates and defines what the governmental data mining initiatives are. This is a crucial step, as the term "data mining" has almost taken on a life of its own, and is applied in several, at times contradictory, ways. The chapter also notes specific unique traits of these practices, while focusing on the distinct roles of humans and machines. These will prove constructive later, when striving to understand how it differs from its alternatives. The next Sect. 5.2 maps out, with a very broad brush, the various concerns data mining generates while drawing from the ongoing literature regarding this issue. The last Sect. 5.3 introduces four alternative strategies of personal data usage and management (or lack thereof) for achieving the governmental objectives of security and law enforcement. It also addresses an additional strategy (contemplated by policy makers and think tanks) for using a specific form of data mining while anonymizing the data. In the second segment of this section, I sharpen the distinctions between the central alternatives, so to promote a better understanding of their advantages and shortcomings.

The discussion of data mining and its alternatives goes beyond the actions of government. Private entities are using similar techniques to distinguish among their actual or prospective clients/customers, while analyzing personal behavior. These practices are applied by advertisers, marketers, management and in even more questionable settings, banks credit card issuers and insurance companies (Scism and Maremont 2011). While this context is important, it is beyond our current scope. It should be noted, however, that the rationales and internal balances discussed in the governmental context cannot be applied directly to the private sector. With private firms, competitive forces (when these indeed exist) might play an important role in achieving some of the needed objectives.[6] However, these differences and their implications must be explored elsewhere.

Finally, although the paper claims to merely make a methodological contribution, I confess to arguing a normative point between the lines. While I do not carry through a full analysis of the pros and cons of the data mining strategies, my sense is that when taking the full scope of alternatives into account, data mining is far less problematic than when considered at first blush. The problems data mining brings to mind persist,

[6] In some instances, the services rendered are not essential, thus allowing for consumer choice—an option which requires rethinking many of the elements to be addressed below. Finally, the obligations and motivations of governmental entities are different than their commercial counterparts, thus altering the internal calculus leading to the final recommendations.

and with greater force, when applying other options. Understanding this point might lead policy makers to reconsider the overall negative treatment data mining options receive in many circles. Furthermore, data mining indeed presents difficult challenges, yet these might not be the ones which intuitively come to mind—an insight which calls for further contemplation and analysis.

5.1.1 Data Mining: In Theory and in Practice

5.1.1.1 Data Mining: Definitions, Processes, and General Terms[7]

The term "data mining" has recently been used in several contexts by policy makers and legal scholars. For the discussion here, I revert to a somewhat technical definition of this term of art. Here, data mining is defined as the "*nontrivial process of identifying valid, novel, potentially useful and ultimately understandable patterns in data*." Even within this definition, there are several intricacies. The term "data mining" refers to both "subject based" and "pattern based" searches (Cate 2008; Slobogin 2008, 323).[8] The former refers to database searches of and for specific individuals, events and predetermined patterns. However, the core of this chapter focuses on the latter forms of analysis (also referred to as "event-based" data mining). These methods provide for a greater level of automation, and the discovery of unintended and previously unknown information. Such methods can potentially generate great utility in the novel scenarios law enforcement and intelligence now face—where a vast amount of data is available, yet there is limited knowledge as to how it could be used and what insights it might provide.

In "pattern based analyses," the analysts engaging in data mining do not predetermine the specific factors the analytical process will use at the end of the day. They do, however, define the broader datasets which will be part of the analysis. Analysts also define general parameters for the patterns and results which they are seeking and could be accepted—such as their acceptable level of error. Thereafter, the analysts let the software sift through the data and point out trends within the relevant datasets, or ways in which the data could be effectively sorted (Zarsky 2002–2003).[9] The data mining process could achieve both descriptive and predictive tasks. Descriptive data mining provides analysts with a better understanding of the information at their disposal, while uncovering hidden traits and trends within the dataset. When applied by law enforcement to vast databases of personal information, such analyses can uncover disturbing behavior patterns, and assist in ongoing investigation to find criminals and terrorists they are already seeking. While these practices generate concerns, this paper focuses on the use of the data mining of personal information for

[7] Since the matters addressed here were drawn out elsewhere, the analysis is brief. For a more in-depth discussion, see DeRosa 2004; Zarsky (2002–2003). See also Taipale (2003).

[8] For a discussion regarding the distinction among the two—see Cate (2008).

[9] For a discussion as to how these data mining techniques are carried out see, Zarsky (2002–2003).

predictive modeling and analysis—an issue which generates far more interest (and subsequent fear).

In a predictive process, the analysts use the data mining application to generate rules based on preexisting data. Thereafter, these rules are applied to newer (while partial) data which is constantly gathered and examined, as the software constantly searches for previously encountered patterns and rules. Based on new information and previously established patterns, the analysis strives to predict outcomes prior to their occurrence (while assuming that the patterns revealed in the past pertain to the current data as well). In the law enforcement and national security context, such insights can prove quite helpful—at times allowing for sufficient reaction time before it is too late.

5.1.1.2 Data Mining: Automation and the Human Touch

When considering the detrimental aspects of data mining, the automated nature of the process quickly comes to mind. Therefore, it is important to address the extent of automation and human influence in this process. Counter to what one might initially believe, even with predictive data mining, the role of the human analyst and her discretion is quite extensive. For example, the dataset must be actively constructed, at times by bringing together data from various sources. The analysts also predefine the parameters of the search.[10] These actions directly impact the outcome of the process, and thus policy.

The extent of human discretion involved in this process is not a factor set in stone. Rather, it is a result of various policy decisions. For instance, it is impacted by whether the process is *interpretable* or *non-interpretable*. As this term is not commonly visited in the literature, I will devote a few lines to address it. With a non-interpretable process in place, the actions premised upon the predictions the data mining process provides are not necessarily explainable to humans. Namely, the software makes its decisions based upon multiple variables that were learned throughout the data analysis. This process is not easily reduced to comprehensible human language. Therefore, applying non-interpretable schemes affects the role and discretion of the analysts. In non-interpretable processes, human discretion is minimized to setting the parameters for generating predictive algorithms ex ante. The subsequent process of sorting objects, events or people is carried out automatically, with minimal human oversight. Yet perhaps the greatest effect on the role of the human comes after the fact. When a process is non-interpretable, it is very difficult to provide an answer as to why a specific result was reached beyond stating that this is what the algorithm found based on previous similar cases in the past.[11]

---

[10] This is done both in advance, and after the fact, by "weeding out" results she might consider as random, wrong or insignificant.

[11] I was told by data mining experts that this is usually the case with face and image recognition software.

The flip side of these processes would be a fully interpretable analysis—one that uses a limited number of factors, which in turn could be reduced to a human-language explanation. With interpretable results, an additional stage could be added in which the analyst works through the patterns and criteria set forth by the computer algorithms for the prediction tasks. These could be indications of higher risk associated with individuals of a certain height, age, specific credit or purchasing history—and, of course, the interaction of all these factors. With an interpretation in hand, the analysts can track and set aside factors and patterns which they find offensive, ridiculous and problematic. In addition, the analyst could provide a response to inquiries as to what initiated special treatment of an event or individual. The interpretation process would no doubt prove costly, both in terms of additional expenses for analysts, efficiency and effectiveness lost in the process. However, it provides advantages in terms of accountability and transparency.

Providing for an interpretable process also enables an additional level of human scrutiny in the predictive data mining dynamic. If analysts have a good grasp of the elements used, they can further seek out a theory of causation. Such a theory would go beyond the mere correlation data mining reveals and seek out explanations as to why these are proper indicators[12] beyond the notion that they merely "work." This step as well can prove helpful in weeding out ridiculous and random findings. It can also block practices which resemble problematic (or even illegal) discrimination.

To summarize, this segment provided a broad overview of the meaning and use of data mining when applied to the analysis of personal information by governments. It also briefly clarifies the extent of human discretion and computer automation. The entire discussion is, however, premised on an underlying assumption that data mining tools are effective in achieving their analytical objectives, while maintaining an acceptably low-level of false positives and negatives. Whether this is indeed true is currently hotly debated (Jonas and Harper 2006; Schneier 2006), and notoriously difficult to measure. The answer to these questions will depend on context, as well as on the costs, consequences and levels of false positives and false negatives. Therefore, prior to engaging in data mining, a relevant authority must conduct an assessment of the effectiveness of the data mining process (TAPAC 2004). If such analysis indicates that data mining schemes are doomed to technical and operational failure, data mining must be abandoned. However, the analysis presented below is premised upon the contrary assumption—that data mining indeed works, and at times even too well.

5.2 The Fears and Challenges of Governmental Data Mining

Data mining presents vast opportunities for bridging the gap between the government's informational needs and the vast datasets of information at its disposal. With data mining, such data could be transformed into knowledge. However, these practices generate a variety of concerns. These concerns, in turn, are now requiring policy

[12] However, "building" a theoretical justification to a statistical correlation is usually easy and merely requires some imagination. Thus, one can easily question the extent of protection from arbitrary results a call for "causation" provides.

makers and courts to engage in an extensive discussion and analysis. A discussion of these matters splinters quickly into a multitude of claims and counterclaims. Fully addressing all these issues is beyond the confines of this (or any) article. For that reason, this chapter focuses on a specific methodological point which must be applied in every one of the data mining contexts—*addressing alternatives*.

Yet, in the interest of providing context, this segment maps out the specific analytical junctures where data mining is challenged. It is at these points where addressing alternatives is crucial. This analytic mapping relies upon scholarship and policy reports addressing such matters in the last few years. For the sake of clarity, I distinguish among the different steps of personal information flow such as the collection and analysis stage and the usage of personal data.[13]

The following description is mostly theoretical and normative, with only limited attention provided to positive law. I chose this analytical path for several reasons: First, temporarily setting aside the positive analysis allows for quickly working through the relevant issues, and leaving room for an in-depth discussion of the alternatives below. As the law greatly differs among jurisdictions, a full-blown analysis of positive law would be lengthy and complex. Second, to a great extent, the legal and policy standing on these issues is still up for grabs. In the United States, most of these issues have not been decided upon in the courts and are waiting for regulation and legislation. They probably do not amount to breaches of constitutional rights—or as Daniel Solove succinctly summarized—"... data mining often falls between the crevices of constitutional doctrine" (Solove 2008, 355). They are also probably permitted according to current privacy laws in view of various exceptions and loopholes. Yet public opinion and various policy groups do not approve these practices (Slobogin 2007, 194)[14]—and thus some changes in the law are inevitable. In Europe, the entire legal structure governing privacy and data protection within government is being revamped as part of the Lisbon Treaty's aftermath. Yet new policy will surely follow as both privacy and data protection are recognized as basic human rights. Therefore, a discussion at a high level of abstraction is still fitting.

Collection and Analysis A data mining process inherently calls for automatically reviewing and analyzing profiles filled with personal information regarding many different individuals. This process will be carried out without their consent to such analyses. The data used was previously collected by either government or commercial entities. It is hard to imagine that individuals' conceded to the data mining process here described at the time of collection, or at a later stage. If the information was collected by government, citizen might not have conceded to data collection at all. Rather, they were forced to provide their data and settle for a basic and vague notice of the collection and future uses provided by the government.[15]

---

[13] Transparency is an additional category which requires scrutiny and discussion, yet it calls for a very different form of analysis. For more on this issue, see Zarsky (2012).

[14] For an empirical study pointing in this direction, see Christopher Slobogin (2007).

[15] In the United States, such rights are governed by the Privacy Act, which call for the publication of SORNs to notify the public of such uses. For more on this, see the Privacy Act Overview of 2010, accessed July 12, 2011, http://www.justice.gov/opcl/1974indrigacc.htm.

Engaging in personal data analysis without the direct consent of relevant data subjects runs counter to several legal concepts in the context of privacy and data protection. First, such actions might constitute searches (Blitz 2011; Slobogin 2010).[16] If so, data mining will be considered an illegal search when carried out without sufficient judicial approval—approval which is not currently sought. According to other privacy theories, which are more central in European thought, data mining without prior consent constitutes a violation of the realm of control individuals have over their personal information (Solove and Schwartz 2006; Westin 1967).[17] The information is also analyzed and used outside the original context in which it was collected, thus violating the principles of "Contextual Integrity" set forth by Nissenbaum to describe proper information uses and flows (Nissenbaum 2009). Currently, under US law at least, such practices are permitted if the data were collected legally and a very general and vague notice is provided (TAPAC 2004).

On a more pragmatic level, these vast analyses projects might generate a "chilling effect" with regard to many important human activities and behaviors; if citizens fear that specific actions will generate additional governmental scrutiny, they will refrain from these actions—such as travel, communications or consumption—even when they are legal and at times socially beneficial (Cate 2008; Solove 2001; Strandburg 2008).[18] From a somewhat different perspective, knowledge of such actions impedes upon the citizens' autonomy; it does not allow them to develop their "self" to the greatest extent possible.

Finally, even if these practices are justifiable in one context, such as that of homeland security, there is the fear that government and its agents will not stop there. Equipped with these powerful tools and immense datasets, they will use them for other, more mundane, objectives. While data mining could be justified to protect citizens from upcoming risks which might lead to devastating outcomes, it probably cannot be justified as a tool for locating deadbeat dads. This is the "Project/Function Creep" concern, which has many commentators and policy makers worrying. This concern might lead to recommendations that data mining projects should be stricken down in their entirely (Slobogin 2008, 326).

---

[16] This is not the classic understanding of a "search," which does not pertain to searches of data which were already collected. However, newer theories reexamining the "search" terminology question such wisdom. Slobogin, for instance, believes the term should be used in the same way the public understands it. According to his empirical studies, that includes data mining. Mark Blitz is also examining whether searches within data or other sources the government obtained lawfully could be considered a "search," nonetheless, while focusing on DNA samples.

[17] The notion of "privacy as control" was set forth by Alan Westin and implemented in various elements of both the OECD Principles and the EU Data Protection Directives. See generally Westin (1967); on the EU Data Protection Directives in general, see Solove and Schwartz (2006).

[18] For a discussion of this argument in the Data Mining context, see Cate (2008) who notes it as perhaps the most powerful one in this context. Strandburg makes a similar argument, while pointing out that in some contexts data mining might impede on US Constitutional First Amendment Rights, such as freedom of speech and association. For a general discussion of privacy and autonomy, see Solove (2001).

Usage Using the knowledge derived from the data mining process for various governmental objectives generates an additional set of problems. A basic concern is that the outcomes will be used to unfairly discriminate among citizens. Discrimination could prove unfair for a variety of reasons: it could be premised (at times, tacitly) upon unacceptable social factors, such as race and nationality. It could also be premised upon partial information, or immutable factors, over which individuals have no control. In addition, some might object to distinguishing among individuals based on mere correlations with others (who might have committed wrongdoings), as opposed to the specific thoughts and actions of the relevant individual. This is the generalized/individualized suspicion distinction some scholars have already considered (Slobogin 2007, 40).[19] I am currently unaware of specific laws addressing discrimination by governmental (Harcourt 2007; Schauer 2006)[20] data mining in the United States (beyond the protection provided through the Equal Protection Clause to all). In the EU, specific rules governing automated searches might apply, and indeed provide individuals with additional rights to learn the internal process used (Korff 2011).[21]

An additional concern often mentioned when addressing the data mining process, is that it is ridden with errors. These errors can be of different forms and come at various stages of the process: they can result from errors in the initial data, in the aggregation process,[22] as part of the statistical modeling and computer programming, in the implementation of the system or in the ability to correctly define the risks and match them to the strategies on the ground. The errors can have devastating outcomes. First, they can render the entire process ineffective and inefficient—unable to identify real risks while leading law enforcement to follow bogus leads. Yet even when setting these concerns aside (and assuming they can be tested), errors can have detrimental effects on specific individuals; these might be subjected to discomfort, additional scrutiny and even castigation and suspicion by others for no real reason.

It should be noted that data mining tools maintain the ability to self-correct errors in the analysis process. As the process rolls on, information regarding success rates, false positives and false negatives becomes available and is "fed" into the process. Analysts can use such data to fine-tune the algorithms they later apply. In addition, data mining techniques could be used to study the datasets and seek out information which does not fit other data patterns. Analysts could then examine whether anomalies in the data result from errors and correct the database accordingly.

Finally, lack of knowledge and understanding of the data mining internal processes might also raise fears related to "due process" (Steinbock 2005)—or lack thereof. Individuals might fear that adverse action was or will be taken against them without

---

[19] For a discussion and critique of this distinction, see Slobogin (2007).

[20] I intentionally emphasize the lack of laws in the governmental realm. In the commercial realm there is some reference to this issue in the Fair Credit Reporting Act. For a critique of this situation and a call for a change, see Harcourt (2007). For a very different perspective, see Schauer (2006).

[21] For a full discussion of this issue in EU law (as well as the law in the various states) see an excellent discussion in Korff (2011)

[22] For a discussion of errors in general and of this context in particular, see Ramasastry (2004).

their ability to examine the reasons or challenge the allegations. The data mining process might be inherently opaque and its inner working hidden from the public for various reasons. Lacking a better understanding of the internal process encumbers the individual's autonomy and compromises the interests "due process" rules are set out to protect.[23]

5.3 Alternatives to Data Mining

> Indeed, it has been said that democracy is the worst form of government except all those other forms that have been tried from time to time
>
> Winston Churchill

5.3.1 Mapping out Alternatives

As the previous segment shows, a policy analysis of the data mining of personal information is an extremely complex matter. A comprehensive analysis calls for addressing all these elements, and more. In addition, however, a policy study of data mining must consider the alternatives to applying data mining analyses. These are the policy strategies of choice, to be set in place if society refrains from applying data mining. As the quote above demonstrates, examining an issue without considering its alternatives is a futile exercise. In this section, I will briefly present the following five alternatives: (1) altogether refraining from the analysis of personal information to identify individuals and events of higher risk and therefore treating all individuals and events equally; (2) differentiating among events and individuals randomly; (3) doing so while relying on the human discretion of field officers, who examine personal information pertaining to the specific individual; (4) relying upon profiles and patterns constructed by experts and (5) applying data mining only to anonymous or anonymized data.

These alternatives are not without overlaps. Solutions might include elements from some or all of these options. Rather than alternatives, these are trajectories for various policy strategies which could be implemented—with every "alternative" pushing a different form of compromise. An understanding of the solutions' pros and cons along these lines, prior to selecting one of them for further implementation, is imperative. The analysis presented here assists in carrying out such balancing.

(1) The first and most obvious alternative to government data mining initiatives is altogether refraining from the analysis of personal information to identify individuals and events of higher risk, and setting them aside for specific treatment. Generally, this is the alternative to data mining usually envisioned. Yet as I will explain here, it is probably the most unlikely strategy to follow.

---

[23] US "due process" doctrine does not apply for various reasons. In some contexts, EU law provides for a right to understand the processes' internal workings. For a discussion of this issue, see Steinbock (2005).

Setting aside technologies and policies that enable selection will lead to treating all individuals (or events) as potentially risky and subjecting everyone to higher scrutiny. When this happens, however, the potential risk transforms into inefficiencies and discomfort, as well as excessive governmental costs. These costs will no doubt come out of resources that could have been used to have a better society (or left in the pockets of the taxpayers). This strategy might also lead to difficult legal questions regarding the authority to subject all individuals to additional burdens when no evidence indicating elevated suspicion against them exists. Finally, such course of action could lead to substantial breaches in security and system failures. The fatigue resulting from applying higher security standards to individuals and events that are clearly of low risk will adversely impact the alertness of the relevant officials. These officials, at the end of the day, might miss or react poorly to an actual threat when it finally comes their way.

Deciding whether to opt for this option, as opposed to using data mining, calls for a difficult balance of interests. It also requires tough decisions as to whether society should adopt an initiative which will risk the inconvenience, harm and even liberty of specific individuals at several junctures. It must note that this alternative leads society in its entirety to be taxed, either financially, in terms of attention, or even raising risks of security. Clearly, liberal and democratic societies should be willing to refrain from any data analysis if balancing indicates this is necessary.[24] Furthermore, society is mandated to do so (Schauer 2006)[25] when important interests of the specific harmed group are at stake. This is the case when governmental practices intentionally discriminate on the basis of race or nationality. Yet in other instances which do not involve the risk of reinforcing very problematic stereotypes, balancing becomes far more difficult and the results far less clear. In many instances, governments will decide that applying some form of selection and focused attention is prudent.

Yet beyond the normative balancing, this first alternative is politically unsustainable. As risk manifests and law enforcement resources are stretched, politicians and policy makers will face great pressures to "do something" with the vast datasets of personal data at their disposal. Thus, they will be pressurized to move away from this alternative. Given the high risks and limited enforcement resources, a form of selection must transpire. The question is, of course, how the selection will take place. This is where data mining and the other options come into play.

(2) Refraining altogether from selective practices in the context of security or law enforcement is unreasonable and unfeasible; the costs might be too high (costs that might lead to compromising basic rights of many citizens) (Slobogin 2007,

---

[24] It would mean that *all* individuals, for instance, would be required to arrive 30 minutes earlier at the airport to go through heightened security checks.

[25] For instance, discrimination on the basis of "sensitive information" such as race is illegal, even when such discrimination is statistically justified. For a partial critique of this outcome, see Schauer (2006).

102)[26] and the fatigue to the system too great. This leads to considering alternatives which enable the selective allocation of resources. This second alternative applies randomness to meet the security risks at hand (Harcourt 2007).[27] Searches, stops and other steps of enforcement would be applied to random individuals by the relevant field officer.

Scholarship points to this option as either a strategy that must complement data mining profiling or replace it entirely (Harcourt 2007). Random allocation and testing is an important measure to be applied in conjunction with data mining analyses (or any other strategy). It is crucial for statistically monitoring the effectiveness of data mining initiatives and examining whether they are justifying the compromises they entail. Here, however, I am referring to a much broader implementation of random allocation and a much narrower role for data mining.

While broadly applying a random scheme when specific personal information is available for analysis might seem as a strange (to be polite) option, in some contexts it certainly might suffice. When carried out in public, random checks might achieve sufficient deterrence of criminals and others fearing to be singled out. It will also allow government to show it is doing something—or in other words create a "security theater" (Schwartz 2008). By doing so, governments will sidestep many of the problems data mining present, while also averting the problems of fatigue and overstretching of resources.

With randomness, as with almost any strategy, there are several crucial details which must be attended to. First, there is the actual chance of being randomly selected. A very low chance due to limited law enforcement resources will probably fail to achieve deterrence.[28] A very high chance will begin generating the problems of alternative (1). Another issue is how "randomness" would be achieved. While this might sound trivial, in fact, it is quite difficult for individuals in the field to engage people randomly. They are quite often affected by internal biases and external factors (notions to be explored in depth below) when striving to make a random selection. This, of course, leads to unfair outcomes on the one hand and the fear of gaming[29] and ineffectiveness on the other hand. For a random search to be truly random, a randomizing tool must be applied—a computerized gadget that will indicate when someone would be selected, stopped or questioned.[30] Training field agents to ignore their judgment and succumb to a random number generator will not be simple. For all these reasons, administrating randomness might not be as easy as one might think.

---

[26] For instance, one might argue that encumbering the ability of all individuals to travel when striving to provide for security might limit their freedom of movement. I will refrain from developing this notion. For more on this point, see Slobogin (2007, 102).

[27] An option promoted by Harcourt (2007).

[28] When the chance for selection is very low, such enforcement loses its teeth, as the penalties inflicted cannot be disproportionate to the specific transgression. See similar dynamics occurring in the music and film industry when striving to enforce their rights online.

[29] Clearly, just selecting every tenth person or a similar strategy will allow easy gaming of the system by interested parties (all they have to do is travel in pairs and one of them will surely be beyond suspicion!).

[30] I thank Kathy Strandburg for making this point.

Yet even if these problems could be resolved, I believe the "random" alternative is unfeasible. Engaging in mere random selection, when a great deal of information which could be of relevance is available, might be hard for the public to swallow. The notion of ignoring information on the one hand, and subjecting individuals who are clearly of a very low risk to a higher level of scrutiny on the other, would be difficult to accept politically and might even make a mockery of the entire process. At times, the public must overcome its aversion of solutions which generate such "intentional blindness" for reasons detailed above (such as avoiding racial discrimination). Yet there is a paucity of strong justifications for applying randomizations broadly.

(3) The third alternative concedes to both the need for specific treatment of individuals and the use of personal information in this process. With this alternative, a decision maker examines specific personal information about an individual and makes an informed, ad hoc, decision. The decision maker might rely on the information she directly collects at the time of a personal encounter (what the individual is carrying, doing, saying). Yet she might also rely upon information in the individual's governmental profile when making this decision (What has he done? Where has she been?). In most cases, the decisions made in this scheme involve a field officer or a lower-level bureaucrat exercising their discretion. Possible examples are tax officers selecting a return for audit, security officers deciding which individuals to subject to additional questioning, or police officers deciding what street to walk or drive by (Slobogin 2007, 23).[31]

To further explain the nature of this alternative, it is important to note what decision makers are not doing. First, they are not running analyses which involve the datasets of the entire public (and thus individuals entirely removed from the relevant context). Second, the process is not automated (in the computerized sense), although the decision maker might use a computer to view personal information about the subject in real time. Third, it does not involve the formulation of factors, representing statistical groupings which indicate a higher or lower level of risk (at least not intentionally or explicitly). In addition, this alternative might have operational advantages. It requires officials to think on their feet, as opposed to some data mining schemes which require individuals to merely apply an algorithm. This latter role might adversely impact official's motivation and performance (although the motivational problem could probably be resolved with alternative measures).

In its most basic form, this alternative is merely hypothetical. Governments no longer operate in this way. Field officers never have full discretion, but are subject to protocols, which are a result of central planning. Allowing full discretion and lack of any protocol is simply unthinkable given the inability to control and regulate the actions of these officers, which might open the door to massive abuses (Slobogin 2007, 123). In addition, opting for this alternative will call for ignoring a great deal

[31] The discussion is intentionally avoiding instances in which the actions resulting from the higher level of scrutiny constitute searches, or other actions which directly impede upon the liberty of the subjects. I am doing so to sidestep the broader discussion about *Terry* stops and other such actions, where "reasonable cause" or other levels of scrutiny are mandated. For a mapping of these contexts, see Slobogin (2007, 23).

of knowledge within the system—knowledge which one field officer cannot integrate effectively. When neglecting to make use of such additional information, existing threats will not be sufficiently met, and potential evil doers will easily circumvent security measures by hiding their intentions.

For these and other reasons, addressing and critiquing this option might resemble attacking a straw man. However, there is still merit in examining this practice, even in its purest form. While this alternative is probably rarely exercised or even advocated, policy choices will no doubt reflect some variation of it. The options which are finally selected will be somewhere along the continuum between this alternative and the next one to be discussed ("4"). In other cases, some balance between this option and a data mining-based system which provides officers with recommendations, will be applied. Therefore, this alternative's pros and cons must be accounted for.

It is also important to point out that these practices are not as distinctively different from the use of profiles (to be addressed below) or even data mining, as they purport to be. The difference between them is one of degree, as in this model greater importance is vested with individual discretion. On its face, this alternative seems to be distinctively different, while treating every individual separately, and reaching conclusions while relying on data pertaining to the relevant subject. It is perhaps the most salient example of "individualized suspicion" (as opposed to generalized one). However, every future-looking statement pertaining to one individual's risk and prospects is actually premised upon a statistical analysis (even if it is an unconscious one) of the behaviors of others (Schauer 2006).[32] The prediction is carried out within the minds of the field officers, who generate it on the basis of behavioral patterns they witnessed or learned of in the past.

In addition, the policy structuring the law enforcement framework which leads to the field officer's discretion is based (at times, quite subtly) upon predictions. These predictions, in turn, were premised on some form of statistical analysis. In some cases, field officers are instructed that relatively minor crimes or actions (such as carrying box cutters) are indicative of other, more serious crimes (such as commandeering aircrafts). This rule is in fact a prediction premised on previous findings and behaviors. In other instances, field officers are required to present specific tests or questions and study the results they receive. Again, these questions and tests were structured with previous encounters in mind and an assumption that similar behavior patterns will reoccur.

To sum up our introduction to this alternative, let us examine two important parameters which were previously introduced: interpretability and correlation/causation. On its face, the process involving this alternative is interpretable. It is possible to learn the reason for any specific decision simply by asking the decision maker (and steps could be taken to assure that decisions would be logged to assure effective retrieval). Thus, this aspect provides an important advantage over the data mining practices which might lack interpretability. Yet the interpretability of this alternative could be called into question; the reasons the officials or field officers report might

---

[32] For instance, if the officer focuses on someone with a gun, it is because he created a mental profile with the category "people with guns," and is focusing his attention on those within that category.

not be the true ones (and there is almost no way to verify them). In addition, if the officer states that he relied on a basic intuition or hunch (which might be the case in many instances), the decision is virtually uninterruptable.

A similar observation could be made regarding the correlation/causation divide mentioned above. On its face, the field officers will refer to theories of causation when applying various decisions and measures. This will provide a safeguard against unfair or erroneous policies. However, when law enforcement decisions are opaque and rely upon intuition, they might be merely premised on assumed correlations the relevant official noted in the past, which have yet to be backed by a relevant theory (or even authenticated empirically). Thus, a closer look at this alternative shows that it is not as promising as we might have originally thought.

(4) The fourth alternative to data mining requires law enforcement to rely upon predetermined profiles for the allocation of resources and risks among individuals and groups (Schauer 2006, 166).[33] This profile is constructed by experts, who apply their common sense, expertise and experience to the task, in a top-down process. Experts will set up parameters for selecting tax returns, individuals at borders or the location of police cars. They will do so while working through datasets of previous actions and perhaps other forms of knowledge sets from the social sciences.

The differences between this alternative and data mining (as well as the former) could be set along three themes. First, the process does not call for "combing" through the entire dataset of personal information available to the government in the same way data mining applications operate (yet surely to a greater extent than the previous alternative). Note, however, that the profiling stage calls for some examining of datasets pertaining to previous problematic acts. In addition, the general parameters of personal datasets will be reviewed, to get a sense of the "normal" levels of the parameters used, so that a profile of deviations from the norm could be constructed.

Second, the process will not be automated but generated by human discretion. As opposed to the previous alternative, this process is triggered by the discretion of experts. Obviously, this option calls for some use of technology—a system will provide the decision maker with relevant facts, perhaps even with recommendations. Yet the final decision would be of the experts. In addition, the focus of discretion in this context is quite different than the one explored in the previous example; discretion is centralized, as opposed to being dispersed on the periphery of the bureaucratic infrastructure which is what the previous alternative called for.

The third difference between this alternative and the previous one (and a theme it shares with data mining) pertains to the notion of relying on statistics and an "actuary model." This model uses "generalizations" while making decisions regarding specific individuals. Here, analysts create groups and subgroups of individuals based on set parameters. These groupings instruct law enforcement to treat those within it differently. Such modeling relies on specific assumption regarding the ability to predict the future behavior of individuals, as well as deduce it from the actions of

[33] As Schauer explains, such practices are wide spread, and applied by customs, as well as by the IRS; see Schauer (2006).

others. It also accepts the risk of wrongfully treating an innocent individual who happens to fit within a problematic group or profile.

I again conclude this segment by returning to the elements of interpretability and causation. As opposed to the options explored thus far, with this alternative, the process will not only be inherently interpretable but will usually rely on various theories of causation for explaining the elements it includes. This will arguably enhance the autonomy of those subject to the analysis; there will always be an understandable answer to explain the singling out of a specific individual. It will also promote transparency in the procedure, which could be easily explained as a logical thought process. Finally, relying on causation will, as explained above, provide a check against problematic forms of discrimination and errors. This is an important benefit of this alternative, although interpretability and causation could be designed into data mining tasks, if deemed important.

(5) The fifth and last alternative already accepts the ability of data mining to achieve the objectives at hand. However, it requires that the analysis is conducted using anonymous (or anonymized) datasets. This recommendation, set forth by several recent policy reports (TAPAC 2004; Markle Foundation 2003), calls upon the government to engage in the analysis through the usage of several cryptographic tools. These tools allow for data matching, warehousing, and even mining, without providing the analyst with actual access to the personal information being mined. Access to personal data could be provided at a later time if suspicion arises, yet safeguards could be set in place to block unrestricted data sharing.

This alternative calls for a different form of balancing. It mitigates only some of the problems of data mining, while leaving others unaffected or even exacerbated. This strategy might reduce some forms of privacy and autonomy-related fears, as the public's concerns of being searched and tracked will be eased by knowing the government cannot connect their personal data to their real identity (Slobogin 2007, 195).[34] However, this alternative increases the chances of errors within the process and the lack of transparency. In addition, concerns regarding the practices which follow from data mining—the generation of patterns which would later be used to unfairly distinguish among individuals and events as parts of groups—will still persist! Finally, applying this alternative comes with non-trivial costs (in terms of both real out-of-pocket costs as well as costs of errors and engaging the system with additional process).

Considering this alternative also requires some rethinking as to the actual protection anonymity provides. Recent studies have indicated (Ohm 2010) that a massive anonymous database of personal information, which includes a multitude of factors about every individual, can be re-identified by sophisticated users if another database of identifiable personal information is at their disposal (Ohm 2010, 1746–48).[35] Thus, the government would probably be able to circumvent the protection measures mentioned here, should it choose to do so. These new findings weaken the attractiveness

---

[34] For empirical findings showing this point, see Slobogin (2007, 195).

[35] This was the case in the Netflix/Imdb fiasco. Such multi-factored datasets are now at the disposal of many public and private entities.

of this alternative. However, in the governmental context at least, these concerns of hacking and circumvention are probably manageable though various technological tools and disciplinary measures which will limit access and control the data.[36] All in all, however, this fifth alternative still requires a great deal of additional consideration.

5.3.2 *Distinguishing between the Field Officer, Profiler and Data Miner*

Three key alternatives enable government to engage in selective enforcement and scrutiny: data mining and alternatives (3) and (4) above. There are key differences between these options—differences which have crucial policy implications. In this segment, I will examine with greater depth the differences among them. The points made here can be used in future analyses of data mining and its alternatives, which must account for these elements and the differences they generate. Of course, in varied contexts, these differences will have different implications—yet a basic understanding of this framework is imperative. I also point out which differences are not crucial to the discussion, but can prove to be a distraction from addressing other important elements.

First, let us take a look at the notion of human *discretion* and the different methods of decision making the models employ. More specifically, the alternatives lead to a choice between various forms of human discretion, as well as a balance between human and automated discretion. Selecting between methods of discretion has several implications. The most central one is the forms of errors it generates.[37] If one form of discretion generates predictable errors (even if those are not substantial) the system would be easily gamed and manipulated. If the errors are systematic, a specific set of the population would be harmed, leading to distributive and other harms (again, even if overall efficiency is maintained). If the errors are both systematic and detrimental towards specific segments of the population, which are either weak or were singled out in the past, this leads to an additional set of problems. When balancing alternatives against each other, the errors resulting from the different forms of discretion must be accounted for. The next few paragraphs briefly map out how that could be done.

Preferring human discretion, as opposed to deferring to the output of a data mining-powered application, leads to at least two shortcomings (which pertain to

---

[36] This option still holds substantial benefits, as it minimizes the risk of illegal abuse of the information by a government executives (such as the many stories occurring every year of tax officials sharing or selling personal information about citizens). Note, however, that this problem could also be mitigated through disciplinary actions.

[37] If one form of discretion generates errors which are frequent, the entire process is compromised. However, let us assume that the threshold of a reasonable level of errors would be attended to as a preliminary matter—and if the level of errors will be unacceptably high, the project would be set aside. Yet as I demonstrated in the text, even with an overall acceptable level of errors, problems can still prevail.

almost all decisions premised on human cognition) that quickly transform to errors in the final outcome: Human decisions: (a) tend to rely upon heuristics and (b) at times employ hidden biases. Both dynamics are systematic and predictable. The latter also generate errors detrimental to specific weaker and vulnerable segments. I now turn to take a closer look at both elements, explain how they generate differences between the models and briefly note the implications of these differences.

A vast psychological literature regarding heuristics clearly indicates that when dealing with complex tasks, the human brain applies various shortcuts which allow it to overcome information overload (Korobkin 2003; Tor 2008). These rules of thumb often lead to correct decisions. However, at times, heuristics lead to predictable errors. This occurs when individuals face the need for quick decisions, with limited attention and vast information to consider. While some errors could be corrected through training and experience, many others cannot.

Considering the alternatives pointed out above quickly leads to recognizing flaws in the third alternative, which relies heavily on the individual discretion of field officials. This alternative will lead to predictable cognitive traps where heuristics will be applied but lead to a wrong result, which adversaries might abuse. Thus, for this reason alone, opting for this (third) alternative will come at a high price in terms of efficiency and fairness. When opting for the forth alternative (expert-driven profiles), this concern is somewhat mitigated. Experts might have greater awareness to these tendencies to err, and focus on empirical findings, rather than mere intuitions. They also need not make quick decisions under pressure. However, this process could be inflicted with heuristic-related errors as well, given the reliance on human-based discretion.

On the other hand, data mining faces the least of these troubles. Computers have no need for shortcuts and heuristics when they have the capacity to address all data. When indeed, for efficiency purposes, only segments of the data are addressed or another analytic shortcut is used, it is a shortcut the operators are well aware of and can take into consideration.

Relying upon discretion also allows for the internal biases of the individual decision makers to impact their actions and decisions, even inadvertently. At times, the discrete decision of the experienced decision maker is a discriminatory one. Such discrimination is premised upon (at times, subconscious) animosity towards specific segments of the population, or other forms of prejudice. This might result in an inefficient outcome (Schauer 2006, 179).[38] Far worse, however, this might lead to unfairness towards the weaker segments of society, or against groups society designated as protected.

Biases can transpire within the frameworks of both the third and forth alternatives. Field officers are most susceptible to generate these distortions. Indeed, a recent review of studies addressing law enforcement field decisions with regard to race

[38] This was exactly, according to Schauer, the case in O'Hara airport, where it was revealed that the percentage of minorities made subject to intrusive cavity searches was very high. When such practices, which were no doubt motivated by racial animosity, were stopped, the success of such searches increased. See Schauer (2006).

shows an alarming and distorted picture (Harcourt and Meares 2010). For this reason, providing full discretion to field officers is unthinkable (Harcourt and Meares 2010).[39] Yet even relying on expert decisions (as in alternative #4) might not resolve many of these concerns. Experts might be plagued with internal biases and generate policies which are unfair to weaker and protected groups. Rather than relying upon strong data findings and expertise, they might be motivated by beliefs and prejudice. Note, however, that the fourth alternative has the advantage of a central process. As opposed to a system where decisions are made at the periphery, the expert profiles could be closely audited and studied in an attempt to identify arbitrary conduct that might lead to unfair discrimination. This, of course, is easier said than done.

With data mining, applying an automated process allows the central planner to retain better control over the actions in the periphery as well. Yet data mining provides an additional benefit; computer modeling is not driven by human assumptions (which might be both hidden and biased) but by the data itself. Therefore, concerns regarding hidden biases premised on prejudice might be sidestepped by applying data mining.

Many will disagree with this last statement. Beyond the fact that data mining has systematic flaws, hidden biases might be a feature of data mining, and lead to even graver concerns. These biases might be put in place at the points of human interaction listed above, which in many cases are hidden from public scrutiny. Thus, data mining allows for the embedding of values as well. The difference between the options here discussed amounts to the ease of embedding values ex ante and the ability to identify these instances ex post. Those arguing against data mining will state that biases can be built into decision-making processes quite easily ex ante, and are difficult to identify, if hidden well, after the fact. For that reason, data mining runs high risks of generating biased conduct.

I believe, however, that the problems mentioned are not inherent features of data mining, and certainly are not beyond repair. If the data mining process is sufficiently transparent, it can effectively overcome these challenges. Adding interpretability and even causation to the data mining process could allow policy makers to assure that biases are averted. In addition, analysts could keep a close eye on the forms of software used, and the protocols applied when using it. Biases in a central computer code, once acknowledged, could be tackled with ease and identified effectively by external review. This task is certainly easier to tackle than achieving this objective with the other alternatives mentioned. Managing and mitigating hidden biases in the actions of numerous field officers vested with a great deal of discretion is a much harder task. This would call for tracking, evaluating and disciplining all actions carried out in the periphery.[40] Even doing so with a group of central experts seems daunting, and will generate numerous painful confrontations. For these reasons, I

---

[39] The authors explain that part of the role of the 4th Amendment is to limit the discretion of law enforcement. Harcourt and Meares (2010).

[40] I acknowledge that even when using a central system, some level of examining of the actions of the periphery operation is needed as well. Yet this would be substantially less than the level required in the third alternative model.

believe this segment of the analysis clearly points to the superiority of data mining initiatives.

A *second* difference between these central alternatives pertains to the use of decisions premised on statistical groupings, as opposed to individualized suspicion. Deciding on the basis of a group's statistical analysis leads to a much broader debate, in which some scholars show great resentment to the "actuary method" (Harcourt 2007). This is the notion that individuals are treated as parts of groups, which have specific predefined traits and parameters, as opposed to actual clinical work to examine the relevant situation. Similar methods are broadly adopted in many stages of modern life (especially in insurance), and generate aversion there as well. Unlike the previous issue, this one should not weigh heavily when balancing alternatives.

While using this form of statistical analysis in data mining might generate negative sentiment, I believe categorically rejecting the "actuary method" is unwise. Merely relying on an individual's record, not only is inefficient, but includes implicit reliance on groupings as well. In addition, the individualized process generates several crucial detriments, especially the lack of interpretability and transparency. Therefore, when opting for "individualized" treatment, the public does not always understand the process's underlying assumptions and inner workings. Options which use statistical analysis (such as alternative (d), or data mining), might indeed be rejected, but it should be for other, more specific, reasons.

The *third* issue is that of automation. Deciding between the alternatives mapped out above is again merely a subset of a broader discussion concerning the role of computer-generated decision making in a technological society (Bamberger 2010).[41] Data mining calls for a great deal of automated decision making, while the other options do not. Philosophically, those fearing automated decision making show disdain for the tyranny of computers, which might make systematic errors and are unable to take into account the delicacy of the human condition. They also fear that society does not easily accept errors made by computers, as opposed to easily accepting that "to err is human." Finally, they might consider the notion of individuals treated by computers (as opposed to fellow humans) as undignified (Korff 2011). These are all powerful arguments against the spreading use of data mining. Yet again I do not believe these distinctions, on their own, should receive a great deal of attention when comparing alternatives.

When addressing this issue, we must always worry that behind many of the arguments stated in the previous paragraph, stands a fear of technology with a neo-Luddite flavor. In other cases, these arguments might result from a tendency to underestimate technology and its ability to match human achievements. However, the analysis of alternatives presented here shows that non-automated decision making features substantial problems as well. Yet it would be wrong to reject the notion of unease with computerized decision making in its entirety. The public's resentment towards data mining could be a result of an irrational fear of automation. Yet, this sentiment might be derived from other strong and meaningful sources which we must diligently seek out and explain—among others the fear of errors, loss of autonomy and the lack of

[41] For a discussion of this matter in the Corporate Risk Management setting.

transparency. After uncovering these concerns, they must be tackled directly. The broader, somewhat vague, notion of automation must be set aside, especially when accounting for the balances alternatives entail.

5.4 Conclusion: Alternatives as Building Blocks for Future Analyses

This chapter strived to illuminate a crucial methodological step which can assist policy makers seeking balances in today's world of global insecurity. Such policy makers are now charged with structuring schemes for using databases of personal information to promote law enforcement and stability. In doing so, policy makers will be called upon to assess the option of data mining. The first step of this process calls for understanding the technology at hand. The second step the analysts face is identifying the variety of problems and questions these methods generate. The third step is introducing alternatives and illuminating the differences between them. These steps formulate a novel methodology for examining data mining practices. Comparing among alternatives will provide for a better sense of the balances and realistic compromises required at every juncture. The comparison must account for all the elements of the discussion. It should account for legal analyses, economic costs, technological abilities, and psychological limitations.

Existing risks call for the use of personal information in an effort to preempt possible harms and attacks. Society will be forced to decide among several non-ideal options. At the end of the day, the solution finally chosen would no doubt be a compromise. The methodological steps presented in this chapter strive to assist in these balancing efforts, while acknowledging that there is still a great deal of work to be done. I hope this small contribution promotes this broader objective.

Acknowledgments This chapter is part of an NWO-funded research project "Data Mining without Discrimination." I thank Kathy Strandburg, Richard Stewart, the participants of the NYU Law School Hauser Research Forum, the NYU Privacy Reading Group and the DePaul Law School CIPLIT presentation for their comments. I also thank Talya Ponchek for her comments and research assistance. For an extended version of the ideas presented here, see: Zarsky, Tal Z. 2012. Data Mining and its Alternatives Penn State Law Review 116(2):101.

References

Ayres, Ian. 2007. *Super crunchers*. New York: Bantam Dell.

Baker, Stephan. 2008. *The numerati*. New York: HMH.

Bamberger, Kenneth A. 2010. Technologies of compliance: Risk and regulation in a digital age. *Texas Law Review* 88 (4): 669–739.

Blitz, Mark. 2011. Warranting a closer look when should the government need probable cause to analyze information it has already acquired? *PLSC 2011 Workshop*. Draft, on file with author.

Cate, Fred H. 2008. Data mining: The need for a legal framework. *Harvard Civil Rights-Civil Liberties Law Review* 43 (2): 435–489.

DeRosa, Mary. 2004. *Data mining and data analysis for counterterrorism.* Center for Strategic and International Studies (CSIS) report, 14. http://csis.org/files/media/csis/pubs/040301_data_mining_report.pdf. Accessed 12 July 2011.

Harcourt, Bernard E. 2007. *Against prediction*. Chicago: University of Chicago Press.

Harcourt, Bernard E., and Tracey L. Meares. 2010. Randomization and the fourth amendment. *University of Chicago Law & Economics, Olin Working Paper No.* 530:3–76.

IBM. 2010. Memphis police department reduces crime rates with IBM predictive analytics software. http://www-03.ibm.com/press/us/en/pressrelease/32169.wss. Accessed 12 July 2011.

Jonas, Jeff, and Harper, Jim. 2006. Effective counterterrorism and the limited role of predictive data mining. *Cato Institute, Policy Analysis* 584: 1–12. www.thebreakingnews.com/files/articles/datamining-cato-report.pdf. Accessed 12 July 2011.

Korff, Douwe. 2011. Data protection laws in the EU: The difficulties in meeting the challenges posed by global social and technical developments. *Working Paper No. 2, European Commission Directorate-General Justice, Freedom and Security* (January 20, 2010), final [extended and re-edited] version. http://ec.europa.eu/justice/policies/privacy/docs/studies/new_privacy_challenges/final_report_working_paper_2_en.pdf. Accessed 12 July 2011.

Korobkin, Russell. 2003. Bounded rationality, standard form contracts, and unconscionability. *University of Chicago Law Review* 70:1203–1295.

Markle Foundation. 2003. Creating a trusted network for homeland security (December 1, 2003). http://www.markle.org/publications/666-creating-trusted-network-homeland-security. Accessed 12 July 2011.

Nissenbaum, Helen. 2009. *Privacy in Context*. California: Stanford University Press.

Ohm, Paul. 2010. Broken promises of privacy: Responding to the surprising failure of anonymization. *UCLA Law Review* 57:1701–1777.

Ramasastry, Anita. 2004. Lost in translation? Data mining, national security and the adverse inference problem. *Santa Clara Computer & High Tech.Law Journal* 22:757–796.

Schauer, Frederick. 2006. *Profiles, probabilities and stereotyping*. Harvard University Press.

Schneier, Bruce. 2006. Why data mining won't stop terror. *Wired* (September 3, 2006). http://www.wired.com/politics/security/commentary/securitymatters/2006/03/70357. Accessed 12 July 2011.

Schwartz, Paul M. 2008. Reviving Telecommunications Surveillance Law. *University of Chicago Law Review* 75:310–311.

Scism, Leslie, and Maremont, Mark. 2011. Insurers test data profiles to identify risky clients. *The Wall Street Journal*. http://online.wsj.com/article/SB10001424052748704648604575620750998072986.html?mod=WSJ_hp_LEADNews-Collection. Accessed 12 July 2011.

Slobogin, Christopher. 2007. *Privacy at risk: The New Government Surveillance and the Fourth Amendment*. Chicago: The University of Chicago Press.

Slobogin, Christopher. 2008. Government data mining and the fourth amendment. *The University of Chicago Law Review* 75:317–341.

Slobogin, Christopher. 2010. Is the fourth amendment relevant in a technological age? *Governance Studies at Brookings* (December 8, 2010). http://www.brookings.edu/~/media/Files/rc/papers/2010/1208_4th_amendment_slobogin/1208_4th_amendment_slobogin.pdf. Accessed July 12, 2011.

Solove, Daniel J. 2001. Privacy and power: Computer databases and metaphors for information privacy. *Stanford Law Review* 53:1393–1462.

Solove, Daniel J. 2008. Data mining and the security-liberty debate. *University of Chicago Law Review* 74:343–362.

Solove, Daniel J., and Schwartz, Paul M. 2006. *Information Privacy Law*. New York: Aspen.

Steinbock, Daniel J. 2005. Data matching, data mining, and due process. *Georgia Law Review* 40:1–86.

Strandburg, Kathrine J. 2008. Freedom of association in a networked world: First amendment regulation of relational surveillance. *Boston College Law Review* 49:741–822.

Taipale, Kim A. 2003. Data mining and domestic security: Connecting the dots to make sense of data. *Columbia Science and Technology Law Review* 5 (2): 1–83.

TAPAC. 2004. The report of the technology and privacy advisory committee, safeguarding privacy in the fight against terrorism. http://epic.org/privacy/profiling/tia/tapac_report.pdf (Hereinafter TAPAC Report). Accessed 12 July 2011.

Tor, Avishalom. 2008. The methodology of the behavioral analysis of law. *Haifa Law Review* 4:237–327.

U.S. General Accounting Office. 2004. Data mining: Federal efforts over a wide range of uses. Report to the ranking minority member, subcommittee on financial management, the budget, and international security, committee on governmental affairs, U.S. senate, GAO-04–548. Washington: 9–54. http://www.gao.gov/new.items/d04548.pdf. Accessed 12 July 2011.

van der Veer, R.C.P., Roos, H.T., and van der Zanden, A. 2009. Data mining for intelligence led policing. Paper presented at the proceedings of the 15th ACM SIGKDD International Conference on Knowledge Discovery and Data Mining, Paris, France (June 28–July 01, 2009). http://www.sentient.nl/docs/data_mining_for_intelligence_led_policing.pdf. Accessed 12 July 2011.

Westin, Alan. 1967. *Privacy and Freedom*. New York: Atheneum.

Zarsky, Tal Z. 2002–2003. Mine your own business!: Making the case for the implications of the data mining of personal information in the forum of public opinion. *Yale Journal of Law & Technology* 5:1–56.

Zarsky, Tal Z. 2012. Transparency in data mining: From theory to practice, in Discrimination and Privacy in the Information Society, (Forthcoming) (Springer)

Chapter 6
Managing Suspicion and Privacy in Police Information Systems

Negotiated Work in Local Police GIS in Romania

Vlad Niculescu-Dinca

> *Not even imagination can limit this revolution [the use of GIS in public administration] as it will, with only a few noticing, change many areas of work and form the basis for other new practices that will be obligatory implemented.*
> *(IT director)*

6.1 Introduction

Over the past several decades and especially in the past decade, a broad trend in the world of policing has been the move toward more preventive and proactive styles. Beginning as a way to increase police legitimacy (Tilley 2003), community policing promotes stronger relations between police and communities to develop a greater flow of information about crime problems and aims to cultivate an ethos of policing as a public service. Although widely adopted, community policing coexists in practice with other managerial strategies (Tilley 2009). One of these, CompStat, is a multilayered approach characterized by both setting an organizational mechanism for officer accountability and placing at its core geographic information systems (GIS) for crime mapping and analysis. In this lineage of policing managerial philosophies, spreading gradually and developing incrementally, a constant has been the increasing employment of data gathering and analysis and the promotion of intelligence sharing as key elements in efficient resource allocation and strategies of crime control (Ratcliffe 2008).

In the European Union, this trend can be seen both in the promotion of new security technologies (i.e. multiple security research projects within EU FP7) as well as in the security strategy policies. On one hand, identifying the cross-sectorial nature of threats calls for strengthened cooperation and coordination between European

V. Niculescu-Dinca (✉)
The DigIDeas Project, Zuyd University & Maastricht University,
Brusselseweg 150, 6419 Maastricht, The Netherlands,
e-mail: vlad.niculescudinca@zuyd.nl

S. Gutwirth et al. (eds.), *European Data Protection: In Good Health?*,
DOI 10.1007/978-94-007-2903-2_6, © Springer Science+Business Media B.V. 2012

agencies, Member States and local authorities as "even seemingly petty crimes such as burglary and car theft, sale of counterfeit and dangerous goods and the actions of itinerant gangs are often local manifestations of global criminal networks" (European Commission 2010a, p. 4). On the other hand, efficient law enforcement in the coming years is seen as facilitated by powerful technological systems that need greater interoperability as well as overcoming obstacles posed by lack of standardization and divergent approaches of Member States.

At the same time, there is concern in the Commission that developments in the area of police and judicial cooperation throughout the EU could also have privacy infringements and other undesired harms. Therefore, recommendations have been made to ensure that "different categories of data should be distinguished in accordance with their degree of accuracy and reliability, that data based on facts should be distinguished from data based on opinions or personal assessments, and that a distinction should be made between different categories of data subjects (criminals, suspects, victims, witnesses, etc.), with specific guarantees laid down for data relating to non-suspects" (European Commission 2010b, p. 14). In general, while taking into account the specific nature of policing activities, the Commission recommends "the extension of the application of the general data protection rules to the areas of police and judicial cooperation in criminal matters, including for processing at the domestic level" (European Commission 2010b, p. 15).

6.1.1 Information Technologies in Local Policing in Romania

In preparation for entering the European Union and afterwards, Romania began a process of reforming and employing information and communication technologies in public administration and government. More recently, accelerated and concerted efforts are being made towards developing integrated information systems of various local and national government agencies (CSAT 2007). As future (national) security threats are envisioned to require integrated operative activities, interoperability and interconnectivity of information systems has become emphasized in strategy documents (CSAT 2010). While there is recognition of insufficiently reformed institutional frameworks,[1] the development and deployment of such technological systems in these organizations is happening at an increasing pace—with design decisions concerning personal data models and flows, categories and processing algorithms largely closed from public debate. Moreover, there are relatively few studies analyzing the appropriation of these technological systems and innovations in day-to-day working routines of practitioners, and their implications for privacy, not only in terms of the protection and security of personal data but concerning the ways data are being classified, aggregated, and acted upon.

[1] According to the Ministry of Interior, Traian Igas, talking in the context of the ongoing police reform and referring to an internal police study in an interview on 26 July 2011, http://www. b1.ro/ stiri/ politica/traian-iga-peste-30-din-angajarile-in-poli-ie-s-au-facut-pe-rela-ii-video-8520.html, a significant percent of the police staff employed in the past 13 years did not receive formal police academy training.

In this context, this chapter focuses on local policing, analyzing the socio-technical practices related to a geospatial information system (and its relations to information systems of other agencies) in the local police station of a major city in Romania referred in this document as M city.[2]

This chapter shows that data processing practices, and in particular the constructions of suspicions, may violate privacy principles through incorrect and ambiguous data and, through data aggregation, cause accumulative informational harm. The responsibility for such outcomes is shown to remain a complex mix of factors including vague procedural provisions, particularities of organizational culture, institutional arrangements, the subjectivity of police agents, as well as technological design. By highlighting these factors, the chapter suggests that such technological innovations in public administration may contribute to make government more efficient but also create privacy risks and informational harm. Therefore, it concludes that processes of digitalization and the interconnection of information systems need to be concerned not only with data protection but also with the ways this data is generated and processed in ways sensitive to possibly affected values.

This chapter first introduces the case study of the local police of M city with its system and associated practices. Then it accounts for the theoretical approach and the methods employed in data gathering during fieldwork. The next section analyzes the data processing practices and how they have been influenced by the introduction and shaping of GIS in the local police. It does that by following the thread of data flows in the organization and beyond, focusing on data concerning suspicions. In the last part, the chapter analyzes the privacy related implications as well as reflects towards a broader value-conscious design of such systems.

6.1.2 *GIS in Local Policing: M City Case Study*

In addition to national policing agencies, coordinated by the Ministry of Administration and Internal Affairs, local police in Romania were established as municipality services under the authority of local councils in an effort to promote the ethos of policing as a public service. These police services are maintaining public order, handling offences until they reach the penal law at which point they are delegated to the national police.

The local police of M city embraced a community policing style and implemented a CompStat police managerial philosophy. Weekly strategy meetings assess staff accountability as well as analyze the previous week's reports and broader spatial and temporal trends. These analyses are enabled by reports and maps afforded by the geographic information system, developed by the local partner of a global GIS corporation. The spatial and temporal analysis of criminality is backed by a geospatial

[2] To protect the anonymity and confidentiality of officials and police staff, who collaborated generously, the name of the city has been turned into M city. The same applied to the names of police staff throughout the chapter.

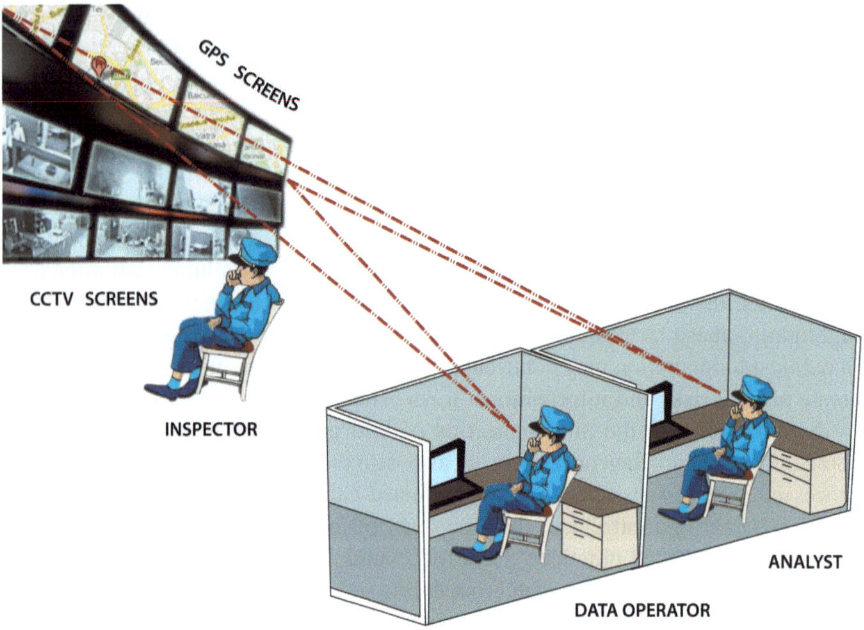

Fig. 6.1 Schematic representation of the control room. (The staff in the control room works separated by cubicles, designed at particular heights, fostering specific lines of sight and division of labour: The GPS screens should be visible from any point in the room while each worker should concentrate on their work)

information system that allows registration, categorisation, geo-localization and retrieval based on a broad range of fields and complex filters. The data introduced goes back several years in time and includes geo-location and temporal data of incidents, identifying attributes of involved goods and persons, etc.

At the same time, the system enables real-time global positioning of police agents in the field (both walking patrols and cars). This feature not only allows efficient dispatch and monitoring of operative activities but also minute logging of police agents' patrol routes. The dispatch inspectors in the control room are able to visualize the agents' GPS-enabled radio units on a screen wall and also generate a history of their movements when required by police management (Fig. 6.1).

At this point, the system and associated practices are regarded by police representatives as having stabilized in their daily routines. For example, concerning the practice of real-time GPS tracking, most answers suggested that the measure presents "no problem" for their privacy. However, this was apparently not so from the beginning.

> In the beginning we did extensive checking of the itinerary of the agents. They did not believe that it is actually possible or that we actually do it. They turned the stations off; some broke the wires in the car units and meddled with the settings. Now we don't check as often as they also understood that it works. Also after some were called on the 'carpet' (i.e. disciplining committee) the acceptance improved.
> (inspector in M city local police)

However, in more informal settings, agents are still expressing a degree of resistance and frustration: *"Of course we were wandering more throughout the city before this system"*, mentions one agent. *"I did find strange the idea of being tracked all the time"*, mentions another under her breath, while another expresses relief when learning his GPS unit is broken: *"Tell me X, are you tomorrow at the base? I hope you are not gonna' work us to death. Oh, anyway, my GPS is broken..."*

Yet it is not only field agents that are surveyed. Dispatch inspectors working in the control room are not only surveyors of cameras and of GPS units of agents but surveyed themselves though the GIS. They are monitored to introduce the dispatch jobs in the system as they occur. That is, the time that an event is inputted in the GIS should be close to the time the event is reported to have happened.

6.1.3 Institutionalizing Suspicion

Local police collaborates with the national police and other law enforcement agencies in various common activities and exchange information on a daily basis on a whole spectrum of criminal occurrences they encounter, including the identification of potential suspects.

While local police in Romania are gaining increasing access to data produced by other organizations, data produced by local police can be further aggregated into the systems of the national police. As the head of M city local police mentions, sometimes this information may help them with new leads in investigations and new links about suspects.

Local police were concerned about the issue of suspicion before the introduction of geographic information system. Still, the term *suspect* in the local police does not refer to a person connected to a committed crime, but to a person raising suspicions due to behaviour and other reasons. Police tactics procedures (Stancu 2006) specify that the measure of interception applies not only to those for which "there are clues to have committed crimes" but also to those assessed as *suspect* by police agents due to "their presence at a particular place and time, their clothing, luggage or behaviour".

One reason for registering suspicion came from community policing strategies. As one member of the police staff mentions, interception and asking for identification documents is practiced as a preventive strategy: *"It may make them think: 'Wait, perhaps it's not a good time since the police has already ID'd me"'*. Upon such a stop and identification, the agents are instructed to report not only the name of the person(s), the location and temporal data of the event, but also what they suspected the person of and what lead them to this assessment.

The reason for registering suspects in the system came also due to protocols of data exchange with the national police. The daily exchange of data, including that of suspicions, is viewed as a useful contribution to solving crimes, providing leads that are otherwise difficult to find. In a typical situation of collaboration, the local police patrol stops and identifies a person considered as having suspicious behaviour or due to his/her presence at a particular place or time and exchanges this information with the national police. The latter may in turn benefit, if that persons turns out to be involved in a crime dealt by them, but who could have hardly been associated with it.

6.1.4 Theoretical Approach

This chapter draws on ideas developed in science and technology studies (STS) and in particular Actor-Network Theory (ANT) (Latour 1988) in order to offer complementary, empirically grounded insights regarding privacy implications emerging from socio-technical practices. This is in part because—in stressing the need to go back and forth constantly between the agency of technological artefacts and that of human beings—ANT avoids the pitfalls of both considering technologies to *determine* human behaviour and social relations (whether for better or worse) or reducing technologies to the neutral tools and clear-cut outcomes of particular social practices. From an ANT perspective, actors—defined as entities that do things—should be conceived in relations with other actors for it is from these networks of relations that they become what they are. The technical and the social cannot be analyzed separately and are part of a *seamless web* of elements that are neither fully technical nor fully social (Bijker and Law 1992).

Recognizing that technological designs and material artefacts are more than mere tools, that they are able to influence user behaviour (Latour 1992) enables the study of technology-society interactions with a vocabulary developing attention to their mediating role (Verbeek 2005). Through this mediation, technologies can be said to be involved in *translating* programs of action, meaning that either human beings translate their intentions, *delegating* them to the functions of artefacts or, symmetrically, artefact's functions can be said to impose *prescriptions* guiding user's behaviour without the need of the designer's presence. Such built-in scripts can be said to be the program of action a technology invites or imposes. While designers can anticipate user behaviour building in *preinscriptions*, technologies can also implicitly supply their program of action, without such script originating in a designer's intention.

But it is not only use of technology that matters. Other social studies of technologies have shown that both use and *non-use* (Star 1991; Oudshoorn and Pinch 2003) are important. Attention should be paid both to actors directly involved in technology use as well as to those excluded in various ways but related to technological applications—and which cannot be meaningfully understood as users (e.g. data about persons classified and acted upon in databases). Exclusions such as these should not be seen as coming from deficiencies of users; rather from the design of technologies—and in a broader sense of socio-technical ensembles—that did not take their interests into account (e.g. various labels categorizing personal data in virtue of criteria possibly ignoring or against the person's interests).

This approach proves fruitful when analyzing a slippery concept such as privacy in the context of evolving socio-technical practices. This is because conceptualizations of privacy as well as privacy expectations have been continually changing also in connection with technological developments and their associated practices. This is attested by the wealth of privacy theories that were developed and refined in the past decades along the emergence and proliferation of various information and communication technologies.

In the face of—and sometimes anticipating—the spread of information and communication technologies raising various privacy concerns, legal scholars and philosophers have defined regulatory and conceptual responses in defence of privacy interests of individuals. This mutual shaping of socio-technical practices and privacy can be seen for example in the emphasis that many of these theories have on informational privacy and data protection, assuming therefore the production and processing of this data by computational means. Such data protection principles and informational privacy concepts, focusing on individual's rights to control their personal information, were moreover translated into privacy regulations (e.g. data protection legislative instruments (European Union 1995) or privacy-enhancing technologies).

However, regulations are in continuous need of adjustment as previous regulation often proves inadequate in face of new technological possibilities and changing social practices (Gutwirth 2002). Many scholars have thus commented on the privacy related issues raised by a wealth of technologies such as databases, networked data processing, data mining, location-based services and the changes they ensued both on related practices as well as in privacy expectations.

For example, in the context of policing and law enforcement, Vedder shows how certain techniques of data processing are able to discover knowledge in data, previously integrated from several sources. Such analysis, that can be used for a wide area of purposes, including analyzing medical data and drug consumption as well as criminal investigations (Vedder 2001), also highlights the inadequacy of certain regulations that concentrate on data protection. With data subjects largely unaware of retention and processing of data about them in multiple interconnected organizations, privacy protecting principles such as consent to information use (Fulda 2000), the value of social forgetfulness (Blanchette and Johnson 2002) or data minimisation (Flaherty 1997), prove difficult to be defended. The very development and refinement of such principles reflecting evolutions in technical affordances attests to the ways in which privacy has been shaped along the co-evolution of technologies and social practices, sometimes this being seen as a process of slow erosion (Koops and Leenes 2005) but never altogether discarded (Koops 2003).

Given this wealth of privacy perspectives and the growing variety of technological means employed also by policing and law enforcement, it is part of the proposed approach of this chapter to allow privacy issues to emerge from empirically grounded field work about socio-technical practices, rather than being prescribed by theory. Of course, privacy conceptualizations developed in tight relation with certain socio-technical ensembles may remain informative and suitable but it could also be possible that they prove inadequate in face of novel socio-technical innovations.

Therefore, this chapter shows how the empirical study of technological appropriation in daily practices, using the vocabulary of ANT, can both inform privacy debates as well as reflect on steps towards the design of these systems in ways sensitive to values.

6.1.5 Fieldwork Data Gathering

Being inspired by these ideas, the analysis of this chapter draws from an ethnography of actors and networks I performed during July/August 2010 in the police station of M city, analyzing the relations within organizational, legal and architectural arrangements and between police staff, screens, cubicles and information systems. Performing ethnographic research in a police station, inspired by the works of Norris and Armstrong (1999) and Dubbeld (2004), allowed for my close observation of work processes of police staff and the relations between them and the technological systems they engage with.

Gathering of material has been achieved in the course of roughly 100 hours of participant observation in various situations, including (night) shifts in the dispatch centre with its screens wall, strategy meetings where police management analysed and took decisions based on spatial-temporal analysis of geo-coded events represented on big screens, data gathering and introduction from paper based reports of field agents towards their introduction in the system, data analysis with map comparisons and filtering, preparation for information sharing with other police agencies, street patrolling with GPS tracked agents and informal meetings with inspectors of various ranks. The analysis draws from several interviews, internal police documents, system requirement documents and field notes.

6.2 Shaping of GIS in the Local Police

The introduction of geospatial information systems in public administration and the police in particular occurs against the background of ambitious modernization efforts towards integrated information systems at local and national level of public administration. The solution analyzed in M city is currently being advertised as a preconfigured product, requiring minimal configuration efforts and easily integrating with other modules and other GIS solutions. Moreover, it is viewed as part of a technological revolution that will be implemented in many areas of public administration, affording aggregation of data from multiple sources:

> This is where the revolution begins [. . .] Just think that you'll be able to define a zone and you'll obtain the [geo-referenced] persons attached to this zone and then you can have all kinds of detailed analysis regarding distributions of unemployment, age groups, you name it. . . .
> (IT director)

Resistance towards these changes is seen as coming from individual employees or from institutional frictions. However, these are not regarded as able to deter the processes of technological development:

> You know, it is not only the GIS developers that are making the whole thing work. It's also the Special Telecommunications Services. And besides they have more departments participating. They were not particularly glad to participate at the implementation of a

> system that was not by default part of their structures. But in the end they had to agree to help as they could have only postponed things. This is the direction, there is no other.
> (M city official)

The reception of the system by police management was also portrayed in enthusiastic terms. As the head of police mentions concerning the inspiration for the system: "*We were inspired by the '24' TV series.* [3]*You can't stop watching them, it's really 24 hours watching*". This inspiration drove them towards a solution enabling easy retrieval and visualization of information in a centralized way and the result was seen a great step forward in their practices:

> We didn't think we'll reach the same level as we have seen there (i.e. in the '24' series), to type a name and get what milk he drank as a child, but setting our standards high got us here. When we first had the system working and saw everything on the big screen, we all said in one voice: Wow! It makes a huge difference to see in one glance a certain crime distribution instead of going through paper reports or even through Excel files.
> (local police official)

The quotes are indicative of a relatively determinist perspective on technology in government services. However, as Van der Ploeg (2003) makes us aware, conceptualizations of technology can be seen as discursive strategies serving different purposes in public debates. In this way, they may also have implications for assigning responsibility, understanding the distribution of human and non-human agency and the space left for possibilities to shape the technologies and related practices. In this case, dominating discourses highlighting technology's ontological stability, easiness in utilization and inevitability in their development and deployment may leave little space for critical analysis of possible vulnerabilities and uncharted issues related to various gathering and processing of personal data in increasingly integrated and interconnected government information systems.

On the other hand, describing technology *in the making* (Latour 1987), with less fixed characteristics, other views may come to the fore highlighting different distributions of agency. An analysis of M city internal police documents shows 144 change requests of various complexities made since the initial deployment by the police towards developers. The system is being modified not only to correct anomalies and fix bugs but also to account for local police working practices.

The spectrum of system requirements shows that the police shaped the system since its initial deployment in as much as its introduction shaped police work. The requirements came from all the layers of the organization, from both management and continuing to emerge out of daily routines. These include requests for specific facilities for local police strategies, specific police categories (i.e. suspect, instigator), functionality to facilitate agent and inspector accountability (i.e. generate reports per each agent, not per patrol, to allow implementation of Compstat), capabilities for interoperability with other police agencies systems, measures to prevent wrong

[3] The action in the "24" TV series is centred in the high-tech hub of a fictional counter terrorism unit, where the staff work surrounded by a multitude of screens and are able to simultaneously access and aggregate information from a multitude of databases.

identification of persons in the system (i.e. checking of duplicate personal identification codes at introduction of entries), more fine grained access control (i.e. blocking of editing fields for certain roles) and also features to maintain local hierarchy and preserve the privacy of management officials (i.e. the impossibility for inspectors to track the GPS stations of management officials).

The next sections follow police workers in their routines while describing distributions of agency between humans and artefacts as they have been influenced by the introduction and shaping of the GIS. The thread guiding the sections is following the flow of data about persons—and in particular managing of suspicion—from the situations in the field towards its classification, aggregation and preparation for analysis and exchange with other police agencies.

6.2.1 Data Introduction

While geospatial solutions in the police are being shaped routinely in police practice, its introduction—along CompStat managerial philosophy—induced significant changes in the organization processes and working routines of agents, inspectors and analysts.

For one, field agents received additional responsibilities on completing the paper-based field reports with previously ignored details. They were initially required to introduce the reports themselves at the end of their shift. To make sure all details are introduced, the designers *inscribed* certain constraints in the introduction procedure. The goals of police management towards standardization and efficiency were *translated* into making certain fields mandatory in the information system. However, this *delegation* of disciplinary goals to the technological system came on the background of specific institutional and culturally shaped attitudes. The relatively low level of training among police agents as well as the little number of available computers in the police section led to long waiting times for data introduction by each agent, while the procedure of geo-coding incidents was considered too elaborate by the agents. As the head of police mentions, due to these factors and the particularly culturally shaped attitudes of the agents, this original arrangement was abandoned shortly. The procedure of introducing field data in the GIS system was then delegated to office staff and not regarded as highly interesting. As this tends to be the case, it is given, when possible, to new staff (Fig. 6.2).

6.2.2 The "susp." Notes

During this research, in several sessions in which field reports were introduced into the system, I noticed that many of the reports arrived incomplete from agents. Some arrived without the precise address and others contained only the note "susp.", with no details regarding the situation, reasons for assigning this label to a person or the

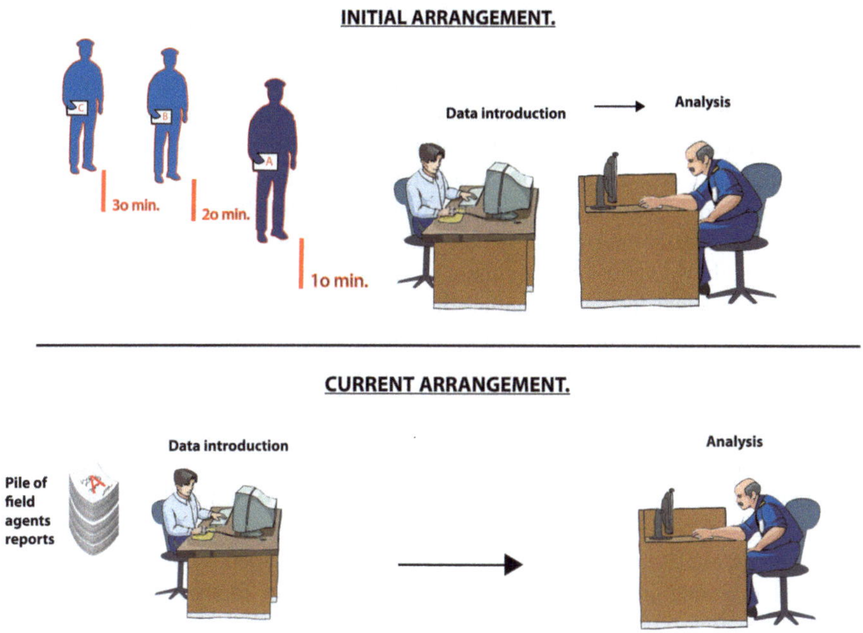

Fig. 6.2 Data introduction arrangements

type of suspicion. During the registration of such a note, after date, time, location and names involved, the system displayed a drop down list containing types of crimes and, after a small moment of looking at the report, the operator chose "Theft". Upon my request, the explanation followed:

> The program asks for an offence to be specified before going to the next step. Probably the suspect was searching through the trash bins as an alibi for steeling, probably bad clothing, kind of walking, hair style. What else could he have done in the parking lot at that hour?
> (local police data operator)

This situation can be analyzed on several accounts and it illustrates the mix of factors contributing to the construction and representation of suspicion in police systems.

The operator's decision to select a type of crime was, for one, induced by the system design. As the data operator mentioned, it prescribed her actions in making a choice on what the person was a suspect of, even if the field report did not provide details. The specific design was conceived anticipating a context of use in which the field agents were the represented users, introducing the field reports into the system themselves. As this task has since been delegated to office workers, the paper reports became the only easily available reference (except in situations in which the operators phone the tired field agents to ask for the missing details).

As pointed out by Akrich (1992), these situations are demonstrative of how designers define anticipated users, endowing them with particular capabilities, motives or aspirations, when *inscribing* predictions about the context of use. The '*script*' resulting from this process defines, like a film script, the framework of action in which the actors are supposed to act. However, even if the technical artefact may *prescribe* certain imperatives or invite dominant patterns of action, it may still be that actors define their own roles or, as the case of the local police system, that the anticipated actors do not come forward to play the roles envisaged by the designers. The arrangement chosen by the local police to both delegate disciplining to system design and to delegate data introduction from agents to office staff yielded a new situation not initially envisioned by designers and police management.

However, the operator's action cannot be said to have been completely determined by the technological design. The choice of selecting a category without having details on the paper report was also fostered by the organizational arrangements of rewards and punishments. During other shifts, I witnessed operators being reproached for leaving fields empty, as this could be noticed by the head of police during strategy meetings, possibly triggering further investigation within the organization. These reproaches were followed by the completion of the empty fields by the superior with data that was probably considered accurate but also not written in the reports.

With regard to the actual category of suspicion chosen by the operator—theft—this could be explained by a combination of culturally shaped opinions and situated knowledge. Sometimes, operative activities are guided by 'themes' of interest looking for particular types of offences. Although the system allows for specifying such themes, this step does not always occur, as several police staff mentioned. Without these additional details, the system does not distinguish for example, between 'suspects' identified as part of a drag net search, 'suspects' identified upon a theme operation or 'suspects' that gave the agent particular motives for identification. However, in the case of the 'susp.' notes, they were not informed by such a theme and, moreover, none of the assumptions regarding hairstyle, walking style, clothing style or behaviour of the identified person had a reference in the paper reports.

Upon my request, another operator retrieved afterwards the data collected about that particular person. We found that he was a teenager, age 14. The system retrieved 5 entries reporting that he was identified several times but, except the last one, none of the entries described theft or any committed offence.

To be sure, assigning the attribute 'suspect' in the local police system does not necessarily entail the starting of the procedure of detaining the person. Moreover, there is awareness among some police officers concerning the sensitivity of the *suspect* category and the *incompleteness of representations* (Gerson and Star 1986, p. 267) in the geographic information system in general. As one inspector mentioned me, "*This issue of suspicion is one of the controversial issues. Why should you be in our databases because you were wandering late and because you were not from this city?*" However, as the head of police notes, information produced by the system can inform and influence the attitudes of police agents when they verify a suspect or address. This in turn may potentially yield later on biased decisions or erroneous interventions on innocent citizens or at wrong addresses.

What the analysis above illustrates is the complex mixture of factors contributing to associating the attribute "suspect" to a person in a police information system, an association which proved erroneous in this case. More than being a mere "social" construct (Norris and Armstrong 1999), suspicion in these examples appears a socio-technical construct. It may be influenced not only by culturally shaped categories and opinions configuring the agent's formal and informal perception and attendance to the situation but also by particular system designs, vaguely defined procedures, and organizational arrangements.

6.2.3 Analysis Work and Strategy Meetings

With data introduced in the GIS, the analyst became responsible for generating all kinds of thematic maps (for example scandals and incidents with cars in a specific day/month/year), comparisons with maps of crime incidents exchanged with the national police (first introducing and geo-coding them in the local GIS), identification of patterns and informing management with suggestions about possible next steps.

In preparation for strategy meetings, the analyst, working in the cubicle-shaped workspace, is generally absorbed into intensive screen interactions, switching very fast between windows with maps afforded by the GIS, often moving with the pen on the screen and looking at each indicator that produces a brief description upon a mouse over:

> Beginning this year [i.e. 2010] we had a boom in thefts probably because of the [economic] crisis. Then we reacted by sending agents in those areas and did identification just as on the time of Ceausescu: Everyone after a certain hour was identified as suspect.
> [...]
> Look how they move after our actions. You can see how they cross the boulevard and move into this neighbourhood. Or look here: Nothing this week, then boom, then nothing, nothing, nothing: They tried this area but left it.
> (local police analyst)

The quotes disclose, for one, a tendency of the analyst to use data being previously introduced as a complete and accurate representation of events. This is suggested both by the vocabulary and the tense of the verbs used to refer to the elements displayed on the screen as if they are a real-time unfolding of events. Moreover, the quotes illustrate the practice of assigning suspicion during theme operations. These are justified by pre-emptive police strategies but they may also result in the registration of persons as suspects simply because of their presence at a particular place and time.

With the maps and reports prepared, the local police management gather weekly in strategy meetings. In attendance are the head of police and the chiefs of departments, all facing the screen wall while the analyst presents the maps from a computer. These contain each type of offence represented by coloured markers (car incidents in red, begging in black, etc.). Moreover, a separate map, on a separate screen, contains the data coming in from the national police each morning. These maps are compared to assess the next steps and allocation of resources, depending on the reports on the situation in the field. During one such meeting, the chief of police asked:

> Why is that whole neighbourhood empty [of incidents], we used to have much more events there? Has it become so quiet?

The response from the analyst follows:

> Rather that we're not there so much... [on top of] lack of motivation since the reductions ... (i.e. 25% salary cuts and the 40% personnel reduction).

The head responds:

> Yes, that's probably it. Next week we will make a special action in this neighbourhood on every offence.

The meeting is focused on the screens, with the head of police asking for reports and maps on certain criteria. At one point he stares for several minutes, mentally being absorbed in the screens, with a silence that got the others stare at each other. "*Show me what happened one week earlier*", said the head of police, breaking the silence, requesting a map which displays the events that were geo-coded one more week in the past, in the previous year. The analyst generates a new map but does not remember any details regarding the displayed representations of events. The head of police looks at the screens displaying the aggregated set of events and decides that all strategic orders stay the same just as on the week of the previous year.

The meeting closes with an order given by the head of police to a chief of department to make a covert action in civilian clothes concerning a new gang observed around a certain location. The head of police suspected them of doing more than begging and of being coordinated from outside the group. However, this plan of action and information was not recorded in the GIS.

For one, these situations described above show that an awareness of ambiguities in representations increases the chance for decisions being informed by local knowledge. When the head of police doubts the apparent lack of offences in a certain neighbourhood, he allows the analyst to remind him of recent personnel reductions and demotivating salary cuts, better explaining what the system displays.

Secondly, when this sort of local knowledge is not present, decisions rely solely on what the system displays. Thus, the system can be said to prescribe a particular way of resolving the situation, inviting a particular kind of use, implicitly co-shaping the use that is made of it. GIS ethical literature (Graeff and Loui 2008; Jenkins and McCauley 2006) identifies multiple ways in which such systems may induce value-laden decisions. GIS can contain inherent inaccuracies, inconsistencies, misrepresentations or alterations of data. Moreover, decisions disadvantaging certain categories of non-users can come from the use of GIS algorithms, either by ignoring or by combining multiple sources of data.

Thirdly, as shown in the last instance, not all plans for actions, contextual information, or information on certain suspicious activities (such as the suspected gangs), are recorded in the GIS. This reinforces the idea of unavoidable inaccuracies of representations. Additionally, the superficiality in data recording shows that users, possibly influenced by culturally shaped attitudes towards discipline, can refuse to use technologies. Selective ways of using technologies can coexist with the designer's inscriptions aimed at shaping user behaviour.

6.3 GIS and Privacy

The analysis so far showed that besides contributing to more efficient police practices, technologically mediated work introduces its specific risks. The analysis elaborated on cases showing the complex intermingling of human and technological factors involved in particular practices of data processing. Citizens, judged as "suspects", sometimes without solid reasons or with these reasons erroneously registered, have data about them classified and geo-coded in information systems, where they are being analyzed and exchanged with other agencies in preventive strategies or to help in criminal investigations.

These socio-technical practices are generally viewed by the police as presenting little risks for innocent citizens' rights and legitimate in balance with the overwhelming benefits (in this case 'suspicion', it is argued, is normally cross-checked with other data sources if acted upon, while erroneous interventions are analyzed as exceptions, compensated for, and tolerable especially in face of greater problems). However, precisely as increased legitimacy of security actions may downplay the implications of practices prone to cause informational privacy infringements, the identified situations provide the reasons for analyzing data processing practices, emerging within socio-technical ensembles.

While the above analyzed data flow is shown to generate inaccuracies, data subjects are largely unaware thus also unable to consent to information use or verify if and at what quality the data about them has been processed. While this may be unavoidable to a certain extent in police practices, it may also cause privacy violations even after the data is no longer needed.

At the same time, technologically mediated work is shown to induce in practitioners a tendency towards reliance on the data displayed on screens. Further on, data about particular persons can be processed in new ways and retrieved in different contexts but its partiality or inaccuracy tends to become 'black-boxed' to those other practitioners. In this way, once a person is classified as "suspect", the combination of this with other classifications may have multiple implications.

Firstly, it may trigger further verification in additional databases, effectively eroding privacy in the process of verifying whether the suspicion is justified or not. Secondly, when retrieving such data in different contexts it may jeopardise the *presumption of innocence* as the practitioner's attitudes relying on displayed data are shown to be influenced by the retrieved information. Thirdly, when processed in integrated information systems, aggregating multiple sources of personally identifiable data from interconnected organizations, may amount to causing, what Noëmi Manders-Huits (2011), referring also to Van den Hoven (1999), calls *accumulative informational harm.*

6.3.1 GIS and Accumulative Information Harm

Borrowing the notion of accumulative harm from Joel Feinberg (1984), Manders-Huits adopts it to e-government identity management systems by indicating that these may inflict harms either by malicious intent or recklessness by the government (e.g. careless implementations) and/or citizens (e.g. supplying of incorrect information). Whereas Feinberg explains accumulative harm as being harm inflicted by a collective through the accumulation of multiple seemingly harmless acts, Manders-Huits suggest that this also can occur in the accumulation of seemingly innocuous bits of information.

However, she distinguishes differences between the two notions. Feinberg notion of accumulative harm is caused when negligible individual acts turn out harmful in retrospect of their collective sense. However, as Manders-Huits observes, Feinberg's examples of individual acts (e.g. one person walking on the grass may not wrack the lawn but if enough people were to follow the exact same action, the grass would be unable to recover) are harmful on their own, only in negligible amounts. In the case of accumulative information harm, it is argued that the individual bits of information may not be considered harmful on their own, but the accumulation of data that may potentially cause harm.

In the case of large-scale identity management systems in e-government, Manders-Huits identifies three ways in which accumulative informational harm can be caused and sees the potential of such harm as inherent in their deployment regardless of practice. Firstly, these can come from incorrect identity related information stored in databases; secondly, from technical hitches of technical infrastructures (e.g. unintentional classifications disadvantaging categories of citizens); thirdly, the presence and accessibility of personal information contributing to a shift in power balance between citizens and government, rendering the latter more vulnerable.

The system analyzed in this chapter focused the potential for these three kinds of informational harm. It shows that (geographic) information systems contain inaccurate information and highlights that the appropriation of technological systems is done in the context of particular organizational cultures. As this is the case, the process is fraught with unanticipated situations. In addition, through the interconnection with other information systems—and even more so in the case of integrated or centralized information systems—the practice of data aggregation may cause accumulative informational harm, shifting the balance of power between citizens and government.

6.4 Concluding Remarks

Identifying rapidly evolving and complex sets of threats, with local incidents mirroring global phenomena, security strategies emphasize the need for integrated, comprehensive approaches. Towards these goals, interoperable and integrated information systems are seen as key factors in increasing the capacity for data gathering, processing and information exchange for common operative activities of local and national policing agencies. In this context, this chapter focused on the geographic information system of a local police station. It followed local police in their practices of data processing and preparation for collaboration with other policing agencies, exchanging data on a daily basis, including identified potential suspects.

Attempts towards regulating and assigning responsibility for the outcomes of these kinds of data processing lead to the identification of multiple factors. However, as the analysis in the chapter showed, blaming individual factors fails to account for the complex intermingling in a seamless web of human and machine relations.

Pointing only towards poorly trained agents or their culturally shaped attitudes fails to account for the ways their behaviour is steered by all kinds of programs of action, human and technological, conditioned by procedures, organizational arrangements as well as technological scripts. Pointing only towards technology for its affordances to quickly classify, aggregate and distribute (possibly erroneous) information or erode values fails to account for its appropriation in particular institutional arrangements and culturally shaped organizational contexts in which practitioners are able to use it in novel ways or not use at all. As one example from this case shows, the privacy of management officials has been preserved, despite the real-time tracking affordances of GIS, as their GPS-enabled units were removed by design from the gaze of police inspectors.

Pointing towards procedures and legal frameworks, fails to account for the agency of practitioners to interpret vague regulation as well as for the technological affordances able to induce new behaviours and render previous regulation inadequate. Therefore, the analyses of the ways privacy could get eroded and the ways informational harms can be generated in socio-technical practices shows the need for privacy protective principles as well as awareness among designers and practitioners regarding the management of identity related information.

Such a series of approaches to design (Friedman 1997; Nissenbaum 1998; Van den Hoven 2007), could provide principles and methodological contributions towards building such systems in ways sensitive to the values that may be affected. However, it needs to be employed and continuously developed in interdisciplinary cooperation (Monasso 2011) just as these systems need also to be *iteratively and incrementally* developed in the social and institutional context in which they feature (Gerson and Star 1986, p. 268).

This study highlighted that not only is informational harm caused by reckless data generation and processing but wrong or partial information coexists with accurate information (Sanders 2006, p. 732). This aspect points towards the need for transparency enabling tools such that data quality can be monitored and increased. At the same time, improving data quality may induce increased reliance on information, which in turn could decrease cautiousness when acting upon it. Therefore, this suggests the need for complementary awareness rising among practitioners regarding inherent ambiguities to prevent erroneous interventions.

As the potential for accumulative informational harm may be inherent in the deployment of interconnected and integrated information systems (Manders-Huits 2011), their development requires the employment of the principle of data minimization. In this way, minimum amount of data should be recorded for clearly defined purposes and stored for no longer than necessary. In turn, the implementation of this principle requires the transparency of categories and algorithms and putting in place advisory boards monitoring the levels at which data aggregation practices can amount to causing accumulative informational harm.

Acknowledgements The research leading to these results has received funding from the European Research Council under the European Union's Seven Framework Programme (FP7 2007–2013)/Grant. No. 201853.

Besides the formal support of the DigIDeas project, the author wants to thank Irma van der Ploeg and Jason Pridmore for their guidance and useful comments as well as police staff, municipality officials and technology developers for their generous collaboration.

References

Akrich, Madeline. 1992. The de-scription of technical objects. In *Shaping technology/building society: Studies in sociotechnical change,* ed. Wiebe Bijker and John Law. Cambridge: MIT Press.

Bijker, Wiebe, and John Law, eds. 1992. *Shaping technology/building society: Studies in sociotechnical change*. Cambridge: MIT Press.

Blanchette, J.-F., and D. Johnson. 2002. Data retention and the panoptic society: The social benefits of forgetfulness. *The Information Society* 18(1):1–13.

Dubbeld, Lynsey. 2004. *The regulation of the observing gaze: Privacy implications of camera surveillance*. Enschede: PrintPartners IpsKamp.

European Commission. 2010a. COM 673, the EU internal security strategy in action: Five steps towards a more secure Europe. Brussels: European Commission.

European Commission. 2010b. COM 609, a comprehensive approach on personal data protection in the European union. Brussels: European Commission.

European Union. 1995. *Directive 95/46/EC of the European parliament and of the Council on the Protection of Individuals with regard to the processing of personal data and on the free movement of such data*, Brussels, OJ No. L281, (The EU Data Protection Directive).

Feinberg, Joel. 1984. *The moral limits of the criminal law, harm to others.* Vol. 1. New York: Oxford University Press.

Flaherty, D. 1997. Controlling surveillance: Can privacy protection be made effective? In *Technology and privacy: The new landscape,* eds. P. Agre and M. Rotenberg, 167–192. Cambridge: MIT Press.

Friedman, B., ed. 1997. *Human values and the design of computer technology.* New York: Cambridge University Press.

Fulda, Joseph S. 2000. Data mining and privacy. *Albany Law Journal of Science and Technology* 11:105–113.

Gerson, E. M., and Susan Leigh Star. 1986. Analyzing due process in the workplace. *ACM Transactions on Information Systems (TOIS)* 4(3):267 (Special issue: selected papers from the conference on office information systems).

Graeff, C., and Michael C. Loui. Ethical implications of technical limitations in geographic information systems. *IEEE International Symposium on Technology and Society.*

Gutwirth, S. 2002. *Privacy and the information age*. Boston: Rowman & Littlefield.

Jenkins, D.G., and L.A. McCauley. 2006. GIS, SINKS, FILL, and disappearing wetlands: Unintended consequences in algorithm development and use. In *Proceedings of the Twenty-First Annual ACM Symposium on Applied Computing*, Dijon, France.

Koops, Bert-Jaap. 2003. The shifting 'balance' between criminal investigation and privacy. A case study of communications interception law in the Netherlands. *Information, Communication & Society* 6(3):380–403.

Koops, Bert-Jaap, and Ronald Leenes. 2005. 'Code' and the slow erosion of privacy. *Michigan Telecommunications & Technology Law Review* 12(1):115.

Latour, Bruno. 1987. *Science in action*. Cambridge: Harvard University Press.

Latour, Bruno. 1988. Mixing humans and nonhumans together: The sociology of a door-closer. *Social Problems* 35(3). (Special issue: The sociology of science and technology).

Latour, Bruno. 1992. Where are the missing masses? Sociology of a few mundane artefacts. In *Shaping technology/building society: Studies in sociotechnical change,* eds. Wiebe Bijker and John Law. Cambridge: MIT Press.

Manders-Huits, Noëmi. 2011. Regulating invisible harms, In *Innovating Government, Information Technology and Law Series,* eds. S. van der Hof and M.M. Groothuis, 20(1):57–73.

Monasso, Ton. 2011. Electronic exchange of signals on youth at risk. In *Innovating government, information technology and law series*, eds. S. van der Hof and M.M. Groothuis, 20(1):41–56.

Nissenbaum, Helen. 1998. Values in the design of computer systems. In *Computers in Society*, 38–39.

Norris, C., and G. Armstrong. 1999. *The maximum surveillance society.* Oxford: Berg Publishers.

Oudshoorn, N., and T. Pinch. 2003. Introduction: How users and non-users matter. *How users matter The CoConstruction of users and technology,* 247–270. London: MIT Press (Print).

Ratcliffe, J.H. 2008. *Intelligence-led policing,* 5. Cullompton: Willan Publishing.

Sanders, Carrie. 2006. Have you been identified? Hidden boundary work in emergency services classifications. *Information, Communication & Society* 9(6):714–736.

Stancu, Şerb, ed. 2006. *Police tactics manual. Ministry of administration and Internal affairs publishing house.*

Star, S.L. 1991. Power, Technologies and the phenomenology of conventions: On being allergic to onions. In *A sociology of monsters? Essays on power, technology and domination, sociological review monograph,* ed. J. Law 38:26–56. London: Routledge.

Supreme Council for National Defence (CSAT). 2007. *National defence strategy of Romania.* Bucharest: Supreme council for national defence (CSAT).

Supreme Council for National Defence (CSAT). 2010. *National security strategy of Romania.* Bucharest: Supreme council for national defence (CSAT).

Tilley, Nick, ed. 2003. Community policing, problem-oriented policing and intelligence-led policing. In *Handbook of policing*, 326, ed. T. Newburn. Cullompton: Willan Publishing.

Tilley, Nick. 2009. *Crime prevention*, 95. Cullompton: Willan Publishing.

Van den Hoven, Jeroen. 1999. The internet and the varieties of moral wrongdoing. In *Internet and Ethics,* ed. D. Langford London: McMillan.

Van den Hoven, Jeroen. 2007. ICT and value sensitive design. *IFIP international federation for information, the information society: Innovations, legitimacy, ethics and democracy processing*, vol. 233.

Van der Ploeg, Irma. 2003. Biometrics and privacy: a note on the politics of theorizing technology. *Information, Communication & Society* 6(1):85–104.

Vedder, A.H. 2001. KDD, privacy, individuality, and fairness. In *Readings in cyberethics,* eds. R.A. Spinello and H.T. Tavani, 404–412. Boston: Jones and Bartlett Publishers.

Verbeek, P.-P. 2005. *What things do: Philosophical reflections on technology, agency, and design.* Pennsylvania: Pennsylvania State University Press.

Part II
Regulation, Enforcement and Security

Chapter 7
The Set Up of Data Protection Authorities as a New Regulatory Approach

Philip Schütz

7.1 Introduction

Embedded in a dissertation project that is dedicated to a comparative analysis of data protection authorities (DPAs; EU Directive 1995),[1] this chapter aims to shed light on the conception of DPAs as a new regulatory approach by the European Union (EU). Since there is little research on DPAs from a political science perspective, the theoretical foundations of and empirical findings about independent regulatory agencies (IRAs) can help to serve as a template model providing a framework for analysis of DPAs.[2]

IRAs represent a crucial instrument of the *regulatory state*, which is characterised by ongoing deregulation, increased delegation and reregulation processes (Majone 1994, 1996, 1997). They comprise a relatively new attempt to cope with societal challenges that elude traditional models of governance. The EU Data Protection Directive makes use of this instrument, stipulating mandatory supervisory authorities, which have to fulfil a variety of different functions. DPAs are not only expected to serve as ombudsmen, auditors, consultants, educators, policy advisers and negotiators, but they should also be able to enforce changes in behaviour, when private or public actors violate data protection legislation (Bennet and Raab 2006, p. 135).

Most importantly, contrary to most IRAs, DPAs are not only assigned to supervise private entities such as companies of various business sectors, but they are also expected to watch over public authorities, i.e. executive, legislative and judicial

[1] Data protection authority refers in the following text to the term of supervisory authority, stipulated in the Directive 95/46/EC.

[2] Since the dissertation project is in its theoretical conceptualisation phase, this work attempts to explore and discuss new theoretical and methodological approaches to the analysis of DPAs rather than to present substantial empirical results. This is also reflected in the structure of the chapter.

P. Schütz (✉)
Fraunhofer Institute for Systems and Innovation Research ISI, Karlsruhe, Germany
e-mail: philip.schuetz@isi.fraunhofer.de

S. Gutwirth et al. (eds.), *European Data Protection: In Good Health?*, 125
DOI 10.1007/978-94-007-2903-2_7, © Springer Science+Business Media B.V. 2012

institutions and bodies.[3] Despite the traditional checks and balances in a democratic and constitutional state, the monitoring of governmental bodies by an authority closely linked to the government is particularly new in the theoretical framework of the *regulatory state* and IRAs. Since "[w]e are not normally accustomed to think of government as 'regulating' itself" (Hood 1999, p. 3), the setting up of DPAs by the political actors in order to exercise control over the same presents an interesting and new aspect in the theory of regulation. Thus, the concept of independence in this context seems to require special attention. However, other essential and problematic features of IRAs and DPAs such as credibility, accountability, democratic deficit, legitimacy and effectiveness will be discussed as well.

As a final remark, the author wants to emphasise that this work displays the starting point of the above mentioned dissertation project. The text should therefore be considered as exploratory and tentative, reflecting new ideas to examine the role of DPAs in society.

7.2 The Development of the Regulatory State

One of the most controversial issues in political economy and public policy research has been the question of how deep the state should penetrate the market economy. Taking a closer look at dominant theories behind public policy-making processes of post-war countries in Western Europe, there seems to be a wavelike motion over time; a pendulum swinging back and forth between the poles of interventionist and free market approaches. While the strategies of stabilisation and consolidation—which included centralised administration, state ownership and planning—mainly shaped the political programmes of the 1950s and 1960s in most West European countries, the privatisation and deregulation dominated the policies in the 1970s and 1980s, strongly influenced by the American model (Majone 1994).

Today's *regulatory state* represents, in many ways, a mélange between the *dirigiste* and the *neoliberal state* of the preceding decades (Mayntz 2009).[4] On the one hand the state continued, throughout the 90s, to retreat from sectors which were once publicly owned, e.g. utilities such as traffic, gas, electricity, water, etc. In the new millennium the trend of privatisation seems to have manifested itself, since once considered core competences of the state such as education, welfare, pensions, police, military, and even whole policy-making processes, are subject to delegation (Kemp 2011). However, critics state that deregulation has not necessarily led to increased efficiency and a decrease in governmental rules and regulatory activities (Vogel

[3] Of course, there are additionally numerous data protection commissioners and officers in private companies and governmental organisations working together with DPAs on the international, national and federal level. However, they will not be part of the analysis in this work.

[4] Regulation refers in the following to "the development and application of rules (e.g. laws, directives, guidelines, standards, codes of conduct etc.) directed at specific target populations, and the—public as well as private—rule-makers involved." Moreover, this work mainly focuses on public regulation, i.e. regulation by the state, neglecting, for example, self-regulation approaches.

1996). On the contrary, privatisation frequently resulted in a massive increase of new state regulations, as happened in the British utilities sector in the 1980s and 1990s (Thatcher 1998).

On the other hand, the state "reregulates the now-liberalized markets [and sectors] with less intrusive instruments" (Gilardi 2002, p. 873). Rather than direct state intervention, reregulation ideally implies the idea of improving "the efficiency of the economy by correcting specific forms of market failure such as monopoly, imperfect information, and negative externalities" (Majone 1994, p. 79). Yet, imposing rules on highly dynamic and innovative sectors has turned out to be highly problematic and extremely difficult.

In fact, the information technology (IT) sector, for example, where the emergence of the personal computer, the Internet as well as mobile computing revolutionised the collection, storage, processing and exchange of information, has mostly evaded public regulation successfully. There are two main reasons for this: First of all the immense velocity of technological development has been constantly leaving state regulation to lag behind. Secondly, the transnational and globalised nature of the sector makes it extremely difficult for policy makers to pass effective regulatory legislation on a national level. Hence, the demand for reliable and trustworthy rules in these heavily under-regulated innovative sectors increases and the call for regulation has become more and more prevalent across national boundaries.

Furthermore, the evolution of the *regulatory state* in Europe was closely linked to the growing importance of the EU. Since the EU neither represents a federal state nor a confederation but rather made it necessary to redefine the term "union" as a new form of federation *sui generis*, based on the idea of a multilevel governance (Hooghe 2001), the EU legislation was designed to be regulatory rather than redistributive (Caporaso 1996, p. 39). Comprehensive empirical research on the *regulatory state* by Majone (1997, p. 139) suggests that "rule making is replacing taxing and spending". Traditionally, taxing and spending powers have always been the most powerful and prevalent instruments of national governments. Member States were therefore cautious in transferring these competences to EU institutions, although, at the same time, they had to provide a budget that ensured the Union's capacity to act.

Whether or not the EU can thus be considered a *regulatory state* remains a controversial subject; also because it is highly contested as to which political level is actually responsible for most of the regulations, as well as the most important ones. Whereas advocates of the Europeanisation thesis argue that EU-regulations and their influence on national legislation have grown exponentially in absolute as well as relative terms (Majone 1996, p. 144), adversaries emphasise the ongoing dominance of national regulatory institutions as well as the incapacity of European regulation.

Eberlein and Grande (2005, p. 98) introduce a third perspective, i.e. the differentiation thesis, which takes the constraints of harmonisation as well as the self-interest of the Member States into account. "Regulation in Europe, whether market creating or market correcting, thus includes both levels: it is national and European".

This governance model (Levi-Faur 1999, p. 201),[5] however, poses a serious dilemma for effective regulation. The two authors identify a so-called *supranational regulatory gap*, which, *inter alia*, originates from the latitude that is given to Member States when implementing EU law into national legislations (Eberlein and Grande 2005, p. 98). One of the key findings is that informal institutions and above all transnational networks serve as a back road to effective regulation in order to circumvent the regulatory lacunae (Eberlein and Grande 2005, p. 91).

In a nutshell, the concept of the *regulatory state* is characterized by an ongoing deregulation combined with (re-) regulation processes in either already liberalized markets or highly dynamic sectors that bring societal challenges. In Europe, public regulation is characterised by a complex, often state-centred, multi-level governance model that primarily comprises the local, regional, and particularly the national as well as European level.

7.3 Independent Regulatory Agencies

A distinguishing attribute of the *regulatory state* is reflected in the concept of delegation, which, contrary to privatisation, describes the transfer of authority and responsibility from the state to another private or public organisation without being completely exempted from accountability. Delegation of sector-specific regulation assignments to IRAs is an essential and prevalent tool of the *regulatory state*. Yet, the historical origins of IRAs in Europe date back to the early post-war years, when Britain and Germany introduced independent competition authorities (Wilks and Bartle 2002). Even prior to that, IRAs in the United States had become an integral part of the governmental agency landscape, typically operating outside the federal executive departments. Today, the model of independent central banks, which has been rapidly spreading throughout Europe and the rest of the world, is considered as one of the main archetypes of IRAs (Quintyn 2009, p. 267).

IRAs can be defined as "a body with its own powers and responsibilities given under public law, which is organisationally separated from ministries and is neither directly elected nor managed by elected officials" (Thatcher 2002, p. 956). As the name already suggests, independence from governmental influence plays an essential role in the conception of IRAs.[6]

Yet, it seems surprising that the state is willing to accept a loss or at least a reduction of control and power in certain regulatory sectors. Gilardi (2005, p. 102) identifies several reasons for this transfer of competencies to IRAs. One of the most important causes involves the objective of governments to reduce their decision-making costs

[5] In this context Levi-Faur coined the even more appropriate term "state-centred multi-level governance".

[6] This contribution solely deals with IRAs/DPAs on the national level, although the international, and particularly the EU level, would be interesting to look at as well.

by e.g. taking advantage of IRAs' specialised expertise in the relevant field of regulation. "Faith in the power of expertise as an engine of social improvement—technical expertise which neither legislators, courts nor bureaucratic generalists presumably possess—has always been an important source of legitimisation for regulators" (Majone 1997, p. 152).

Moreover, due to their independence, IRAs are not only more flexible in adjusting regulations to changing conditions, but they also work more effectively and efficiently, presenting better regulatory outputs than traditional bureaucracies. They tend additionally to organise their decision-making processes in a more open and transparent way and eventually policy makers are able to profit from shifting blame to IRAs when regulation fails to succeed.

However, the most convincing argument as to why governments delegate power to IRAs is offered by the *credibility hypothesis*. Since "politicians have few incentives to develop policies whose success, if at all, will come after the next election [. . .], it is difficult for political executives to credibly commit themselves to a long-term strategy" (Majone 1997, p. 153). Being exempt from elections and the associated political short-term thinking, IRAs are able to fill this credibility vacuum. They can provide a certain time consistency in their policies leading to a more "stable and predictable regulatory environment" (Gilardi 2005, p. 102). Fearing rapid changes in popular support, governments have, after all, an interest in preserving their policy achievements through IRAs, in order to prevent future parties in power from altering them too easily.

7.3.1 The Model of Regulatory Governance

Whereas the motivation for reducing decision-making costs does not necessarily rely on the agency's independence, most of the other reasons for the delegation of authority to IRAs are founded on this distinguishing attribute. The *credibility hypothesis* especially, is linked interdependently to the feature of independence.

Yet, other important characteristics of IRAs should not be left unconsidered. According to Quintyn (2009, p. 272), an International Monetary Fund (IMF) economist, who has extensively dealt with the concept of independent central banks,there has been too much focus laid upon independence. Being of the opinion that there is a permanent independence-bias, he suggests concentrating instead on the entire **governance model of IRAs** (Fig. 7.1), including, besides independence, accountability, transparency and integrity. In the case of central banks, the author argues that independence is a necessary, but not sufficient, condition for the main objective of price stability (Quintyn 2009, p. 274). He notes that independence, which only represents one pillar of his regulatory governance model comprises a principle that is not absolute and never has an end in itself. On the contrary, there is interdependence between the four pillars of the model.[7]

[7] However, due to the relative newness of the attempt to apply the theoretical concept of IRAs to the analysis of DPAs this work will mainly concentrate on the feature of independence and

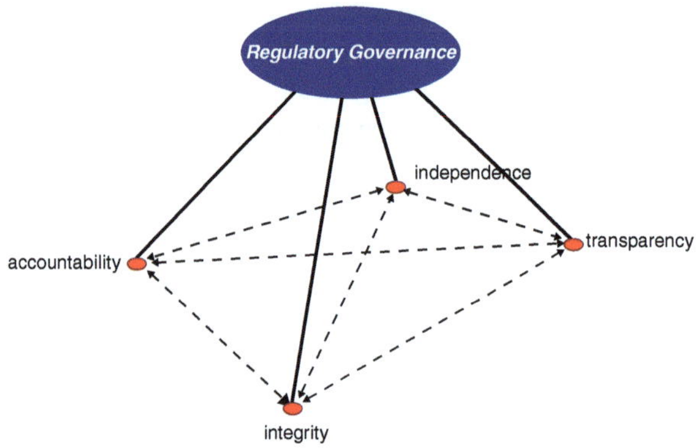

Fig. 7.1 Regulatory governance. (Source: Qintyn 2009, p. 282)

7.3.1.1 Independence of IRAs

Even though the concept of IRAs' independence seems to be rather straight forward, it is in fact highly complex, implying different nuances and dimensions. In order to exemplify the challenges in measuring the independence of IRAs, two empirical studies are briefly discussed.

Though being confronted with operationalisation and quantification problems, Gilardi has developed an independence index concentrating on formal, i.e. legally stipulated, independence. The comparative analysis embraces 33 IRAs from 5 different regulatory sectors in 7 European countries. Originally developed by Cukierman et al. (1992) in a comprehensive study focused on measuring the independence of central banks, the index involves five groups of indicators: The agency's head status, the management board members' status, the general frame of the relationships with the government and the parliament, financial and organisational autonomy, and the extent of delegated regulatory competencies (Gilardi 2002, p. 880). In order to avoid subjective valuation, he attributes the same weight to each of his 21 indicators.[8]

In a second step Gilardi (2002, p. 884) tries to explain why the independence of regulatory agencies varies from sector to sector and country to country. The results of his quantitative analysis, deploying multivariate regression models, reveal a significantly positive correlation between the country's degree of market openness and independence. Moreover, national veto players are negatively associated to agency independence, and eventually economic interdependence has no measurable impact on the formal independence of IRAs. He also finds an important difference between

accountability. Principles such as transparency and also integrity will mostly be neglected, although they comprise crucial elements of a good governance model, which will be subject to a more comprehensive assessment within the dissertation project.

[8] As an example, the indicator "term of office" can have six different parameters: "Over 8 years", "6–8 years", "5 years", "4 years", "fixed term under 4 years or at the discretion of the appointer", and eventually "no fixed term". Each parameter is assigned a value evenly spread between 1 (= complete independent) and 0 (= not independent). Since there are six parameters, the assigned values are accordingly: 1, 0.8, 0.6, 0.4, 0.2, 0.

economic and social regulation, namely that only the first is likely to be carried out by IRAs.

One of the main problems of Gilardi's quantitative analysis is that it only captures legally stipulated independence. Although pointing to that problem, he does not consider informal dependences, which are likely in an agency created by government. A combination of quantitative and qualitative research would have been interesting in that respect.

Thatcher (2002, p. 959), in contrast, includes informal aspects into his comparative analysis of 23 IRAs from 8 different sectors in 4 countries, using 5 groups of indicators: Party politicisation of regulators, departures of IRA members before the end of their term, official tenure of IRA members, financial and staffing resources as well as the use of powers by elected politicians to overturn the decisions of IRAs. The author not only integrates informal aspects of independence such as politicization of regulators into his models, but also expands his analysis to possible dependencies on regulatees, i.e. large companies, which is operationalised by the percentage of IRA members formerly or subsequently active in the private sector (*revolving doors*). Eventually, the influence of IRAs on decision-making processes is scrutinised.

Thatcher (2002, p. 966) concludes: "Having created IRAs, governments do not use their most visible formal powers to control them, with the exception of limiting IRA's resources (and the partial exception of politicization in Italy)". In addition, IRAs seem separated from business by and large, meaning that the *revolving door* phenomenon occurs rather seldom. Besides, sharp conflicts are frequently carried out between IRAs and the private sector, including legal action. Britain, however, represents an exception in both cases. In the end, he is of the opinion that IRAs have significantly contributed to making decision-making processes more open and transparent (Thatcher 2002, p. 969).

Both of these studies show the complexity of independence in the context of IRAs. Seemingly simple questions such as "Who is independent (the agency or single officials)?" and "Independence from whom (public or private actors)?" turn out to be difficult to answer. There is a formal (statutory) as well as informal dimension of independence and particularly the latter needs to be qualitatively examined, e.g. by conducting interviews with IRA officials.

In sum, it seems crucial that public as well as private sources of influence are separately identified in the analysis of IRAs, and accordingly, DPAs. Though stressing the importance of independence, private and public policy-makers are, in fact, confronted with a conflict of interests when an agency is enabled to escape their spheres of influence. Thus, it is most likely that IRAs are subject to attempts at severe manipulation. A comprehensive assessment and evaluation of independence should furthermore include a combination of quantitative and qualitative methods in order to grasp the informal dimension of potential dependencies.

7.3.1.2 The Importance of the Accountability Principle

Accountability is one of the most neglected, yet immensely important elements of IRAs, because, according to Bovens (2005, p. 192), it fulfils, above all, five key functions: Democratic control, legitimacy, enhanced integrity, improved performance and public catharsis.[9]

The first is clearly linked to the idea of *providing public oversight* in order to "give account" of IRAs' actions. In this context Majone (1997, p. 160) emphasises that "the simplest and most basic means of improving agency transparency and accountability is to require regulators to give reasons for their decisions". In this way IRAs are open to external checking mechanisms such as judicial review, public participation, peer review and policy analysis.

Secondly, one of the most important functions comprises *maintaining and enhancing legitimacy*. Since non-elected institutions such as IRAs face the problem of a democratic deficit inherent in their conceptual structure, the formerly supposed advantage of not being part of the electoral process turns out to be disadvantageous when it comes to legitimacy. Hence, the concept of accountability becomes even more relevant when considering that IRAs should avoid giving the impression of being a fourth branch of government (Meier and Bohte 2006).

Disagreeing with this view, Majone (1997, p. 159) argues that IRA's democratic deficit is mainly deduced from the assumption that "the only standard of democratic legitimacy is direct responsibility to the voters or to the government expressing the current parliamentary majority". Being an alternative to the common majoritarian archetype, the Madisonian democracy model primarily aims to share, disperse, delegate and limit power, in order to avoid Tocqueville's *tyranny of the majority* (Hamilton et al. 2008, p. 48; Tocqueville 2000). Consequently, the criticism that accuses IRAs of a serious democratic deficit is rather unfounded when following the Madisonian idea of democracy.

Thirdly, accountability helps to *enhance the integrity of regulatory governance*, since giving public account about one's actions reduces the risk of self-interest capture, i.e. individual staff pursuing their own self-interest by manipulating or subverting regulation (Quintyn 2009, p. 279).

Finally, there is a continuous *improvement of agency performance* mainly referring to individual as well as institutional learning processes. "Norms are (re)produced, internalized, and, where necessary, adjusted through accountability" (Bovens 2005, p. 193).

However, the principle of accountability remains problematic. Regarding the model of regulatory governance, one of the main difficulties lies in the interdependent relationship between accountability and independence, often presented as a trade-off. While doing research on the independence of central banks, Eijffinger et al. (2000) and Bibow (2004) objected to this view. The first has developed a model showing

[9] The fifth function refers to accountability in cases of tragedies or fiascos, which is less important in regards to the topic of this contribution and will therefore be neglected.

that independence and accountability complement each other in the case of monetary policy, while the latter emphasised the need to balance the two principles.

Even though both features do comprise counterparts, accountability should be seen as complementary to independence rather than antagonistic. "The concept of a 'trade-off' is flawed to the extent that it assumes that stronger accountability mechanisms must necessarily mean a less independent regulatory agency" (Quintyn and Taylor 2007, p. 18). Accountability and independence do not have to be mutually exclusive. In fact, it is even possible that they profit from each other, meaning that accountability underpins the autonomous status of IRAs by providing e.g. legitimacy. Yet, it should be noted that "the political principal can transfer his powers to the independent delegate, but not his legitimacy, [. . .] [which is why] IRAs must rely on other external sources of legitimacy [such as output-oriented legitimacy]" (Maggetti 2010, p. 3).

In order to reconcile IRAs' autonomy and the accountability claim of the political principal, Quintyn et al. (2005, p. 19) suggest a set of practical accountability arrangements: Ex ante and ex post accountability mechanisms refer to the obligation of reporting before or after actions are taken. Whereas these mechanisms follow an explanatory rationale, amendatory accountability implies the obligation to resolve deficiencies in policy or regulatory rule making. Furthermore, procedural accountability describes the legally binding procedures that must be followed when regulatory actions are taken and substantive accountability is supposed to ensure the alignment of supervisory actions with the IRA's official objectives. Though rarely applied, personal accountability corresponds to the sanctioning of individual top officials such as the head of an IRA. Financial accountability eventually creates the need to present proper financial statements, while performance accountability would emphasise the duty for IRAs to work effectively.

Despite these rather clearly structured arrangements, one major problem of accountability involves the confusion with control, causing profound problems for IRAs in exercising their tasks properly. Accountability should therefore only be enforced by a combination of monitoring arrangements and instruments (Quintyn 2009, p. 280), which should normally abstain from any governmental control mechanisms such as the exclusive right of appointment, dismissal, budgeting, etc. Therefore, while it remains tremendously important to keep an eye on the degree of influence from governmental and parliamentary actors, it should not be forgotten that accountability represents a vehicle for the construction of legitimacy as well as other important features.[10]

[10] As seen in the preceding paragraphs, the concepts of transparency as well as integrity permeate the accountability principle. Whereas the first is an important driver to enforce accountability, the latter serves as a guarantor of legitimacy and credibility. As mentioned before, however, transparency and integrity are mainly left out of the analysis due to the focus on independence and accountability.

7.4 The EU Data Protection Directive

Directive 95/46/EC represents the most comprehensive and influential legislative framework regarding the protection of personal data. From the start of the Directive's development it became clear that "data protection had ceased to be merely a human rights issue; it was also intrinsically linked to the operation of international trade" (Bennett and Raab 2006, p. 93).

Since the development of national data protection legislation in Europe had started almost 25 years prior to the Directive (Bennett 1992, p. 77), a patchwork of different data protection acts marked the European legal landscape. According to Mayer-Schönberger (1997, p. 221) national data protection legislation in Western Europe predominantly followed four patterns until the emergence of the EU Directive: The first laws in the 1970s were directed towards restrictions of huge centralised data banks and storage facilities run by governments and large companies; a second generation of legal provisions in the late 1970s aimed furthermore at the regulation of new emerging and rapidly spreading decentralised state and business sector databases. The third phase was dominated by the practical question of how the individual should actually exercise control over his/her information. In 1983, the seminal ruling of the German Constitutional Court, overturning the national census law and establishing the right to *informational self-determination*, provided a legal answer to this question. Thus, a fourth wave of legislations and amendments followed, incorporating the newly created norm and strengthening its status as an individual right.

Although the legal situation of citizens willing to defend their privacy rights in court had improved in many European states, the different data protection laws created a complicated patchwork of provisions associated with legal uncertainty, which caused, especially in the case of transferring personal data from one European country to another, enormous impediments for the private sector. That is why the most convincing argument in favour of approving the European Data Protection Directive was the objective of eliminating these impediments in order to harmonise the market (Gutwirth 2002, p. 91; Simitis 1994).

Aiming to create a more homogenous legal and economic sphere to guarantee the free flow of data across European borders, the Directive, which took over 5 years of tough negotiations from first draft to adoption, was therefore mainly developed in the Internal Market portfolio of the EU Commission (Bennett and Raab 2006, p. 93). Furthermore, the economic argument not only helped to persuade stakeholders of the Directive's benefits, but also provided a legal basis, namely Article 100a of the EC Treaty, which seeks, *inter alia*, "to ensure the establishment and functioning of the Internal Market".

Since then the Directive has not only facilitated the transborder flow of personal data within the EU, but has also contributed to higher minimum standards of data protection in quite a few Member States. Due to the fact that EU directives are supposed to be binding, each Member State has, at least formally, implemented comprehensive data protection provisions. Although these provisions have not been able to keep pace with technological developments and new emerging threats to privacy, the Directive, which is currently under revision, still constitutes one of the most advanced legal frameworks in the field of data protection worldwide.

7.4.1 *The Role of DPAs*

The European Data Protection Directive also stipulates the mandatory set up of data protection authorities, the so-called supervisory authorities (EU Directive 1995, article 28). This has mainly been done to harmonise the responsibilities of DPAs, which were quite different in data protection provisions of the Member States developed prior to the EU Directive. Sweden, for example, relied on a licensing model, whereas the French government chose a more selective approach and Germany provided for subsequent monitoring as well as recommendations rather than binding decisions (Hustinx 2009, p. 134).

After the Directive had been implemented, functions of DPAs changed in the following ways: First of all, DPAs are expected to monitor the application of the Directive's provisions in their country (EU Directive 1995, article 28, para. 1). Second, they should serve as policy-advisors concerning administrative measures or regulations associated with the processing of personal data (EU Directive 1995, article 28, para. 2). Third, the Directive provides DPAs with investigative powers, effective powers of intervention and the power to engage in legal proceedings (EU Directive 1995, article 28, para. 3). Finally, they have to fulfil the role of ombudsmen, being obliged to hear claims lodged by any person (EU Directive 1995, article 28, para. 4).

The effective powers of intervention include the provision of "delivering opinions before [potentially privacy threatening] processing operations are carried out" (EU Directive 1995, article 28, para. 3, subparagraph 2), which results in a consultative function. DPAs are therefore supposed to advise not only public but also private actors when, for example, new policies or technologies are developed that might have privacy-invasive implications.[11] In addition, Bennett and Raab (2006, p. 139) have identified education as a rather informal but nonetheless important mission of DPAs, which relates to the objective of raising awareness and supporting data protection by the individual him or herself ("Selbstdatenschutz").

Considering all these assignments, it is not surprising that a comprehensive comparative study of the European Commission comes to the conclusion that DPAs are frequently overwhelmed by an enormous workload and sometimes incompatible functions. The authors note: "We feel that this is too much to ask of any single body"

[11] However, these formal provisions of the EU Data Protection Directive do not mean that national DPAs are actually endowed with the same powers and tasks. As already mentioned, Member States are granted some latitude in the transposition of EU law into national legislation, which often results in quite a different legal set up of DPAs.

(Korff and Brown 2010, p. 44), and in comparison to other IRAs, DPAs are in fact confronted with performing an incredible variety of different tasks.

One of the most distinctive features of DPAs is their assignment to watch over private as well as public actors. This is contrary to the work of most IRAs, monitoring (financial) markets or the utilities sector. Here, DPAs comprise an interesting aspect, namely the idea of an authority that was created by the state in order to control the same. Hood (1999, p. 223) deal with this new facet of regulation, identifying huge deficits in the actual transposition of the concept. Hence, the issue of independence especially from governmental influence requires extra attention.

7.4.2 *Independence of DPAs*

Although the majority of IRAs face similar conflicts concerning their continuous struggle for independence, DPAs are particularly under threat of being held in check by public authorities. Since the state not only delegates power to DPAs, but could additionally be subject to harsh criticism and potentially strict regulations itself, public actors have an increased interest in manipulating the output and outcome of DPAs' actions.[12] Maybe that is why the Directive has specifically stated that DPAs "shall act with **complete independence** in exercising the functions entrusted to them" (EU Directive (1995), article 28, para. 1).[13]

Nonetheless, the legal and political reality in EU countries shows that extremely different interpretations of the term "complete independence" are prevalent. Even though the goals stipulated in the Directive are supposed to be binding, the Member States are granted some latitude in working out the details of the finally implemented national legislation (Simitis 1994). That is why the legal set up and status of DPAs varies from country to country or in a federal state like Germany even from land to land. These legal details, however, can determine what kind of powers and tasks are delegated as well as whether or not DPAs are able to work independently and hence effectively.[14]

In the following section Germany has been chosen as an example to demonstrate how much room for interpretation of the term "complete independence" sometimes remains. At the same time, the German case serves as a starting point to conduct a more comprehensive analysis of DPAs in EU Member States.

---

[12] Furthermore, DPAs are traditionally closely linked to certain ministries.

[13] Even though "complete independence" is the term used in the EU Directive, there is no institution, organisation or individual who can claim to be complete independent. However, the wording refers to the increased relevance the European Union put into the autonomous status of DPAs.

[14] As already pointed out in Sect. 3.1.1, it is important to consider the difference between the formal and informal dimension of independence. This work will only focus on legally stipulated independence features of DPAs.

7.4.2.1 The German Case

As we have shown, the concept of independence is hard to define and can be stretched at will. It also seems that the notion of independence varies over time. In Germany, where DPAs have a long tradition at the national (since 1978) as well as regional levels (in Hesse since 1970), independence was always an essential element of data protection institutions (Hessian Data Protection Act 1970). Supporting this point, German DPAs even served as a role model in the process of developing the EU Data Protection Directive.

Lately, however, the governmental interpretation of independence in Germany has come under closer scrutiny. Because Germany is a federal state, regulatory powers were given to the national DPA, i.e. the Federal Commissioner for Data Protection and Freedom of Information (FfDF),[15] as well as to his regional equivalents on the Länder level. Over the years, a differentiation process has been taking place between the two levels. Today, the FfDF is responsible for data protection issues concerning federal public institutions and their policies, whereas DPAs of the Länder monitor the public sector on the regional level as well as, most remarkably, non-public bodies, *inter alia* private enterprises, within their territory.

Consequently, some Länder governments launched specific governmental agencies, which were put in charge of supervising non-public corporations in regards to their compliance with data protection law. The close relationship between the government and the regulating agencies especially, caused the EU Commission as well as the European Data Protection Supervisor (EDPS) to file a suit against the Federal Republic of Germany for infringement of the "complete independence" principle.

In March 2010 the European Court of Justice (ECJ) ruled that Germany "failed to fulfil its obligations under the second subparagraph of Article 28(1) of Directive 95/46/EC", i.e. the assurance of "complete independence" (Judgment of the Court 2010). Indeed, the ECJ confirmed that some governments of the German Länder had appointed specific "authorities [to be] responsible for monitoring the processing of personal data by non-public bodies and undertakings governed by public law which compete on the market (öffentlich-rechtliche Wettbewerbsunternehmen)" (Judgment of the Court 2010, para. 56).

Furthermore, the court devoted much attention to the clarification of the meaning of "complete independence". The ECJ stated "that a supervising authority must be free from any influence, whether that influence is exercised by other authorities or outside the administration. The fact that these DPAs are subject to State scrutiny in Germany constitutes an infringement of that requirement" (Judgment of the Court 2010, para. 15). Apparently, some Länder governments had a narrower interpretation of "complete independence" in mind, proposing the concept of "functional

---

[15] The FfDF in Germany represents not only the head of the national DPA but also the institution itself. DPA officials are directly working for him.

independence" in the sense that DPAs must be primarily independent of regulatees from the private sector (Judgment of the Court 2010, para. 16).[16]

Despite the judicial decision of the ECJ, the independence of Germany's FfDF seems jeopardised since there are a significant number of gateways and possibilities of governmental influence. Most remarkably, the FfDF is organizationally attached to the Federal Ministry of the Interior, which has several problematic consequences. Even though the commissioner remains, in general, independent from any instructions or orders (functional supervision) by the government, which is specifically stipulated in the national data protection act (Federal Data Protection Act 2009 (1990), article 22, para. 4), he/she is exposed to administrative supervision by the ministry.

According to Dammann (2011, p. 1057), a legal scholar and former top official working for the FfDF, the administrative supervision could not only offer ways to seriously hamper the DPA's work, but also result in a so-called "anticipatory obedience" by the commissioner. Dammann (2011, p. 1058), furthermore, points to the fact that the national DPA is often only a way station for public servants of the Ministry of the Interior, where they will normally continue their career later on. This is highly problematic in terms of the staffs' commitment, orientation and willingness-to-comply.

In addition, the FfDF is not in a position to decide on his/her personnel policy independently (Federal Data Protection Act 2009 (2009) article 22, para. 5, cl. 5), since the ministry also has a say in it. In cases of a promotion or "voluntary" transfer of an employee, the authority even lies exclusively with the ministry (Dammann 2011, p. 1057). Finally, the commissioner is subject to statutory supervision by the government (Federal Data Protection Act 2009 (1990), article 22, para. 5.), which constitutes another potential source of governmental influence. All of this is particularly critical, since the Ministry of the Interior is traditionally in charge of often privacy-invasive national security policies.

All in all, the institution of Germany's FfDF does not seem to fulfil the "complete independence" requirements stated by the decision of the ECJ. Eventually, it should be noted that the "functional independence" approach, presented in the trial of the EU Commission against the Federal Republic of Germany, illustrates the common confusion of accountability with control.

7.4.3 Accountability and Legitimacy of DPAs

When talking about the accountability of DPAs, one would normally think of arguments in favour of more governmental control and against the far-reaching discretion of DPAs. However, this is clearly not the case, if accountability is seen as a complement to independence providing greater legitimacy for DPAs.

[16] Although specific DPAs of the Länder will be scrutinised more thoroughly within the dissertation project, this work will not deal with the regional level in more detail.

Regarding Quintyn et al.'s (2005) practical accountability arrangements, DPAs are already subject to quite a few accountability mechanisms. For example, whereas ex ante accountability is ensured by consultations with stakeholders before audits are undertaken, various publications alongside the mandatory annual report as well as large public relation and awareness raising campaigns represent classical ex post accountability instruments. These tools are rather of an explanatory character, fulfilling the task of shedding light on DPAs' actions. Procedural and substantive accountability mechanisms can be found as well. Since data protection commissioners cannot be sanctioned for regulatory failure, there is no such thing as personal accountability.[17] While financial accountability is reflected by the DPAs' obligation to regularly report on their expenses, performance does not appear as an accountability principle, at least not in the legal context.

However, the performance of DPAs plays a crucial role when it comes to output-oriented legitimacy. Although some data protection commissioners are elected by parliament,[18] the democratic legitimacy of DPAs remains scarce. Therefore, the other sources of legitimacy such as performance and effectiveness become crucial.

Yet, DPAs' performance is problematic, especially when it comes to the effectiveness of their work. This is supported by the results of several comprehensive studies. According to a comparative legal study by the Fundamental Rights Agency of the EU, the reasons for the often-poor effectiveness of DPAs lies predominantly in "the lack of independence, adequate resources and sufficient powers" (EU Report 2009, para. 8). Remarkably, Korff and Brown (2010, p. 44), who come to similar conclusions in their comparative survey on privacy challenges, point to the fact that "weak enforcement in many countries was already noted in a much earlier study [referring to an EC study on case-law on compliance from 1998 by Douwe Korff], and [it] does not appear to have improved much".

In general, DPAs seem to be overwhelmed by their workload, facing an incredible variety of different tasks and additionally being forced to operate on two regulatory fronts. On top of this, they have to deal with the complex functionalities of modern technologies as well as lacunae in data protection law, which makes their work even more complicated. Thus, DPAs can often not live up to the high expectations placed upon them.

In a nutshell, DPAs are confronted with several accountability arrangements. The emphasis lies on explanatory accountability, which is linked to the concept of transparency: DPAs provide public oversight over their actions on a regular basis and are therefore subject to a variety of external checking mechanisms. The lack of effectiveness, however, has in turn severe negative impacts on the legitimacy of DPAs in the long-term. In order to enhance effectiveness, public policy-makers could reduce the workload by defining fewer, and more specific, tasks, as well as increase DPAs' budgets and personnel resources significantly.

---

[17] Yet, in cases of serious misdemeanours DPAs are, of course, subject to statutory supervision by the executive, legislative or judiciary.

[18] In order to obtain additional democratic legitimacy, the German Bundestag elects the FfDF at the suggestion of the Federal Government, following an amendment to the Federal Data Protection Act in 1990.

7.5 Conclusion

Marking the starting point of a dissertation project that deals with a comparative analysis of DPAs, this chapter presents an exploratory perspective on DPAs in the EU, drawing on theories of regulation such as the concept of the *regulatory state* and IRAs. Since there is no clear framework for a political science analysis of DPAs, theoretical and methodological approaches to IRAs could provide a template for the research on DPAs.

Central to the analysis of this work are features of IRAs such as their independence, credibility, accountability, democratic deficit, legitimacy and effectiveness. These aspects are also valid and relevant for the set up of DPAs. Therefore, the lessons learned from research about IRAs could represent a valuable asset in the assessment of DPAs. For example, Thatcher as well as Gilardi have demonstrated interesting ideas as to how to measure formal independence using quantitative methods. Furthermore, Quintyn has emphasized the relative value of independence pointing to principles such as accountability, transparency and integrity as equally important in his regulatory governance model. Although these authors mainly concentrate on IRAs in the financial and economic sector, their hypotheses and methods have been proven useful for the analysis of DPAs. That is why they will not only be applied, but also further developed within the dissertation project.

The most pressing topics for DPAs appear to be their lack of independence, adequate resources and sufficient powers, as several comprehensive studies concluded. Independence is particularly threatened, since DPAs face two fronts of regulatees, i.e. private and public actors, which both have a potentially strong interest in avoiding or manipulating regulatory actions.

Although the European Data Protection Directive stipulates "complete independence", DPAs need continuously to struggle for autonomy, recently exemplified by the ruling of the ECJ on the case of the European Commission against the Federal Republic of Germany in 2010. But also the administrative incorporation of the German FfDF into the Federal Ministry of the Interior poses serious problems for the independence of the DPA.

The "functional independence" approach, presented during the trial by German representatives, comprises a striking example of the common confusion between control and accountability. Hence, clearly defined accountability arrangements that involve a combination of monitoring instruments as well as accountability relationships towards multiple stakeholders are of the utmost importance. However, the traditionally close link between DPAs and the government (certain ministries in particular) has resulted in a rather one-dimensional accountability relationship, i.e. the often-exclusive right by the government to appoint and dismiss the commissioner (head of DPA) as well as to finance the agency.

Yet, accountability should not be seen as a trade-off in regards to independence. If appropriately structured and arranged, accountability actually serves as a complement to independence, providing not only transparency, but also greater legitimacy. Regarding the first, DPAs have clearly succeeded in making extensive information about their actions available to the public. However, the latter suffers from the lack of effectiveness of DPAs' work. Facing a broad spectrum of activities as well as two regulatory fronts, DPAs appear to be overwhelmed by the enormous workload and the technological dimension of the tasks.

All in all, it seems tremendously important, not only to strengthen DPAs' independence, particularly from public policy-makers, but also to increase their financial and personnel resources significantly in order to enhance their effectiveness. The dissertation project will eventually try to answer the question as to how this can be achieved.

References

Bennett, C. 1992. *Regulating privacy: Data protection and public policy in Europe and the United States*. Ithaca: Cornell University Press.

Bennett, C., and C. Raab. 2006. *The governance of privacy: Policy instruments in global perspective*. Cambridge: MIT Press.

Bibow, J. 2004. Reflections on the current fashion for central bank independence. *Cambridge Journal of Economics* 28 (4): 549–576.

Bovens, M. 2005. Public accountability. In *The Oxford handbook of public management*, eds. Ewan Ferlie, Laurence E. Lynn and Christopher Pollitt, 182–208. Oxford: Oxford University Press.

Caporaso, J. A. 1996. The European Union and forms of state: Westphalian, regulatory or post modern? *JCMS: Journal of Common Market Studies* 34 (1): 29–52.

Cukierman, A., S. B. Web, and B. Neyapti. 1992. Measuring the independence of central banks and its effect on policy outcomes. *The World Bank Economic Review* 6 (3): 353–398.

Dammann, Ulrich. 2011. Bundesbeauftragter für den Datenschutz und die Informationsfreiheit. In *Bundesdatenschutzgesetz—Kommentar*, ed. S. Simitis. Baden-Baden: Nomos.

Eberlein, B., and E. Grande. 2005. Beyond delegation: Transnational regulatory regimes and the EU regulatory state. *Journal of European Public Policy* 12 (1): 89–112.

Eijffinger, S. C., M. M. Hoeberichts, and E. Schaling. 2000. A theory of central bank accountability. CEPR Discussion Paper.

EU Directive. 1995. Directive 95/46/EC of the European parliament and of the council of 24 October 1995 on the protection of individuals with regard to the processing of personal data and on the free movement of such data. *Official Journal of the European Communities* L 281.

EU Report. 2009. Executive summary of the final draft of the comparative legal study on assessment of data protection measures and relevant institutions. In *Report: Fundamental Rights Agency (FRA) of the European Union*.

Federal Data Protection Act, *Germany*, 2009. 1990.

Gilardi, F. 2002. Policy credibility and delegation to independent regulatory agencies: A comparative empirical analysis. *Journal of European Public Policy* 9 (6): 873–893.

Gilardi, F. 2005. Evaluating independent regulators. Paper presented at the organization for economic cooperation and development, designing independent and accountable: Regulatory authorities for high quality regulation, working party on regulatory management and reform, proceedings of an expert meeting, London, United Kingdom.

Gutwirth, S. 2002. *Privacy and the information age*. New York: Rowman & Littlefield.

Hamilton, A., J. Madison, J. Jay, and L. Goldman. 2008. *The federalist papers*. USA: Oxford University Press.

Hessian Data Protection Act, *Hesse (Germany)*. 1970.

Hood, C. 1999. *Regulation inside government: Waste watchers, quality police, and sleaze-busters*. USA: Oxford University Press.

Hooghe, L., and G. Marks. 2001. *Multi-level governance and European integration*. Lanham: Rowman & Littlefield.

Hustinx, P. 2009. The role of data protection authorities, eds. Serge Gutwirth, Yves Poullet, Paul Hert, Cécile Terwangne and Sjaak Nouwt, 131–137. Netherlands: Springer.

Judgment of the Court (Grand Chamber) of 9 March 2010. 2010. European Commission v Federal Republic of Germany. Failure of a member state to fulfil obligations—Directive 95/46/EC—Protection of individuals with regard to the processing of personal data and the free movement of such data—Article 28(1)—National supervisory authorities—Independence—Administrative scrutiny of those authorities. Case C-518/07.

Kemp, J. 2011. The slow death of the regulatory state. http://blogs.reuters.com/great-debate/2010/06/04/the-slow-death-of-the-regulatory-state/. Accessed 3 Aug 2011.

Korff, D., and I. Brown. 2010. Final report: comparative study on different approaches to privacy challenges, in particular in the light of technological developments. European commission—directorate-general justice, freedom and security.

Levi-Faur, D. 1999. The governance of competition: the interplay of technology, economics, and politics in European Union electricity and telecom regimes. *Journal of Public Policy* 19 (2): 175–207.

Maggetti, M. 2010. Legitimacy and accountability of independent regulatory agencies: A critical review. *Living Reviews in Democracy* 2:1–9.

Majone, G. 1994. The rise of the regulatory state in Europe. *West European Politics* 17 (3): 77–101.

Majone, G. 1996. *Regulating Europe*. London: Routledge.

Majone, G. 1997. From the positive to the regulatory state: Causes and consequences of changes in the mode of governance. *Journal of Public Policy* 17 (2): 139–167.

Mayer-Schönberger, V. 1997. Generational development of data protection in Europe. In *Technology and privacy: The new landscape,* eds. Philip Agre and Marc Rotenberg, 219–241. Cambridge: MIT Press.

Mayntz, R. 2009. The changing governance of large technical infrastructure systems. In *Über governance: institutionen und prozesse politischer regelung,* ed. Renate Mayntz, 121–150. Frankfurt: Campus.

Meier, K. J., and J. Bohte. 2006. *Politics and the bureaucracy: Policymaking in the fourth branch of government*. Belmont: Wadsworth.

Quintyn, M. 2009. Independent agencies: More than a cheap copy of independent central banks? *Constitutional Political Economy* 20 (3): 267–295.

Quintyn, M., and M. W. Taylor. 2007. Robust regulators and their political masters: Independence and accountability in theory. In *Designing financial supervision institutions: Independence, accountability and governance*, eds. D. Masciandaro and M. Quintyn, 3–40. Cheltenham: Elgar.

Quintyn, M., E. Huepkes, and M. Taylor. 2005. The accountability of financial sector supervisors: Principles and practice. IMF Working Paper.

Simitis, S. 1994. From the market to the polis: The EU directive on the protection of personal data. *Iowa Law Review* 80:445–469.

Thatcher, M. 1998. Institutions, regulation, and change: New regulatory agencies in the British privatised utilities. *West European Politics* 21 (1): 120–147.

Thatcher, M. 2002. Regulation after delegation: Independent regulatory agencies in Europe. *Journal of European Public Policy* 9 (6): 954–972.

Tocqueville, Alexis de. 2000. *Democracy in America,* vol. 1. New York: Bantam Books.

Vogel, S. 1996. *Freer markets, more rules: Regulatory reform in advanced industrial countries.* Ithaca: Cornell University Press.

Wilks, S., and I. Bartle. 2002. The unanticipated consequences of creating independent competition agencies. *West European Politics* 25 (1): 148–172.

Chapter 8
Information Sharing in the Area of Freedom, Security and Justice—Towards a Common Standard for Data Exchange Between Agencies and EU Information Systems

Franziska Boehm

8.1 Introduction

In the Area of Freedom, Security and Justice[1] (AFSJ), the process of European integration has considerably supported the establishment of Union bodies, agencies and information systems in recent years. Horizontal information sharing, including the exchange of personal data between these bodies, has become an essential tool in the internal security policy of the European Union (EU). Inter-agency cooperation between AFSJ actors, such as Europol, Eurojust or Frontex as well as the Commission's anti-fraud unit, OLAF, led to the conclusion of agreements providing for mutual information exchange. In addition, the access of law enforcement and judicial agencies

This contribution is based on my PhD research carried out during the last years. It provides a brief overview of some of the results of the research. The complete thesis with the title: "Information sharing and data protection in the Area of Freedom, Security and Justice" is published by Springer.

[1] The term AFSJ is a political notion describing several policies brought together under the umbrella of an overarching concept. Introduced by the Treaty of Amsterdam and further developed in the Lisbon Treaty, this policy aims at offering "its citizens an area of freedom, security and justice without internal frontiers, in which the free movement of persons is ensured in conjunction with appropriate measures with respect to external border controls, asylum, immigration and the prevention and combating of crime" (Article 3 (2) TEU). These political goals are practically enforced by the adoption of multi-annual work programmes (the Vienna (1998), the Tampere (1999), the Hague (2004) and the Stockholm programme (2009)), which establish general priorities and political objectives in this area. Although multi-annual work programmes are not as such binding instruments, these programmes set different political goals, which are subsequently legally implemented by the instruments available to the European legislator, primarily by way of Directives, Regulations and Council Decisions. As a result thereof, these programmes have a substantial effect on the future institutional policy and often directly influence legislative actions in this area.

F. Boehm (✉)
University of Luxembourg, Luxembourg
e-mail: franziska.boehm@uni.lu

S. Gutwirth et al. (eds.), *European Data Protection: In Good Health?*,
DOI 10.1007/978-94-007-2903-2_8,

to data stored in the European information systems, such as the Customs- (CIS), the Schengen- (SIS) or the Visa Information System (VIS) and Eurodac, occupies an increasingly important place in this area.

Post-9/11 policy concepts, such as "the Hague" and "the Stockholm programmes", promote an enhanced cooperation and coordination of law enforcement agencies and other agencies in the AFSJ.[2] Under their influence, formerly not related policy areas, such as the prevention of crime and immigration, are swiftly linked and lead to an intensive cooperation between AFSJ actors of a completely different legal nature, vested with different powers (Mitsilegas 2009). Without being limited by the former pillar constraints and, above all, in absence of a unified approach to data protection in judicial and criminal matters[3], legal and structurally different bodies, equipped with different tasks, exchange and transfer personal data within and outside the EU. The result is that data collected for one specific purpose may be transferred and used for other purposes completely unrelated to the original collection. This fast increasing cooperation at multiple levels necessarily touches upon different data protection regimes. Title V TFEU specifies the policies of the AFSJ.[4] They are a mix of former first as well as former third pillar policies.[5] While on the one hand information and personal data exchange is identified as a priority in this field, on the other hand data protection concerns risk to be undermined.

---

[2] The Hague programme adopted in 2004, for instance, promoted the enforced cooperation of the actors in the AFSJ and introduced the "availability principle", which should govern law enforcement-related data exchange from then on. Bilateral agreements between EU bodies and provisions in secondary legislation were foreseen intending to exchange data and leading, amongst others, to a reinforced inter-agency cooperation. Other measures aimed to allow mutual access to databases or their common use. National databases were supposed to become "interoperable" and direct access to central EU databases such as the SIS should have been established whereby nevertheless data protection standards should have been "strictly observed" (The Hague Programme: Council doc. 16054/04 from 13 December 2004, point 2.1, pp. 18–19). As a main consequence of this instrument, which covered the period from 2005 to the end of 2009, more and more data were shared and the actors in the AFSJ worked closer together than before. The period after 2009 is now covered by the Stockholm programme valid from 2010 to 2014 endorsing the availability principle while repeating the data protection pleas (The Stockholm Programme, Council doc. 17024/09 from 2 December 2009, point 4.2.2, pp. 37–38), Compare also note from the General Secretariat to the Standing Committee on operational cooperation on internal security (COSI), final report on the cooperation between JHA agencies, Council doc. 8387/10 from 9 April 2010.

[3] Council Framework Decision 2008/977/JHA on the protection of personal data processed in the framework of police and judicial cooperation in criminal matters, OJ 2008, L-350/60, in the following FDPJ, OJ 2008, L-350/60 represents a first step towards a comprehensive framework in this area; the FDPJ is, however, very restricted in scope as it is, for instance, not applicable to the data processing of most of the AFSJ law enforcement agencies, such as Europol and Eurojust, as well as at other AFSJ exchange systems, that is, the Schengen or the Customs Information Systems; moreover, excluded from the scope is also the internal processing of the Member States in police and criminal matters.

[4] Four main areas stand out: policies on border checks, asylum and immigration, judicial cooperation in civil as well as in criminal matters and police cooperation (Title V Chapters 2–5 TFEU).

[5] The Provision on police and judicial cooperation in criminal matters (former Title VI EU Treaty) are former third pillar policies whereas the provisions on asylum and immigration were regulated under former first pillar Community law (Title IV EC Treaty).

Questions relating to the coherency and the respect of data protection rules within this cooperation network of the AFSJ actors seem to be pushed into the background. This unbalanced situation can have a profound impact on the rights of the individuals. It is worth pointing out that, even though the context in which information is used is changing rapidly, no evaluation or overview of the existing data collection, processing and data-sharing systems, including a thorough assessment of their effectiveness, their possible overlapping effects, proportionality and their respect of data protection rights have been carried out so far.[6]

In the light of these considerations, this chapter first, in Sect. 2, briefly illustrates the legal background of data protection rules in the AFSJ. Section 3 focuses on the organisation of the existing and the planned instruments governing AFSJ data exchange as well as their compliance with the data protection rules mentioned in Sect. 1. Inconsistencies in the AFSJ data exchange network relating, among others, to gaps of protection, transparency issues and incoherent access procedures and conditions are disclosed. In the respective subsections, comments and criticism are offered and problems are highlighted. Section 4 suggests some basic data protection standards, which follow from the respect of Article 8 ECHR and would improve the respect of data protection rules in the field of internal AFSJ information sharing.

8.2 Legal Background

Before analysing the instruments governing AFSJ information exchange, the data protection rules applicable in this area need to be briefly identified.

8.2.1 Data Protection Before Lisbon

Due to the former pillar structure, data processing in third pillar security-related matters was not included in the relatively comprehensive data protection framework of the first pillar. While, since 1995, the Data Protection Directive 95/46[7] accompanied by sector-specific first pillar instruments[8] has established a wide-ranging data and privacy protection for individuals in an economic-related first pillar context,

---

[6] Communication from the Commission to the European Parliament, the Council, the European Economic and Social Committee and the Committee of the Regions—Delivering and area of freedom, security and justice for European's citizens—Action Plan implementing the Stockholm Programme, COM(2010) 171 final, in particular p. 6.

[7] Directive 95/46/EC of the European Parliament and of the Council of 24 October 1995 on the protection of individuals with regard to the processing of personal data and on the free movement of such data, OJ 1995, L-281/31.

[8] For instance: Directive 97/66/EC of the European Parliament and of the Council of 15 December 1997 concerning the processing of personal data and the protection of privacy in the telecommunications sector, OJ 1998, L-24/1.

data processing for security purposes carried out by governmental law enforcement agencies was excluded from the scope of Directive 95/46.[9]

For a long time, data protection in the framework of former third pillar matters was therefore covered by public international law instruments instead of EU law, most notably by the instruments of the Council of Europe (Siemen 2006).[10] Article 8 of the ECHR and its interpretation by the Strasbourg Court as well as Convention No. 108[11], its respective additional protocols[12] and Recommendation (87) 15[13] built the reference instruments for security-related data processing in the EU.[14]

8.2.2 Guarantees for Security-Related Data Processing in Article 8 ECHR

Although it seems to be difficult to derive principles of general application from the case law tailored to a specific situation, the ECtHR succeeds, nonetheless, in developing a quite comprehensive data protection framework in this specific area (Siemen 2006; De Schutter 2008).[15] The main principles are briefly summarised in the following.

---

[9] Article 3 (2) Directive 95/46/EC of the European Parliament and of the Council of 24 October 1995 on the protection of individuals with regard to the processing of personal data and on the free movement of such data, OJ 1995, L-281/31. This statement was clarified by the ECJ in the famous PNR case: joined cases C-317/04 and C-318/04, *Parliament v. Council*, [2006], ECR I-4721.

[10] Compare for a profound analysis of the instruments of the Council of Europe.

[11] Convention No. 108 of the Council of Europe for the protection of individuals with regard to automatic processing of personal data from 28 January 1981.

[12] In particular the additional protocol to Convention for the protection of individuals with regard to automatic processing of personal data regarding supervisory authorities and trans-border data flows, which entered into force in 2004.

[13] Recommendation R (87) 15 of the Committee of Ministers to the Member States regulating the use of personal data in the police sector, adopted 17 September 1987.

[14] However, since the adoption of the Framework Decision "on the protection of personal data in the framework of police and judicial cooperation in criminal matters" (DPFD) in 2008, OJ 2008, L-350/60, certain minimum requirements also apply in the field of security-related data processing at the EU level.

[15] See: Siemen (2006). Admittedly, it does not cover all difficulties arising in an EU law enforcement context and is the lowest common denominator as the guarantees of the ECHR apply in a public international law context, but the interpretations of the ECtHR have attained a far-reaching significance for the EU over the years and cooperation between the EU and the Council of Europe in fundamental rights matters continually improves. Compare also: De Schutter (2008). See also: joint declaration on cooperation and partnership between the Council of Europe and the European Commission from 3 April 2001, accessed July 12, 2011, http://www.jp.coe.int/Upload/91_Joint_Declaration_EF.pdf; Memorandum of Understanding between the Council of Europe and the European Union from 10 May 2007, CM(2007)74, accessed July 12, 2011, https://wcd.coe.int/ViewDoc.jsp?Ref=CM(2007)74&Language=lanEnglish.

The Strasbourg Court refers to the right to private life of article 8 ECHR when data protection infringements are at stake.[16] Even if personal data are not expressly protected by this article, the ECtHR insists that "the protection of personal data" is of "fundamental importance" to a person's enjoyment of his or her right to respect for private and family life.[17]

The jurisprudence of the ECtHR clearly illustrates that governmental data collection and retention interferes with the right to private life as protected by article 8 ECHR.[18] Every transmission of personal data from one authority to another, including the subsequent use of such data, constitutes another separate interference with individual rights under article 8 ECHR. The transmission enlarges the group of individuals with knowledge of the personal data and can therefore lead to investigations being instituted against the persons concerned.[19] The indented AFSJ data exchange therefore undoubtedly interferes with article 8 ECHR.

After the interference has been established, the ECtHR examines whether the measure in question may be justified. In this context, one has to consider three conditions: the act in question must be "in accordance with the law", pursue one of the legitimate aims listed in article 8 (2) ECHR and must additionally be necessary in a democratic society, which means principally that the interfering law must be proportionate to the aim pursued. Whereby in general the ECtHR admits a wide margin of discretion to the Member States when national security is at stake, the interests of the parties, however, have to be reasonably balanced. Moreover, to be in accordance with the law, the measure in question must be "foreseeable", which means formulated with sufficient precision to enable an individual to regulate his conduct and to predict the consequences a given action might entail.[20]

---

[16] Compare for instance: ECtHR, *Leander v. Sweden*, Application no. 9248/81 from 26 March 1987; ECtHR, *Amann v. Switzerland*, Application no. 27798/95 from 16 February 2000; ECtHR, *Rotaru against Romania*, Application no. 28341/95 from 4 May 2000; ECtHR, *Panteleyenko v. Ukraine*, Application no. 11901/02 from 29 June 2006; ECtHR, *S. and Marper v the United Kingdom*, Application nos. 30562/04 and 30566/04 from 4 December 2008; ECtHR *Weber and Saravia v. Germany*, Application no. 54934/00 Admissibility Decision from 29 June 2006; ECtHR, *C.G. and others v. Bulgaria*, Application no. 1365/07 from 24 April 2008; ECtHR, *Association for European Integration and Human Rights and Ekimdzhiev v. Bulgaria*, Application no. 62540/00 from 28 June 2007; ECtHR, *Malone v. the United Kingdom*, Application no. 8691/79 from 2 August 1984; ECtHR, *Valenzuela v. Spain*, Application no. 27671/95 from 30 July 1998.

[17] ECtHR, *Z. v Finland*, Application no. 22009/93, from 25 February 1997, para 95; ECtHR, *Peck v. United Kingdom*, Application no. 44647/98 from 28 January 2003, para 78; ECtHR, *L.L. v France* Application no. 7508/02 from 10 October 2006, para 43; ECtHR, *Biriuk v Lithuania*, Application no. 23373/03 from 25 November 2008, para 39; ECtHR, *I v Finland* Application no. 20511/03 from 17 July 2008, para 38; ECtHR, *S. and Marper v the United Kingdom*, Application nos. 30562/04 and 30566/04 from 4 December 2008, para 103; ECtHR, *C.C. v. Spain*, Application no. 1425/06 from 6 October 2009, para 31.

[18] ECtHR, *Amann v. Switzerland*, Application no. 27798/95 from 16 February 2000, paras 65–67.

[19] ECtHR, *Weber and Saravia v. Germany*, Application no. 54934/00 Admissibility Decision from 29 June 2006, para 79.

[20] ECtHR, *Sunday Times v. the United Kingdom*, Application no. 6538/74, para 49 from 26 April 1979; ECtHR, *Liberty and others v. the United Kingdom*, Application no. 58234/00 from 1 July

To be more precise, in judgments related to governmental data collection and the implementation of surveillance measures in the framework of article 8 ECHR, certain criteria must be fulfilled to guarantee proportionality and in this way the balance of powers between the interests at stake. These criteria include the limitation on the categories of individuals against whom surveillance measures may be taken as well as the clear definition of the circumstances and limits of the storing and the use of the information before the processing.[21] Time limits for storing are essential and the age of the person concerned must be taken into account to avoid indiscriminate storing of personal data in governmental databases.[22]

Prior to surveillance measures and the collection of data in security-related data processing, it is crucial to determine which kind of data are to be stored and for which purposes the data should be used afterwards (purpose limitation principle).[23] Independent review and adequate and effective safeguards against abuse, including effective remedies, must exist to assure compliance with the rule of law.[24]

With regard to the subsequent notification of individuals subjected to surveillance measures, the ECtHR emphasises that this question is closely linked to the effectiveness of remedies before the courts and therefore to the existence of effective safeguards against the abuse of monitoring powers.[25] In the case *Weber and Saravia v. Germany*, the Strasbourg Court adds: "As soon as notification can be carried out without jeopardizing the purpose of the restriction after the termination of the surveillance measure, [. . .], information should be provided to the persons concerned".[26]

2008, para 68; ECtHR *Silver v. the United Kingdom*, Application no. 5947/72 and others from 25 March 1983, paras 85–88.

[21] ECtHR, *Segerstedt-Wiberg and others v. Sweden*, Application no. 62332/00 from 6 June 2006, paras 88–92; ECtHR, *Liberty and others v. the United Kingdom*, Application no. 58234/00 from 1 July 2008, para 68; ECtHR, *Rotaru v. Romania*, Application no. 28341/954 from 4 May 2000, para 57; ECtHR, *Weber and Saravia v. Germany*, Application no. 54934/00 Admissibility Decision from 29 June 2006, paras 116 and 127.

[22] ECtHR, *S. and Marper v. the United Kingdom*, Application nos. 30562/04 and 30566/04 from 4 December 2008, para 119; ECtHR, *Segerstedt-Wiberg and others v. Sweden*, Application no. 62332/00 from 6 June 2006, paras 89–92.

[23] ECtHR, *Weber and Saravia v. Germany*, Application no. 54934/00 Admissibility Decision, para 116 from 29 June 2006, ECtHR, *Rotaru v. Romania*, Application no. 28341/954, para 57 from 4 May 2000; see also: ECtHR, *Association for European Integration and Human Rights and Ekimdzhiev v. Bulgaria*, Application no. 62540/00 from 28 June 2007.

[24] ECtHR, *Rotaru against Romania*, Application no. 28341/95 from 4 May 2000, paras 55–63; ECtHR, *Segerstedt-Wilberg and others v. Sweden*, Application no. 62332/00 from 6 June 2006, para 121.

[25] ECtHR, *Weber and Saravia v. Germany*, Application no. 54934/00 Admissibility Decision from 29 June 2006, para 135: "since there is in principle little scope for recourse to the courts by the individual concerned unless the latter is advised of the measures taken without his or her knowledge and thus able to challenge their legality retrospectively".

[26] ECtHR, *Weber and Saravia v. Germany*, Application no. 54934/00 Admissibility Decision, para 135 from 29 June 2006.

8.2.3 Data Protection After Lisbon

The entry into force of the Lisbon Treaty influenced the aforementioned EU data protection framework in several ways. One of the major changes relates to the abolition of the pillar structure putting an end to the structural separation between "European Community" actions and "European Union" activities, a development, which will largely influence data protection policy in the AFSJ.

The protection of personal data in the AFSJ is strengthened in three ways: its Article 16 (TFEU) guarantees the right to the protection of personal data to "everyone" and Article 6(3) TEU stipulates that the Charter of Fundamental Rights, which shall have the same legal value as the EU treaties, is additionally applicable when it comes to fundamental rights protection in the EU.[27] Its Article 8 includes the right to the protection of personal data. Important improvements are additionally offered by the intended accession of the EU to the ECHR provided for in Article 6(2) TEU. Particular attention is thereby paid to the ECtHR's interpretation of article 8 ECHR, mentioned above. Improved decision making by the introduction of the ordinary legislative procedure in the AFSJ, where Parliament and Council act as co-legislators[28] in data protection matters, upgrades democratic control. Although transitional provisions delay the effects of the full enforcement of Article 16 TFEU in the AFSJ (Hijmans and Scirocco 2009)[29], the exclusive competence of the Council vanishes and the Parliament has co-decision rights in every question concerning the necessary changes in the legal frameworks of the AFSJ actors.[30] With a view to this fundamental change in the upcoming legislative processes, it is important to propose

---

[27] Article 6 (3) TFEU.

[28] Replacing Article 251 EC, which lays down the current co-decision procedure, the ordinary legislative procedure in Article 294 TFEU assures compulsory participation of the European Parliament, additionally the Council's acting by a qualified majority in the legislative process.

[29] For an excellent overview of the situation of data protection after the Lisbon Treaty, see: Hijmans and Scirocco (2009). Article 9 of the Protocol No. 36 annexed to the Lisbon Treaty provides that the legal effects of the acts adopted before the entry into force of the Lisbon Treaty shall be preserved *until those acts are repealed, annulled or amended.* A deadline to adapt the old instruments to the new Treaty provisions, for instance, in case they do not comply with Article 16 TFEU, is not given. With respect to acts in the field of police cooperation and judicial cooperation in criminal matters adopted before the entry into force of the Treaty of Lisbon, the powers of the Commission under Article 258 TFEU (the Commission's right to enact infringement proceedings) as well as the limited powers of the ECJ under Title VI of the former TEU shall remain the same. In this case, the transitional measure shall cease to have effect 5 years after the date of entry into force of the Treaty of Lisbon. Declaration 20 and 21 provide for the possibility to enact other data protection rules in the ASFJ than those being possibly applicable to former first pillar matters as regards national security as well as in police and judicial cooperation. Moreover, certain Member States (United Kingdom, Ireland, Denmark) complicatedly exclude the application of Article 16 TFEU in specific cases.

[30] The European Parliament and the Council will "lay down the rules relating to the protection of individuals with regard to the processing of personal data by Union institutions, bodies, offices and agencies, and by the Member States when carrying out activities, which fall within the scope of Union law, and the rules relating to the free movement of such data" (Article 16 (2) TFEU).

improvement in terms of data protection in the AFSJ, which could then be used by the parliament in future negotiations.

Finally, even though Article 16 TFEU constitutes an enormous step towards the recognition of essential data protection principles in the AFSJ, its guarantees have to be specified to help enforcing the rights of the individuals in the AFSJ. The interpretation of such broad principles, as carried out by the ECtHR in recent years with regard to data protection principles for security-related data processing, could support this process in a valuable way. However, before proposing improvements, it is important to describe the organisation of the AFSJ data exchange and its shortcomings.

8.3 Organisation of AFSJ Data-Exchange

Information exchange in the AFSJ is on the one hand taking place between the AFSJ agencies (Europol, Eurojust, Frontex) and the Commission's anti-fraud unit OLAF[31] (para 2.1) and on the other hand the law enforcement and the judicial agency, Europol and Eurojust, have access to the information systems such as SIS, CIS, VIS and/or Eurodac[32] (para 2.2). In view of the data protection rules described in the first part, this section not only analyses the organisational structure of AFSJ data exchange, but also criticises the legal shortcomings arising in the current exchange network(s).

8.3.1 Inter-Agency AFSJ Data Exchange and OLAF

Inter-agency data exchange is carried out in two situations: data are exchanged during Joint Investigation Teams (JITs) operations or transferred between the actors based on bilateral agreements.

8.3.1.1 Information Exchange in JITs: Europol, Eurojust and OLAF

The idea of JITs was introduced in 2000 by the Convention on Mutual Assistance in Criminal Matters and later reaffirmed by a Framework Decision on JITs.[33]

[31] Europol and Eurojust are Europe's law enforcement agencies, which collect personal data of criminals, but also of suspects, victims and witnesses. Frontex assures the control of the external borders of the EU and collects data of third state nationals trying to pass the border. OLAF is the Commission's anti-fraud unit carrying out internal investigations within the EU institutions, bodies and agencies. The unit mainly collects personal data of individuals suspected of fraud.

[32] The SIS is a database in the framework of law enforcement and immigration control, which contains data of third state nationals, but also EU nationals. The CIS serves customs control purposes and contains personal data of individuals suspected of illicit trafficking activities. The VIS serves the purpose of the exchange of visa data and entails information of third state nationals who apply for a visa to enter the EU. Eurodac stores fingerprint data of asylum seekers and should prevent that asylum seekers make multiple asylum applications in different Member States of the EU.

[33] Council Act of 29 May 2000 establishing in accordance with Article 34 of the Treaty on European Union the Convention on Mutual Assistance in Criminal Matters between the Member States of the

Legal Concept

The concept of JITs involves the "coordination, organisation and implementation of investigative and operational action carried out jointly with the Member States' competent authorities [. . .]".[34] In recent years, Europol's and Eurojust's main responsibilities relating to JITs were rather of organising and supportive nature[35], acting on basis of their establishing Council Decisions.[36]

However, the role of both agencies in JITs continually evolved in the last years.[37] Europol's, Eurojust's as well as OLAF's current function in JITs is described in a JITs Manual from 23 September 2009 (JIT manual). According to it, only Eurojust's national members acting on the basis of their national law can be a member of the JIT, officials from Europol, Eurojust and OLAF may participate but are not allowed to be a member of the JIT (Lopes da Mota 2009).[38] Article 6 Europol Decision and the JIT manual restrict their function to the involvement in the operation of the JIT, but exclude the participation in any coercive measures.[39] These general rules may be, however, subject to further specific arrangements in forming a particular agreement between the participating Member States and the bodies concerned annexed to

---

European Union, OJ 2000 C 197/1, Article 13; to the initiation of the JIT project, see: Horvatis and Bart De Buck (2007) and Rijken and Vermeulen (2006).

[34] Article 88 (2) (b) TFEU.

[35] Compare recital 9 and Articles 5 (1) (d), 5 (5), 6, 8 (7) c and 54 Council Decision of 6 April 2009 establishing the European Police Office, OJ 2009, L121/37 as well as Articles 6 (b) (iv), 9 (f), 12 (2) (d), 13 (2) (5) and 25 (a) (2) Eurojust Decision.

[36] Article 6 Europol Decision and Article 7 (4) Eurojust Decision.

[37] The Framework Decision on JITs (Article 1 and recital (9) of Council Framework Decision of 13 June 2002 on JITs, OJ 2002 L 162/1 and Article 13 Council Act of 29 May 2000 establishing in accordance with Article 34 Treaty on European Union the Convention on Mutual Assistance in Criminal Matters between the Member States of the European Union, OJ 2000, C 197/1) specifies that two or more Member States can set up a JIT for a specific purpose and a limited period of time to carry out investigations while Eurojust and Europol may participate in the JITs. For this purpose, participating Member States conclude mutual agreements and Europol and Eurojust organise information events and publish manuals on the concept of JITs. In their aforementioned joint JIT manual from 2009, both agencies encourage Member States to set up JITs to better coordinate cases involving several Member States. A JIT consists of law enforcement officers, prosecutors, judges or other law enforcement-related personnel and is established in the Member State in which investigations are supposed to be principally carried out. Other European Union bodies, particularly the Commission (OLAF) as well as law enforcement bodies from third states such as the FBI may additionally be involved, however, just as Europol and Eurojust, they may participate in the operation of a JIT, although they cannot lead or be a member of it. They are associated by an agreement between the agency/administration of a Member State as a party to the agreement and the relevant European Union or third state body; compare: Explanatory report on the Convention on Mutual Assistance in Criminal Matters between the Member States of the European Union, OJ 2000, C 379/7 and JITs Manual from 23 September 2009, Council Doc. 13598/09.

[38] JITs Manual from 23 September 2009, Council Doc. 13598/09, p. 10 and Eurojust Decision, Article 9 (f).

[39] JITs Manual from 23 September 2009, Council Doc. 13598/09, p. 10, see also: Article 6 (1) Council Decision of 6 April 2009 establishing the European Police Office, OJ 2009, L121/37.

the initial agreement setting up the JIT, which may confer more rights to Europol, Eurojust or OLAF.[40]

Considering the formulations in the JIT manual, in practice it seems to be hard to distinguish between the "participation in the operation of the JIT" on the one hand and the exclusion of coercive measures on the other, in particular when taking Article 6(2) Europol Decision into account, which stipulates that Europol staff should "assist in *all* activities and exchange information with all the members" of the JIT (De Buck 2007).[41]

Information Exchange in JITs

Rules on information exchange in the JITs follow a local solution and are generally attached to the national law[42] and stipulate that information could be shared within the limits of the national law of the national members seconded to the JIT.[43]

Further details regarding the exchange of information and data protection issues are entailed in the specific arrangements of the agreements setting up the JIT[44], but the specifics of these arrangements are not published and depend on the agreed compromise between the Member State and the relevant European actor in a particular case. Rules of general application regulating this nevertheless rather informal data exchange do not exist, but would definitely lead to more legal certainty and transparency in this context (Rijken and Vermeulen 2006; Mitsilegas 2009).[45]

Despite this rather non-transparent practice, Europol's role in JITs is of great importance: it may provide the JIT members with information stemming from its databases (the EIS or from an analysis work file).[46] Europol can grant access to both systems "by means of a Europol mobile office located where the JIT is operating" (De Buck 2007). JIT members are allowed to have direct access to Europol's information systems, which enables them to have access to information of Member States, which

---

[40] JITs Manual from 23 September 2009, Council Doc. 13598/09, pp. 26 and 27 suggesting a model agreement for the participation of Europol, Eurojust or OLAF.

[41] Emphasis added, Article 6 (1) Council Decision of 6 April 2009 establishing the European Police Office, OJ 2009, L121/37, with regard to this problem, see: De Buck (2007).

[42] They are vaguely mentioned in Article 6 (4) and (5) Europol Decision and Article 13 (9) and (10) Convention on Mutual Assistance in Criminal Matters as well as Article 1 (9) and (10) Framework Decision on JITs (which literally repeats the aforementioned Articles of the Convention).

[43] Usually, the use of this information is restricted to the purpose for which the JIT has been set up and subject to the prior consent of the Member State where the information became available. Information can further be used for preventing an immediate and serious threat to public security and if subsequently a criminal investigation is opened as well as for other purposes to the extent that this is agreed between Member States setting up the team, Article 1 (10) (a)—(d) of Council Framework Decision of 13 June 2002 on JITs, OJ 2002 L 162/1.

[44] See example of a model agreement in: JITs Manual from 23 September 2009, Council Doc. 13598/09, p. 24.

[45] To this problem, see: Rijken and Vermeulen (2006); Mitsilegas (2009).

[46] Council Decision of 6 April 2009 establishing the European Police Office, OJ 2009, L121/37, Article 6 (4).

do not participate in the JIT or to information of third States cooperating with Europol (De Buck 2007).[47] When a Europol staff member during its participation in a JIT obtains information, he can include the information in Europol's data processing systems, after having obtained the prior consent of the relevant Member State.[48]

The active participation of Europol at the information exchange in the JIT nevertheless risks conflicting with the aforementioned local approach chosen in the JIT cooperation when considering that the information could only be shared within the boundaries of the national law of the national members seconded to the JIT. As a result, different domestic rules on data exchange and data protection may conflict with each other and additionally with the Europol rules, which could finally lead to a considerable lack of legal certainty.

Whereas the Europol Decision entails rules allowing for the exchange between its data processing systems and the JITs, Eurojust's or OLAF's data exchange with the JITs is not regulated. Although, for instance, Article 7(a) (iv) Eurojust Decision reinforces Eurojust's participation in JITs and clearly speaks of a participation of Eurojust's officials in JIT operations (Lopes da Mota 2009; Vervaele 2008)[49], information exchange or data protection rules in this regard are missing. The redraft of the Eurojust Decision in 2009 could have closed this regulatory gap, but either it was not detected or intentionally not regulated (Gusy 2008).[50] Rules comparable to the Europol Decision, which clarify the transfer of data between Eurojust and the JITs as well as the specifics of the information entered in the Case Management System are necessary to regulate this specific problem.

Moreover, OLAF's various legal bases do not even give an indication of its inclusion in JITs.[51] While OLAF officials proceed on the assumption that the second

---

[47] Information from third States can be obtained by using the so-called Virtual Private Network (VPN) connecting Europol's national units and offering encrypted lines with third States, see: De Buck (2007). Compare Council Decision of 6 April 2009 establishing the European Police Office, OJ 2009, L121/37, Article 6 (4) and (5).

[48] Compare Council Decision of 6 April 2009 establishing the European Police Office, OJ 2009, L121/37, Article 6 (4) and (5).

[49] JITs Manual from 23 September 2009, Council Doc. 13598/09, p. 10. It is worth mentioning that Eurojust's function is not any longer restricted to a mere "interface" between national authorities, limited to horizontal cooperation given that the Eurojust Decision 2009 visibly extended its operational tasks and Eurojust's role in JITs. For instance, Eurojust's national members are allowed to participate in JITs and the Secretariat of the JIT Experts Network shall form part of the Eurojust's staff, compare: Lopes da Mota (2009) and Vervaele (2008).

[50] It seems also possible that information obtained in course of JITs is entered by the Eurojust's national Members acting on the basis of national law and not by Eurojust officials in Eurojust's Case Management System. This possibility would also lead to a non-regulated transfer of data from the Case Management System to the other JIT members considering that national law does not apply in this rather European context. In addition, if only Eurojust's national members supply Case Management Information to the JIT or information stemming from Eurojust's own analysis, the questions of information transfer from Eurojust's Case Management System to the JIT through a member acting on behalf of Eurojust involved in the JIT is left unanswered, compare to the general data protection problems arising out of JITs: Gusy (2008).

[51] Compare Commission Decision 1999/352/EC of 28 April 1999 establishing the European Anti-Fraud Office (OLAF) OJ 1999 L136/20 and Regulation (EC) No. 1073/1999 of the European

protocol from 1999 to the Convention on the protection of the EC's financial interests[52]—broadly dealing with the cooperation between the Member states and the Commission in fraud-related matters, active and passive corruption and money laundering—taken together with the Convention on Mutual assistance in Criminal Matters enables OLAF to participate in JITs (De Moor 2009; Ralf 2009), none of these instruments explicitly refers to this sensitive subject matter. On the contrary, OLAF is not even mentioned.[53]

Keeping in mind Europol's extensive data exchange possibilities in the JITs, particularly the inclusion of information obtained in the JIT framework in its databases and vice versa, OLAF participation in JITs in absence of a clear legal basis, is legally doubtful.

Therefore, OLAF's role within the JIT structure certainly has to be clarified. In this context, special attention has to be paid to the fact that the cooperation of these two bodies is so far based on an agreement not allowing for personal data exchange (see Sect. 8.3.1.2). The participation of OLAF and Europol in common JITs unavoidably leads to personal data exchange and would therefore contradict OLAF's existing legal bases as well as the cooperation agreement between Europol and OLAF, discussed hereafter.

The question of joint participation in JITs of Eurojust and OLAF is, however, integrated in their cooperation agreement (analysed in Sect. 8.3.1.2).[54] However, details regarding the JITs cooperation, including the applicable data protection rules, are subject to the JIT agreement concluded between the participating parties.

8.3.1.2 Agreements Providing for Mutual Information Exchange

In addition to the cooperation in JITs, information exchange between the AFSJ actors is provided for in the agreement concluded between the relevant parties.

Parliament and the Council of 25 May 1999 concerning investigation conducted by the European Anti-Fraud Office (OLAF), OJ 1999 L136/31; Article 2 (6) Commission Decision 199/352 broadly regulates that "the office shall be in direct contact with the police and judicial authorities" and Article 1 (2) Regulation 1073/1999 only refers to "assistance" from the Commission to the Member States in organising close cooperation between the competent authorities of the Member States.

[52] Second Protocol, drawn up on the basis of Article K.3 of the treaty on European Union, to the Convention on the protection of the European Communities' financial interests—Joint Declaration on Article 13 (2)—Commission Declaration on Article 7, OJ 1997, C-221/12.

[53] Indeed, the Convention provides for "operational assistance" including exchange of personal data in fraud-related offences between the Commission and the Member States, but it does not specify at all the instruments to be used in this context.

[54] If one party is associated to a JIT related to fraud, corruption or criminal offences affecting the EU's financial interest, it shall inform the other party about its participation and propose the Member States setting up the JIT to consider inviting the other party, Practical Agreement on arrangements of cooperation between Eurojust and OLAF from 24 September 2008, point 9 (1).

Europol-Eurojust

The new Europol-Eurojust Agreement from January 2010[55] mainly regulates Eurojust participation at Europol's analysis work files, which is a new development linking the legal framework of the two bodies, hence affecting data protection questions related to the opening of the files to another agency. Problems regarding the accountability of processing as well as the supervision of it might arise.[56]

The agreement stipulates that both, Europol as well as Eurojust, shall "of its own motion" or upon request, provide each other with analysis results including interim analysis results.[57] When the information communicated matches the information stored in the respective processing systems, Europol or Eurojust shall additionally provide each other with data linked to the information provided.[58] This evidently leads to merging of the data yet stored separately either in the Europol or in the Eurojust databases. Article 8(3) Europol-Eurojust Agreement further provides for a regularly transmission of relevant data stored at Eurojust for the purpose of using them in Europol's analysis work files. The same applies to other information, in particular to information on cases provided that they fall within Europol's competence.[59] It is worth mentioning here that both actors are principally competent to deal with the same criminal offences.[60]

In addition to the exchange of information as regards the analysis work files, there is a further profound and important change as regards Eurojust's possibilities to play a part in Europol's analysis work files.

Whereby direct access by Eurojust to Europol's analysis work files was excluded under the former cooperation agreement from 2004, according to the new Europol-Eurojust Agreement, Eurojust has the right to take the initiative to open an analysis work file or even to establish a target group, if Eurojust is associated with the analysis work file concerned.[61]

---

[55] Agreement between Europol and Eurojust, which entered into force the 1 January 2010, Articles 7 (2) and 8 (2), in the following Europol-Eurojust Agreement; this Agreement replaced the Agreement between Europol and Eurojust from 9 June 2004.

[56] The EDPS in its opinion to the amendment of the Eurojust Decision rightly points to the questions of "who will be the processor?" and "who will be the controller?" within this new collaboration structure. Details to these questions are unfortunately not regulated in the Agreement as it indeed provides for the mutual association, but it does neither clarify questions of supervision in case of Eurojust's participation in Europol's analysis work files, nor regarding the transmission of personal data, compare: EDPS opinion on the Council Decision concerning the strengthening of Eurojust and amending Decision 2002/187/JHA from 5 December 2008, OJ 2008, C 310/1, p. 6, para 34.

[57] Articles 7 and 8 Europol-Eurojust Agreement.

[58] Articles 7 (2) and 8 (2) Europol-Eurojust Agreement.

[59] Article 8 (3) Europol-Eurojust Agreement.

[60] Eurojust's mandate refers to list of crimes for which Europol is responsible and which is laid down in Article 3 Europol Decision, compare Article 4 (1) Eurojust Decision.

[61] Article 9 (2) Europol-Eurojust Agreement. Article 11 (1) and (2) of the Europol-Eurojust Agreement 2010 clarifies that: Europol shall associate experts of Eurojust to participate within the activities of Europol's analysis work files, in particular when Eurojust initiated the opening of the respective file. Eurojust may also request to be associated with the activities of a particular analysis group.

The participation of Eurojust in the activities of an analysis work file and an analysis group at Europol is, however, astonishing, in particular with regard to Article 14(2) Europol Decision whereupon the access to analysis work files is strictly restricted to analyst, designated Europol staff, liaison officers or experts from the Member States. This Article moreover provides that only analysts are authorised to enter data into the file and modify such data. Taking into account that Article 13 Europol-Eurojust Agreement stipulates that the transmission shall be in accordance with the establishing act of the parties and additionally considering the enormous variety (information about criminals, victims, witnesses, contacts, etc.) as well as amount of personal data (up to 69 data elements), which can be stored in Europol's analysis work files, each widening of the circle of persons having access to the relevant information should be accompanied with additional safeguards against abuse as well as effective tools of supervision (compare ECtHR case *Weber and Saravia v. Germany*[62]).

It is worth noting that the Europol-Eurojust Agreement, however, lays down access as well as correction and deletion rights.[63] Disappointingly, although the participation of Eurojust at Europol's work files was newly introduced in the 2010 agreement, the data protection provisions introduced in the former 2004 agreement, were not adapted to the new circumstances. Rules requiring information of witnesses, victims or persons requesting access about the transfer of their data as well as rules relating to the information of Europol's or Eurojust's Joint Supervisory Body (JSB) about the transfer, are missing. Provisions regulating the competence for access request once Eurojust's data are included in Europol's analysis work files are additionally not provided for in the agreement, not to mention provisions relating to the supervision of the data generated in this way.

All in all, Eurojust's participation at Europol's analysis work files demands further protections for individuals, in particular regarding the rights of victims or witnesses to know whether and to whom their data are transferred. The JSB and the data protection officers of both agencies should be informed in any case to guarantee at least a minimum supervision. In addition, when taking the enormous amount of

[62] The transmission of personal data to other authorities was only allowed when it was particularly supervised and restricted to the transmission of data arousing the suspicion that specific facts, as opposed to mere factual indications, pointing to the fact that this person has committed a crime, compare: *Weber and Saravia v. Germany*, Application no. 54934/00 Admissibility Decision, paras 42–43 and 123–129 from 29 June 2006; Article 14 (4) Europol-Eurojust Agreement, however, lays down that the transmission of data revealing racial origin, political opinions or religious or other beliefs, or concerning health and sexual life shall be restricted to absolutely necessary cases and that such data shall only be transmitted in addition to other data.

[63] According to Article 18 (3) Europol-Eurojust Agreement, transmitted data shall be deleted when they are no longer necessary for the purpose for which they were transferred or when they are not necessary for the tasks of the receiving party or when no decision has been taken within 3 months after receipt (Article 16 (4)); a retention review must take place within the first 3 years of storage and when the storage exceeds 3 years, an annual review has to be implemented, see Article 18 (5) Europol-Eurojust Agreement.

data into account with which both agencies are dealing[64], it is worth considering the establishment of an independent authority only for the purpose of monitoring the data transfer between them.

Europol-OLAF Cooperation

Europol's and OLAF's cooperation is based on an administrative agreement restricted to the exchange of strategic information signed in 2004.[65] Currently, negotiations are taking place discussing an administrative arrangement similar to that concluded with Eurojust, which allows for personal data exchange.[66]

However, it is worth noting that, after the entry into force of the new Europol Decision on 1 January 2010, Article 22(3) Europol Decision permits Europol to directly receive, use and transmit information, including personal data from OLAF even prior to the conclusion of a formal exchange agreement "in so far as it is necessary for the legitimate performance of Europol's or OLAF's tasks". In case the transmitted data were originally introduced by a Member State, Europol has to ask the Member State for prior consent.[67]

Taking into account the different existing provisions, on the one hand, a valid agreement not allowing for personal data exchange and on the other, the rules stipulated in the Europol Decision, the legal basis for personal data exchange between OLAF and Europol is far from being clear. Theoretically, according to its legal basis, Europol could transmit and receive personal data stored in OLAF's databases, although it has to be taken into account that OLAF's legal framework lags considerably behind. Apart from the fact that data processing must be generally in accordance with the provisions of Regulation 45/2001[68], none of OLAFs legal bases include transfer provisions regulating the personal data exchange with EU agencies such as Europol.[69]

---

[64] Eurojust registered 1,372 new cases in 2009, compare Eurojust annual report 2009, p. 50 and Europol had 88,419 objects stored in the EIS and initiated 8,377 cases in 2008, compare Europol annual report 2008, pp. 33–35.

[65] Administrative Arrangement between the European Police Office (Europol) and the European Anti-Fraud Office (OLAF) from 8 April 2004, accessed July 12, 2011, https://www.europol.europa.eu/sites/default/files/flags/european_anti-fraud_office_olaf_.pdf.

[66] OLAF annual report 2009, ninth activity report for the period 1 January 2008–31 December 2008, section 4.6.2, p. 59.

[67] Article 24 (1) Europol Decision.

[68] Regulation 45/2001 is restricted in scope and refers only to personal data transfer between Community bodies, which represent bodies established under the former first pillar and does not include Europol or Eurojust.

[69] Regrettably, neither Commission Decision 1999/352/EC establishing OLAF nor Regulation 1073/1999 includes transfer provisions regulating the personal data exchange with third states or agencies such as Europol. Article 10 Regulation 1073/1999 refers to the forwarding obtained in course of internal investigations to the bodies, offices and agencies concerned by the investigation, however, this provision does not take the data exchange in the framework of criminal or judicial cooperation into account. Rules on the transfer to agencies are nowhere to be found in OLAF's instruments.

Europol-Frontex

Frontex and Europol cooperate based on a "strategic agreement" concluded in 2008.[70] The agreement is limited to the exchange of strategic and technical information[71] prohibiting the exchange of personal data, more precisely the transfer of "data related to an identified individual".[72]

Astonishing, however, are the provisions regulating the exchange of information. They are remarkably detailed and seem rather to make sense when personal data shall be exchanged.[73]

Such specified provisions are exceptional and not included in similar strategic agreements Europol has concluded with other EU bodies.[74] The existence of such provisions casts doubts on the complete exclusion of personal data exchange from the cooperation between the two actors.

In addition, the agreement's exclusion of personal data exchange seems to be rather obsolete, yet disconnected to a great extent from Europol's and Frontex's cooperation in reality, also in the light of Europol's new Council decision, which provides for personal data exchange even in absence of an agreement allowing for the latter.[75]

The mysterious wording of the agreement seems, however, to make sense when taking the practical cooperation between the Europol and Frontex into account: a

[70] Strategic cooperation agreement between the European Agency for the Management of Operational Cooperation at the External Borders of the Member States of the European Union and the European Police Office from 28 March 2008; in the following: Europol-Frontex Agreement from 28 March 2008.

[71] According to Article 2 Europol-Frontex Agreement: 1. "Strategic information" includes, but is not limited to: (a) enforcement actions that might be useful to suppress offences and improve the integrated border management of the Member States of the European Union; (b) new methods used in committing offences, in particular, those threatening the security of external borders or facilitating illegal immigration; (c) trends and developments in the methods used to commit offences; (d) observations and findings resulting from the successful application of new enforcement aids and techniques; (e) routes and changes in routes used by smugglers, illegal immigrants or those involved in illicit trafficking offences covered by this agreement; (f) prevention strategies and methods for management to select law enforcement priorities and (g) threat assessments, risk analysis and crime situation reports. 2. "Technical information" includes, but is not limited to: (a) means of strengthening administrative and enforcement structures in the fields covered by this agreement; (b) police working methods as well as investigative procedures and results; (c) methods of training the officials concerned; (d) criminal intelligence analytical methods and (e) identification of law enforcement expertise.

[72] Article 1 Europol-Frontex Agreement from 28 March 2008.

[73] For instance: conditions on the further use and transfer of the transmitted information may be imposed on the receiving party, just as Europol shall only supply information to Frontex ", which was collected, stored and transmitted in accordance with the relevant provisions of the Europol Convention and its implementing regulations" though the latter apparently deals with personal data. Compare: Article 5 para 3 et 8 Europol-Frontex agreement.

[74] For instance with: the Central Bank, Commission, Monitoring Centre for Drugs and Drug Addiction, OLAF.

[75] Pursuant to its Article 22 (3).

House of Lords report reveals that Europol has worked "informally" with Frontex since 2006.[76] An external report evaluating Frontex's work and published on Frontex's webpage sheds light on this issue and discloses further problems. According to the report, Frontex collects data in the framework of joint operations in order to send them to other agencies, such as Europol for threat analysis (Holzenberger 2006).[77] Pursuant to the report, 10% of the detained persons during a joint operation are interviewed by Frontex staff[78], which finally means that Frontex itself also collects personal data notwithstanding its restrictive legal framework at present, which does not allow for personal data processing. Consequently, Frontex acts in absence of a legal basis allowing for the collection and processing as well as the transfer of personal data.[79]

Above, we have seen two important facts relating to data processing at Frontex: while neither the Frontex Regulation 2007/2004 nor the Europol-Frontex agreement permit personal data processing or transfer, the reality seems to tell another story. The exchange and in particular Frontex's collection of personal data is neither covered by the Europol-Frontex agreement, nor by Frontex's current legal basis.

For this reason, clarifications in Frontex's legal framework were long overdue and have resulted in 2010 in the Commission's and the Council's Frontex proposal to amend the Frontex regulation 2007/2004[80] by, amongst others, now including two important changes concerning the question of data processing at Frontex: on

[76] House of Lords Europol report, European Union Committee, 29th report of session 2007–2008, "Europol: coordinating the fight against serious and organised crime", published 12 November 2008, p. 80.

[77] Final report of COWI (European consulting group) from January 2009 preparing an external evaluation of Frontex provided for in Article 33 of the Council Regulation (EC) No. 2007/2004 of 26 October 2004 establishing Frontex, p. 48, accessed July 12, 2011, http://www.frontex.europa.eu/specific_documents/other/, joint operations are described as a "good example of integrated analyses by Europol and Frontex" and are regarded as a working practice in which intelligence and operations are brought together as closely as possible". To the details of the cooperation between Europol and Frontex.

[78] Final report of COWI (European consulting group) from January 2009 preparing an external evaluation of Frontex provided for in Article 33 of the Council Regulation (EC) No. 2007/2004 of 26 October 2004 establishing Frontex, p. 48, accessed July 12, 2011, http://www.frontex.europa.eu/specific_documents/other/.

[79] The proposal to amend the Frontex regulation should eventually put this exchange on a legal basis. Nevertheless, even if the proposal enters into force, personal data exchange with Europol or other Union agencies or bodies would generally require the conclusion of a new cooperation agreement. Compare: Proposal for a Regulation of the European Parliament and the Council amending Council Regulation (EC) No. 2007/2004 of 26 October 2004 establishing a European Agency for the Management of Operational Cooperation at the External Borders of the Member States of the European Union (FRONTEX) from 24 February 2010, COM (2010) 61 final.

[80] Proposal for a Regulation of the European Parliament and the Council amending Council Regulation (EC) No. 2007/2004 of 26 October 2004 establishing a European Agency for the Management of Operational Cooperation at the External Borders of the Member States of the European Union (FRONTEX) from 24 February 2010, COM (2010) 61 final and Council document 2010/0039 (COD), 8121/10, proposal for a regulation of the European Parliament and the Council amending Council Regulation (EC) No. 2007/2004 establishing a European Agency for the Management

the one hand, the Frontex proposal allows to collect, process and exchange personal data concerning the detection of criminal networks organising illegal immigration[81] and on the other hand, it supports the use and the possibility of carrying out risks analysis.[82]

While on the one hand the widening of Frontex's mandate in this regard would connect two not directly linked remits (border control and serious crime prevention), on the other hand, the possibility to carry out risks analysis would considerably overlap with Europol's mandate. Regrettably, the proposal does neither specify the details of data processing at Frontex nor the cooperation with EU agencies. Individual rights, such as data protection rights, are not (yet) included in the proposal.[83] According to recent developments, provisions on the cooperation with Europol as well as on data protection issues should be added to the Frontex proposal.[84] However the details of these provisions are not yet published.[85]

Moreover, it is very important that, in contrast to Europol, Frontex's mandate does not (and will not) cover the collection of data related to serious crime or organised immigration crime, which means that the data of Europol and Frontex are definitely not collected for the same purpose. The possible exchange of the data could eventually lead to the connection of data of potential immigrants with data included in Europol's databases, the latter dealing for the most part with data related to persons associated to crimes. Linking these two subjects while disregarding any distinction between data of criminals and data of (possible) immigrants, contravenes the purpose limitation principle and blurs the border between criminals and immigrants. Clear rules respecting the protection of personal data of the individuals concerned in

of operational cooperation at the external borders of the Member States of the European Union (Frontex) 29 March 2010.

[81] Proposal for a Regulation of the European Parliament and the Council amending Council Regulation (EC) No. 2007/2004 of 26 October 2004 establishing a European Agency for the Management of Operational Cooperation at the External Borders of the Member States of the European Union (FRONTEX) from 24 February 2010, COM (2010) 61 final, Article 2; Eurosur is the planned European Border Surveillance System, for more details, see: Commission staff working paper, report on progress made in developing the European Border Surveillance System (EUROSUR) from 24 September 2009, Sec (2009), 1265 final and analysis of the Commission communications on future development of Frontex and the creation of a EUROSUR, briefing paper from policy department C, citizens' rights and constitutional affairs, civil liberties, justice and home affairs, Directorate General internal policies of the Union from June 2008.

[82] Impact assessment accompanying the proposal for a Regulation of the European Parliament and the Council amending Council Regulation (EC) No. 2007/2004 of 26 October 2004 establishing a European Agency for the Management of Operational Cooperation at the External Borders of the Member States of the European Union (FRONTEX) from 24 February 2010, p. 34.

[83] Compare for more details: opinion of the European Data Protection Supervisor (EDPS) on the proposal for a Regulation of the European Parliament and the Council amending Council Regulation (EC) No. 2007/2004 establishing a European Agency for the Management of Operational Cooperation at the External Borders of the Member States of the European Union (FRONTEX) from 17 May 2010.

[84] Compare press release 11916/11, Presse 192 from 23 June 2011, accessed July 12, 2011, http://www.consilium.europa.eu/uedocs/cms_Data/docs/pressdata/en/jha/122983.pdf.

[85] Last verified on 30 June 2011.

the Frontex proposal would help to prevent the criminalisation of this specific group and should accompany the Council's and the Commission's ambitions to extend Frontex's possibilities to exchange data.

Eurojust-OLAF Cooperation

The practical agreement on arrangements of cooperation between Eurojust and OLAF from 2008 provides for the collaboration in operational and strategic meetings as well as the exchange of information including personal data in specific cases.[86] Restrictions, the conditions on the use of the data or the time of storage of the transmitted data are regrettably not given.

Individual rights are not directly mentioned, although OLAF's data processing must usually comply with Regulation 45/2001. The misleading title "rights of data subjects" of point 14 of the agreement only reveals a consultation duty for the requested party towards the other party before deciding about a request by an individual to have access to, to demand correction, blocking or deletion of its personal data transmitted under the agreement.[87] Apart from that provision, the agreement makes reference to the relevant data protection rights of the parties.

However, the mere reference to the applicable rules of the parties does not automatically assure compliance with them. Considering that the motivation to exchange personal data represents one of the main reasons for the amendment of first cooperation agreement from 2003, additional safeguards taking into account the specific risks of data transfer would have illustrated the "good will" of the parties to acknowledge the importance of data protection rights in this context. The indication of an authority exercising, for example, independent supervision of the agreement would have, for instance, emphasised the submission under an efficient data protection regime.[88]

A further important point concerns the different time limit of storage—20 years at OLAF and as long as it is necessary at Eurojust—which is not taken into account by the text of the agreement. Questions relating to restrictions and the conditions on the use of OLAF's data in Eurojust's Case Management System arise.[89]

---

[86] Practical Agreement on arrangements of cooperation between Eurojust and OLAF from 24 September 2008, point 6.

[87] Practical Agreement on arrangements of cooperation between Eurojust and OLAF from 24 September 2008, point 14.

[88] Theoretically, the EDPS and possibly Eurojust's JSB are responsible for this task, it would not do any harm to the parties to mention them in the agreement. A particular problem in this context relates to the fact that the responsibility for personal data transfer from Eurojust to OLAF lies only with the national member and not with Eurojust, having for consequence that supervision is becoming increasingly difficult and can usually not be exercised by Eurojust's JSB.

[89] Mutual information duties apply and include the notification duty of the other party about corrections or deletions made, including the reasons therefore. In addition, regarding cases in which one of the parties assumes that information received is not accurate, not up to date or should not have been transmitted, the other party has to be warned. A further provision consists of the requirement to inform a third party, to which transmitted data have been transferred, about any deletions or

8.3.2 *Europol's and Eurojust's Access to Information Systems*

Personal data exchange is not only limited to AFSJ agencies, it is also taking place between European information systems and the AFSJ agencies. The information systems include the databases SIS (II), CIS, VIS and Eurodac. The increasing data exchange between the mentioned actors considerably enlarges the authorities and bodies having access to personal data originally entered in only one of the databases. Therefore, attention should be paid to the rather limited purpose for which the databases were established[90] and which is continually broadened when allowing various actors, not necessarily connected to this original purpose, to access. In the light of the foregoing considerations, it is therefore interesting to briefly analyse the relation and the data exchange possibilities in the framework of ASFJ agencies and European information systems in order to understand the data protection impact of the access from the AFSJ agencies to the mentioned databases.

8.3.2.1 Europol's and Eurojust's Access to the SIS II

Europol as well as Eurojust have access to the SIS (II).

Europol gained access to information relating to important categories of data contained in the SIS already in February 2005.[91] In the meanwhile, Europol's and Eurojust's tasks as well as the scope of the new SIS II have been evolved continually and the data entered in the respective databases are getting more and more extensive. Europol's tasks and functions remain nevertheless more comprehensive and the data processed in its databases entail much more elements than those stored in the SIS II.[92]

Despite the access, the Europol Decision does not directly mention the SIS II. Article 21 Europol Decision, however, permits wide-ranging access to data of Union databases to the extent "that is necessary for the performance of its tasks". The SIS II Decision 2007/533 mirrors this provision by stipulating that Europol and Eurojust have the right "within its mandate" to access and search data directly in the SIS II.[93]

---

corrections made concerning this data. Finally, the time limits of the storage bases on the respective rules of the parties, compare practical Agreement on arrangements of cooperation between Eurojust and OLAF from 24 September 2008, point 15.

[90] The SIS for security purposes with regard to EU as well as to third state nationals, CIS for customs control, VIS for the exchange of visa data and Eurodac for the exchange of fingerprint data of asylum seekers.

[91] Council Decision 2005/211/JHA of 24 February 2005 concerning the introduction of some new functions for the Schengen Information System, including in the fight against terrorism, OJ 2005 L-68/44, Article 1 referring to Articles 95, 99 and 100 Schengen Convention, OJ 2000, L-239/19 (persons wanted for extradition, persons or vehicles placed under surveillance or subjected to specific checks as well as to objects sought for the purpose of seizure or use in criminal proceedings).

[92] Up to 69 data elements can be, for instance, stored in an analysis work file at Europol.

[93] Articles, 41, 42 and 43 SIS II Decision 2007/533; the scope of the access relates to persons wanted for arrest or surrender purposes, persons and objects for discreet checks or specific checks

Whereas Europol's legal basis mentions the mandate of the access, *Eurojust's access* to other databases, is neither referred to in the new Eurojust Decision, nor in any of its predecessors. Only Article 42 SIS II Decision 2007/533 refers to the possibility of Eurojust's national Members, not including Eurojust staff, to access and search data in the SIS II.[94]

The absence of Eurojust's mandate is particularly striking when looking at the remarks of the House of Lords, already made in 2003, which clearly point to the lacking provisions allowing Eurojust's access.[95] The amendment of the Eurojust Decision in 2009 could have been an opportunity to define the conditions of Eurojust's access to the SIS II as well as the details regarding the use of the data. The non-inclusion of this topic in the instrument leaves strong doubts on the political will to concretely identify Eurojust's mandate regarding the SIS II data and opens the way for a non-regulated data use at Eurojust.

As regards the processing of the data, both agencies may use the SIS II data. The handling of the data is left to the respective legal bases of the accessing actors.[96] Questions relating to the inclusion of data from other information systems in Europol's or Eurojust's databases are left, however, unanswered. Neither the Europol or the Eurojust Decision nor the SIS II Decision 2007/533 provide for clarifications.[97]

Provisions relating to the protection of the information at Europol and Eurojust are limited.[98] Although both agencies must introduce a recording duty of every access and search made by them as well as a provision interdicting the connection, the transfer, the download and the copying of the SIS II data to another computer system for data collection and processing operated by or at Europol or Eurojust, they may introduce SIS II information in their own database (either, by asking the relevant Member State after a hit in the SIS II to introduce the same information in

as well as to objects for seizure or use as evidence in criminal proceedings Eurojust has additionally access to data of missing persons.

[94] Articles 42 (1) and (6) SIS II Decision 2007/533. This might be partially due to the fact that only national members of Eurojust can access the SIS II database, then integrating the data in the Eurojust system, but it does not explain why a reference is entirely lacking.

[95] "The only provision that enables Eurojust access to SIS data appears to be an unpublished non-legally binding declaration annexed to the Eurojust Decision (which we have asked to see but have never received)", compare: House of Lords, Select Committee on European Union Written Evidence Sub-Committee F (Social Affairs, Education and Home Affairs), letter from the Chairman to Bob Ainsworth, MP, Under-Secretary of State, Home Office, Schengen Information System: new functions, (9407/02 and 9408/02) from 9 April 2003.

[96] Article 41 (3) SIS II Decision 2007/533.

[97] Europol's legal basis, for instance, limits further clarifications to the simple provision that the legal instruments of the relevant partner databases shall govern Europol's use of the data as well as its access conditions, "in so far as they provide for stricter rules on access and use" than those of the Europol Decision. Compare Article 21 Council Decision of 6 April 2009 establishing the European Police Office, OJ 2009, L121/37.

[98] Articles 41 (5) and 42 (4), (5) and (7) SIS II Decision 2007/533.

the Europol or Eurojust database or by asking the Member State for consent to use the information in their own databases).[99]

This possibility also influences the following restrictions of Article 41(5) (c) and (d) SIS II Decision 2007/533 pursuant to which Europol must adopt security and confidentiality rules as well as limit access to data entered in the SIS II to specifically authorised staff. Even if the access is initially restricted to certain persons, which is generally a welcomed provision, if the data are later introduced by a Member State in Europol's databases EIS, the initially restricted access only exists on paper.

Article 41(3) SIS II Decision 2007/533 additionally provides for the possibility to transfer the obtained SIS II information to third states (Member State's consent provided), circumventing the initial restriction of Article 54 SIS II Decision 2007/533 whereupon SIS II data should not be made available to third countries.

8.3.2.2 Europol's Access to the VIS

Access to the VIS is limited to Europol. It is briefly mentioned in the VIS Regulation 767/2008 and further detailed in Council Decision 2008/633 concerning access for consultation of the VIS by designated authorities of Member States and by Europol for the purposes of the prevention, detection and investigation of terrorist offences and of other serious criminal offences (VIS access Decision 2008/633).[100]

As in the case of the SIS II, Europol's access depends on its mandate restricted to "the performance of its tasks".[101]

Due to the influence exerted by the European Parliament during the negotiations[102] and compared to the SIS II instruments, the VIS access Decision 2008/633 requires a more sophisticated, if not necessarily always sufficient, data protection framework briefly analysed hereinafter.

As the VIS Regulation 767/2008 does not specify Europol's access conditions, VIS access Decision 2008/633 does not succeed in reaching comprehensive clarification in this regard either.

The purpose of Europol's access remains vague and generally refers to the purpose of prevention, detection and investigation of terrorist offences and of other serious

[99] Compare Article 41 (3) SIS II Decision 2007/533.

[100] Council Decision 2008/633/JHA of 23 June 2008 concerning access for consultation of the VIS by designated authorities of Member States and by Europol for the purposes of the prevention, detection and investigation of terrorist offences and of other serious criminal offences, OJ 2008, L-218/129 (in the following: VIS access Decision 2008/633).

[101] Article 3 (1) VIS Regulation 767/2008, OJ 2008, L-218/60.

[102] The VIS access Decision 2008/633 entered into force in September 2008 and was not, contrary to VIS Regulation 767/2008, which is a former first pillar instrument, adopted by using the co-decision procedure, but formed part of the "VIS legislative package" agreed between the European Parliament and the Council in 2007 after two and a half years of negotiations. The reason therefore can be found in the legal basis of the instrument, which is governed by Title VI of the EU Treaty dealing with police and judicial cooperation in criminal matters, more specifically the Decision bases on Article 30 (1) (b) and 34 (2) (c) EU. Treaty; thus the Council alone could decide about the adoption of the instrument.

crime.[103] Article 7 VIS access Decision 2008/633 refers to the access for the purpose of the performance of Europol's tasks[104] as well as for analysis purposes according to Article 10 Europol Convention.[105]

Similar criticism as mentioned in the SIS II discussion applies also in the framework of the VIS. In both cases, access depends on a variable factor, namely the performance of Europol's tasks, which are subjected to modifications at any time. A good example is the last amendment of the Europol Convention, the Europol Decision entering into force in January 2010, which completely reversed Europol's legal framework and considerably enlarged its tasks.

A further important, although regrettable, aspect in context of the access of Europol to the VIS, is the fact that important requirements restricting the access conditions of national "designated authorities" do apply to Europol.[106] As a result, Europol's access is significantly wider than the access of the national authorities and does not require that the data are necessary for a specific case or that the consultation substantially contributes to the purpose of the access.[107]

In the light of the foregoing, it is interesting to note that both the Commission as well as the European Parliament stressed during the decisions' negotiations that a "routine access" of Europol should be prevented.[108]

---

[103] The offences are further detailed in two Framework Decisions, which list a range of different crimes, not always corresponding to those of the Europol Decision. Terrorist offences means the offences under national law corresponding or being equivalent to the offences listed in Article 1–4 Framework Decision 2002/475 on combating terrorism (OJ 2002, L-164/3) and serious criminal offences embraces the forms of crimes corresponding or being equivalent to those referred to in Article 2 (2) Framework Decision 2002/584 on the European Arrest Warrant (OJ 2002, L-190/1).

[104] Europol's tasks are described in Article 5 (1) (a) Europol Decision and mentions that Europol has the task to "obtain, collate and analyse information and intelligence".

[105] Mainly corresponding to Article 14 Europol Decision, which stipulates the conditions for collection, processing and utilisation of personal data in analysis work files.

[106] Article 5 (1) VIS access Decision 2008/633 dictates three cumulative access conditions for the national law enforcement and intelligence authorities: first, the access must be necessary for the purpose of prevention, detection and investigation of terrorist offences or other serious crime, second, necessary in a specific case and third, consultation must substantially contribute to the mentioned purposes. Once the national authorities comply with these requirements, a two-step access to the VIS data is stipulated in Article 5 (2) and (3) VIS access Decision 2008/633, which, at this stage of the procedure, also applies to Europol. The two-step access limits the initial search in the VIS to 11 data elements, including fingerprints. Only in the event of a hit, the other data from the visa application form, as well as photographs and the data entered in respect of any visa issued, annulled, revoked, refused or extended are open to access. Whereas the Member States have to fulfill all of the conditions of Article 5 VIS access Decision 2008/633, Europol's access seems to be regarded as less intrusive.

[107] However, Member States as well as Europol have to establish a list with the operating units, which are allowed to access the VIS. These units play an important role in the access procedure as they must submit a reasoned written and electronic request to the central access point established in each Member State or, respectively, at Europol to coordinate the VIS access, compare Articles 3 (3), 4 (1) and 7 (3) VIS access Decision 2008/633, OJ 2008, L-218/129.

[108] Report from 21 May 2007 of the European Parliament on the on the proposal for a Council Decision concerning access for consultation of the VIS by the authorities of the Member States

In the current state of play, Europol's rather wide access to the VIS is worrying. The exceptional aspect of allowing a law enforcement authority access to a database dealing with individuals not suspected of any crime should be at least compensated through very rigid access conditions to avoid the transformation of the VIS into a general crime fighting database, disregarding the fundamental rights of individuals. The introduction of stricter access conditions would have been an important step in this direction.[109]

In context of the enlargement of authorities having access to the VIS data, it is worth noting that not only Europol and the participating Member States may access the VIS data, but also Member States to which the VIS Regulation 767/2008 does not apply.[110] It is exercised via a participating Member State in the way that Member States not yet participating at the VIS shall make its visa data available to the participating Member States, on basis of a "duly reasoned written or electronic request".[111] The question arises whether it makes sense to limit the participation in the VIS Regulation 767/2008 to the Schengen Member States when the non-participating Member States eventually could get access to the VIS data pursuant to Article 6 VIS Regulation 767/2008.

Data protection provisions in the framework of the VIS access orientate on the level of protection of Convention No. 108 and its subsequent amendments[112], the case law pursuant to Article 8 ECHR[113], Recommendation R (87) 15 and on the third pillar data protection Framework Decision 2008/977.[114] If the data are transferred to Europol, the general rules of the Europol Decision apply.

The VIS access Decision 2008/633 nevertheless entails an important provision prohibiting the onward transfer of the VIS data at Europol.[115] In "exceptional cases of

responsible for internal security and by Europol for the purposes of the prevention, detection and investigation of terrorist offences and of other serious criminal offences (COM(2005)600final—2005/0323(CNS)), Committee on Civil Liberties, Justice and Home affairs, rapporteur: Sarah Ludford, pp. 7–8, para (7) and proposal for a Council Decision from 24 November 2005 concerning access for consultation of the VIS by the authorities of the Member States responsible for internal security and by Europol for the purposes of the prevention, detection and investigation of terrorist offences and of other serious criminal offences (COM(2005)600final—2005/0323(CNS)), p. 5.

109 A welcomed provision, however, relates to the requirement to designate a specialised unit for the VIS access within Europol, allowing for better supervision while concentrating the request accesses at one specific entity. Such as in the SIS II, Europol's use of the data is subject to the consent of the Member States entering the data in the VIS, Article 7 (4) VIS access Decision 2008/633, OJ 2008, L-218/129.

110 Due to their limited participation in the Schengen cooperation, certain Member States, such as the United Kingdom, are usually not allowed to access the VIS.

111 Article 6 VIS access Decision 2008/633, OJ 2008, L-218/129.

112 For those Member States, which have ratified it, the Additional Protocol of 8 November 2001 to Convention No. 108 should also be taken into account.

113 Recital (9) VIS access Decision 2008/633/JHA, OJ 2008, L-218/129.

114 Article 8 (1) and recital (9) VIS access Decision 2008/633, OJ 2008, L-218/129.

115 Article 8 (4) VIS access Decision 2008/633, OJ 2008, L-218/129.

urgency", third states may nonetheless receive the VIS data.[116] A provision similar to Article 13(1) (d) third pillar data protection Framework Decision 2008/977 according to which the level of data protection of the third party must be adequate for the intended data processing does regrettably not exist.[117] While the rules on third party data transfer apply to the Member States as well as to Europol, the provisions on data security, liability and claims for compensation are governed by national law and are only addressed to the Member States. Europol relies on its own data security rules whose implementation is subjected to a very unconvincing necessity criterion.[118] The right of access, correction and deletion depends on the law of the Member State in which an applicant invokes that right.[119]

8.3.2.3 Europol's and Eurojust's Access to the CIS

In contrast to the VIS access, an agreement regulating the details of the access from Europol or Eurojust to the CIS data does not exist. Therefore, only Article 11 CIS Council Decision 2009/917 on the use of information technology for customs

---

[116] Article 8 (4) VIS access Decision 2008/633, OJ 2008, L-218/129. There is no definition of such an exceptional case, but there are three additional criteria to be fulfilled to transfer the VIS data to third parties: the data must be necessary in a specific case, the consultation must substantially contribute to the mentioned purposes and the Member States having entered the data into the VIS must have given its consent.

[117] Although, as the third pillar data protection Framework Decision 2008/977 is applicable to the VIS access Decision 2008/633, the latter rules must comply with those of the former one.

[118] Article 35 Europol Decision stipulates specific rules relating to data security involving the "necessary technical and organisational measures to ensure the implementation of this Decision". As the wording of this first paragraph of Article 35 Europol Decision suggests, the implementation of data security measures depends on the necessity of these measures. The latter are considered as "necessary where the effort they involve is proportionate to the objective they are designed to achieve in terms of protection". Thus, data security rules are subjected to a necessity criterion whose content leaves open certain questions. Which body within Europol decides about the effort to be made and about the proportionality of this effort? Europol's JSB is not mentioned in this context, but Article 10 (3) Europol Decision refers to the Management Board, which shall ensure that the measures and principles referred to in Article 35 Europol Decision are properly implemented. Consequently, the Management Board decides about the implementation of data security rules and in this way about the question to what extent the effort appears to be proportionate and as a result about the effort to be made to adopt a specific security measure. The internal Data Protection Officer or the JSB are not involved.

[119] Article 14 VIS access Decision 2008/633/JHA, OJ 2008, L-218/129. Individuals interested in knowing whether their VIS data have been transferred to Europol are merely informed in the framework of the information right provided for in Article 37 VIS Regulation 767/2008. According to this Article, the notification of the applicant is broadly restricted to the fact that Europol may receive the data. There is no information duty provided for in VIS Regulation 767/2008 in the very likely case that the data are transferred to Europol after the visa applicant or the person issuing an invitation or liable to pay the applicant's subsistence cost, has been initially informed about Europol's possibility to access the VIS data. Consequently, information about the actual transfer of the information is not given.

purposes, provides for, at first glance, almost unfettered access to the data entered into the third pillar CIS.[120]

The CIS Council Decision 2009/917 uses the general wording within its respective "mandate and the fulfilment of Europol's or Eurojust's tasks"[121], when describing the limits of the right of access of the two agencies to the CIS.[122]

Recital (5) of Council Decision 2009/917 specifies the reason for Europol's access in this way as it "should allow Europol to cross-check information obtained through other means with the information available in those databases, to identify new links that were so far not detectable and thus to produce a more comprehensive analysis".[123] Finally, access should enable Europol to "uncover connections between cases of criminal investigations, so far unknown to Europol that have a dimension in and outside the European Union".[124]

Eurojust's access refers to the need "to obtain immediate information required for an accurate initial overview enabling to identify and overcome legal obstacles and to achieve better prosecution results" as well as "to receive information of ongoing and closed investigations in different Members States and thus to enhance the support of judicial authorities in the Member States".[125]

Regrettably, no further specifications as regards the subsequent processing of the CIS data at Europol or Eurojust can be found in the CIS Council Decision 2009/917, apart from the obligation to ask the Member State originally entering the data for consent when using and transferring the data to third countries.[126]

After having obtained the consent, in case of Europol, the rules of the Europol Decision apply, which do not regulate the use or the processing of data from the other European databases within the databases of Europol.[127]

Comparable to the situation regarding the SIS II, the Eurojust Decision remains silent on the topic of Eurojust's access to the CIS.

More details on Eurojust's access to the CIS are not codified, which reveals a significant lack of legal rules resulting in the complete absence of Eurojust's mandate to access the CIS data in its own legal basis, the lack of provisions regulating both, the individual rights when the data are transferred as well as the technical details concerning the practical implementation of the access.

Moreover, a legally very doubtful provision is Article 8(1) CIS Council Decision 2009/917, which allows Europol and Eurojust to use the CIS data for any other purposes as long as they are vaguely connected to policing purposes.

---

[120] Article 11 CIS Council Decision 2009/917/JHA of 30 November 2009 on the use of information technology for customs purposes, OJ 2009, L-32320 (in the following referred to as Council Decision 2009917, OJ 2009, 323/20).

[121] Articles 11 (1) and 12 (1) Council Decision 2009/917, OJ 2009, L-323/20.

[122] Article 11 (1) Council Decision 2009/917, OJ 2009, L-323/20.

[123] Recital (5) Council Decision 2009/917, OJ 2009, L-323/20.

[124] Recital (5) Council Decision 2009/917, OJ 2009, L-323/20.

[125] Recital (6) Council Decision 2009/917, OJ 2009, L-323/20.

[126] Articles 11 (3) and 12 (2) Council Decision 2009/917, OJ 2009, L-323/20.

[127] Article 11 (3) Council Decision 2009/917, OJ 2009, L-323/20.

The only provision slightly referring to an access restriction relates to the usual interdiction to directly connect parts of the CIS to Europol's or Eurojust's own data processing systems and to transfer, download or copy the CIS data to its systems, although Europol may also request further information from the Member State.[128]

The persons having access to the CIS shall be limited to "duly authorised" Europol staff and the national members of Eurojust. In case of Europol, reminiscent of the SIS II and the VIS access rules, Europol's JSB shall additionally monitor Europol's activities in this regard. As only national members of Eurojust access the CIS, the monitoring of Eurojust's JSB is curtailed.[129]

All in all, the conditions dealing with Europol's and Eurojust's access to the CIS, compared to the SIS II and the VIS, are even more far reaching. Provisions restricting the access cannot be found which leads to almost unrestrained access of Eurojust and Europol to the CIS data.

8.3.2.4 Common Problems with Regard to the Access of Europol and Eurojust to the European Information Systems

Taking the aforementioned examples into account, it is remarkable that the purpose of the use of the transmitted data to Europol or Eurojust, which should usually be defined explicitly and restrictively when transferring personal data[130], is not further explained. The fact that the use of the data for Europol's or Eurojust's purposes considerably varies from a rather restricted use in the SIS II, the VIS or the CIS is not particularly mentioned. Taking Europol's and Eurojust's different tasks into consideration, the possible processing of SIS II, VIS or CIS data, for instance, at Europol, could have serious consequences for the social and legal situation of an individual.

Allowing Europol and Eurojust access to the extent that is necessary "for the performance of its tasks" without restricting the use afterwards is much too far reaching and should be clarified by specifying the purpose of the access and linking it to the purpose of the subsequent use. This has also to be seen in the light of the continually evolving tasks of Europol and Eurojust. A concrete factor not susceptible to change over time should be used to define Europol's and Eurojust's access conditions and the subsequent use of the data. It is, for instance, regrettable that the relatively strict access conditions applying to the law enforcement authorities of the Member States in case of the VIS do not affect Europol's access.

---

[128] Articles 11 (4) and (5) and 12 (4) Council Decision 2009/917, OJ 2009, L-323/20.

[129] However, a responsibility to inform the supplying Member State if Europol or Eurojust have evidence to suggest that an item of data is factually inaccurate or was entered contrary to the CIS Council Decision 2009/917, applies to the body as well as the obligation to introduce security measures, compare Articles 13 (3) and 28 Council Decision 2009/917, OJ 2009, L-323/20.

[130] Compare *Weber and Saravia v. Germany*, Application no. 54934/00 Admissibility Decision, from 29 June 2006.

With regard to the CIS it is important to mention that although the CIS processes various personal data elements[131], Europol's and Eurojust's access and whose subsequent processing, including a specification of the purpose of the processing of the received data, are not regulated. Individual rights, applicable to the transferred data, are limited to the standard Europol or Eurojust rules and not specifically tailored to the received data. It seems that the transfer of CIS data to Europol and Eurojust was found not important enough to be accompanied by the necessary safeguards, which are to be introduced when transferring personal data from a (customs) database to a law enforcement or judicial agency, such as Europol and Eurojust, which tasks significantly vary from the CIS and whose actions might have a serious impact on the situation of an individual.

The entire or, in Europol's case, partial lack of provisions regulating the subsequent use of the SIS II or CIS data at Eurojust and Europol, for instance, produces the situation that the responsibility of the use of the data is to a great part not clarified. Even though this might be the "heritage" of the former third pillar structures, provisions assuring that the decision of the Member States regarding the transfer of the data is supervised should have been included.[132] Otherwise, supervision at

---

[131] According to the CIS Convention, the CIS comprises data necessary to achieve the CIS's aim previously mentioned, such as commodities, means of transport, businesses, persons, fraud trends, availability of expertise. The new CIS Decision 2009/917 added two new categories: items detained, seized or confiscated and cash detained, seized or confiscated. The Member States determine the items to be included relating to the each of the mentioned categories whereby the data elements, which can be entered, relate to a closed list of personal data and are divided into two groups depending on the aforementioned categories. With regard to the four first categories (commodities, means of transport, businesses and persons), 11 data elements can be stored including: names, date and place of birth, nationality, sex, number and place and data of issue of the identity papers, address, any particular objective and permanent physical characteristics, reasons for entering the data, suggested action, a warning code indicating any history of being armed, violent or of escaping, registration number of the means of transport. Data elements relating to the newly introduced last two categories (items detained, seized or confiscated and cash detained, seized or confiscated) refer to names, date and place of birth, nationality, sex and address. Personal data revealing racial or ethnic origin, political opinions, religious or philosophical beliefs or trade-union membership or data concerning health or sex life are excluded in any case from processing (compare Articles 3 and 4 Council Decision 2009/917, OJ 2009, L-323/20).

[132] Such provisions could, for instance, provide for a notification of the relevant national DPA about the access and transfer of the data by Europol or Eurojust. So far, in the case of Europol, in addition to its already exhaustive tasks (it issues opinions and is responsible for various other tasks: additionally to the review of compliance with individual data protection rights at Europol; it should monitor the permissibility of the transmission of data to third bodies as well as it should review the activities of Europol in its exercise of its rights to access and search data in other databases, such as the SIS II or the VIS; the JSB must also produce a report after having carried out an annual inspection at Europol; Whereby, the JSB describes inspection as a key part of its work, it also functions as an appeal committee; additionally, the JSB also interprets and examines the implementation of the Europol Decision; compare: Article 34 Council Decision of 6 April 2009 establishing the European Police Office, OJ 2009, L121/37) Europol's JSB shall also review the activities of Europol in the exercise of its access to SIS II data.

this kind of stage seems to be difficult to exercise and raises concern.[133] A further possibility could be a duty to inform the individual concerned as soon as possible about the access of other authorities to the SIS II or the CIS data or the transfer of them. This is currently left to the Member States and depends on the national data protection systems.[134]

Inconsistencies further concern in particular the general supervision of Europol's or Eurojust's access to the SIS II, CIS or VIS data. There is no coordinated approach such as it is exercised, for instance, by the European Data Protection Supervisor (EDPS) and the national DPAs in context with the central VIS.[135] Meetings between the EDPS and Europol's or Eurojust's JSB should regularly take place to guarantee a minimum of supervision. Although, in case of the VIS, one may even go further and suggest that the EDPS, which supervises the VIS, should become responsible for the supervision of the data transfer from the VIS to Europol, including regular checks on the compliance with the provisions of VIS access Decision 2008/633 during the processing of the VIS data in Europols' databases. This argument should be kept in mind, especially when considering that the VIS data contain data of innocent individuals, which are at no point suspected of a crime. When already allowing wide-ranging access conditions for Europol, the supervision of this access should at least be effective, independent and equipped with the necessary personal resources.

Also regrettably is the fact that no words are made about Europol's and Eurojust's need to access the SIS II or the CIS data, neither about the possibility to obtain the data by other less intrusive means.[136] It is particularly striking that Eurojust does not even have a legal basis to access the CIS data (apart from the CIS Council Decision 2009/917). The deficiencies in context with the CIS are fundamental and clearly need to be corrected as soon as possible to be in accordance with basic legal requirements.

A further important question arises out of the fact that neither the SIS II Decision 2007/533 nor the CIS Council Decision 2009/917 clarifies by whom and in which of Europol's databases the SIS II or the CIS data are to be included. Are they introduced by Europol or by a Member States in the EIS or used in context of an analysis work file? What happens to the data after they were included in one of Europol's databases? Are they marked and remain connected to the purposes, which had justified their collection just as the ECtHR has considered it as appropriate in *Weber and Saravia v. Germany*?[137]

---

[133] Once the consent is given, formerly SIS II data can be entered in Eurojust's and Europol's databases or transferred to third states.

[134] Compare Article 16 Council Framework Decision 2008/977/JHA of 27 November 2008 on the protection of personal data processed in the framework of police and judicial cooperation in criminal matters; OJ 2008, L-350/6.

[135] Compare VIS Regulation 767/2008.

[136] Opinion of the EDPS on the SIS II proposals [2006] OJ C91/38, point 4.2.3.

[137] *Weber and Saravia v. Germany*, Application no. 54934/00 Admissibility Decision, para 121 from 29 June 2006.

Moreover, when requesting further information from the Member States[138] or when introducing the SIS II, CIS or VIS data in, for instance, Europol's databases EIS, it is very likely that the time limit for storing originally provided for in the SIS II, CIS or the VIS starts to run again, then subject to Europol's rules. This would bypass any possible effects of the provisions providing for a time limit, such as in the SIS II (3 years), in particular in cases in which the data are transferred shortly before the original time limit expires.

Another important issue relates to the circle of accessing actors: the SIS II, for instance, prohibits access from states not participating in the Schengen Cooperation, but, Europol allows for access of a much wider range of actors, such as liaison officers from third states or international organisations, invited "experts" from the third states or other European actors such as OLAF.[139] In consequence, the circle of persons and authorities having access to the data is significantly enlarged when transferring (even if indirectly) the data in Europol's databases and could lead to investigations being instituted against the persons concerned.[140] The proposal of the EDPS and the Joint Supervisory Authority (JSA) Schengen to limit searches to the individuals whose name are already contained in Europol's files, was regrettably not considered.[141]

To conclude, in addition to the aforementioned shortcomings in context of Europol's and/or Eurojust's access to the SIS II, CIS and the VIS, it is worth noting that Europol should additionally be allowed to access the Eurodac database in the near future. If the proposal on law enforcement access to Eurodac[142] enters into force, Europol would be granted access to a database concerning exclusively the data of individuals very likely never to be convicted or suspected of a crime. As a result, law enforcement agencies of 30 countries[143] as well as Europol would have access to the data of persons who were never involved in any criminal procedure.

Serious concerns going far beyond data protection concerns arise out of the planned measures. They are among others outlined by the Meijers Committee[144],

---

[138] According to Article 41 (4) SIS II Decision 2007/533.

[139] Compare Articles 9, 22 and 23 Europol Decision.

[140] Compare *Weber and Saravia v. Germany*, Application no. 54934/00 Admissibility Decision, para 79 from 29 June 2006.

[141] Opinion of the EDPS on the SIS II proposals [2006] OJ C91/38, point 4.2.2.

[142] Proposal for a Council Decision on requesting comparisons with Eurodac data by Member States' law enforcement authorities and Europol for law enforcement purposes, COM (2009) 344 final from 10 September 2009, in the following: Proposal on law enforcement access to Eurodac, COM (2009) 344 final from 10 September 2009.

[143] 27 Member States plus Norway, Iceland and Switzerland.

[144] Meijers Committee, standing committee of experts on international immigration, refugee and criminal law, Utrecht/The Netherlands, letter from 30 December 2009 to the European Parliament, Civil Liberties, Justice and Home Affairs Committee on the Proposal on law enforcement access to Eurodac, COM (2009) 344 final.

the EDPS[145] and the Working Party on Police and Justice[146] and can be summarised as follows: the proposals seriously challenge proportionality as well as purpose limitation, compliance with the ECtHR case law is extremely doubtful, the principle of non-discrimination risks to be undermined and the right to asylum and protection against torture and inhuman treatment seems to be disregarded. Data protection questions relating to the storage and the treatment of fingerprint data of not convicted individuals entitled to the presumption of innocence, the reasons for access, the extension of the purpose of processing, the evaluation of existing system (e.g. the Prüm Decision) and the different time limits of storage of Europol and Eurodac data arise and need to be further discussed before the adoption of the proposal.

8.4 Perspectives and Suggestions for Improvement

As follows from the foregoing considerations, information sharing in the AFSJ has become an essential tool in recent years to contribute to EU internal security policy. The Hague as well as the Stockholm programme call for an increasing inter-operability of the AFSJ databases, which in some cases leads to a questionable connection of systems established for different purposes. In view of the authors of the Stockholm programme, inter-operability constitutes a precondition for the efficiency of police and judicial cooperation in the AFSJ, whereby the interpretation of inter-operability is limited to a technical understanding. The legal dimension of inter-operability is not touched upon. Data protection rules are currently (re)negotiated for each new instrument (cf. De Hert and Vandamme 2004). Moreover, the language used in the programmes tends to understate the crucial influence the increasing cooperation has on the fundamental rights of the individuals concerned. Implicitly linked to the technical considerations is therefore the harmonisation of the individual rights standard. Otherwise, inter-operability may be reached at the cost of a week fundamental rights framework.

As a result, in addition to questions relating to the lawfulness of the ever extending functionalities of Europol and Eurojust and the limits of law enforcement access to data originally collected for a different purpose, which have to be answered elsewhere, the growing tendency to exchange data between the different AFSJ actors

[145] Opinion of the EDPS on the amended proposal for a Regulation of the European Parliament and of the Council concerning the establishment of 'Eurodac' for the comparison of fingerprints for the effective application of Regulation (EC) No (.../...) (establishing the criteria and mechanisms for determining the Member State responsible for examining an application for international protection lodged in one of the Member States by a third country national or a stateless person), and on the proposal for a Council Decision on requesting comparisons with Eurodac data by Member States' law enforcement authorities and Europol for law enforcement purposes, OJ 2010, C-92/1, in the following EDPS opinion on the proposal of law enforcement access to Eurodac, OJ 2010, C-92/1.

[146] The Working Party on Police and Justice (WPPJ) is a working party composed of experts from national DPA's and works together with the Article 29 Working Party, compare: Draft Annual Report for the Year 2009, p. 4.

makes it relevant to embed safeguards governing this transfer to compensate for the increased risks caused by the exchange of personal data.

Certainly, as the AFSJ still is a mix of former public international law and intergovernmental structures as well as of supranational EU structures, the data processing and protection framework is necessarily not entirely harmonised. However, the cooperation and the personal data transfer between the analysed systems already goes far beyond the former limited (legal) possibilities. So far, due to the "tendency to agree new functions before deciding the legal or technical limitations required" (Garside 2011), data protection rights could not keep up with the steady extension of the possibilities to exchange data among the AFSJ actors. In some cases, the legal instruments allowing for data exchange have a low level of individual rights protection. In others, data exchange is entirely carried out without a legal basis (e.g. Eurojust-CIS). The need for a coherent and general legal instrument on the exchange of personal data between AFSJ actors respecting the data protection rights of the persons concerned is obvious and should be urgently developed to better comply with fundamental rights in the AFSJ.

The first essential criterion, following from the respect for the rule of law, is, however, first and foremost, a clear legal basis to allow for security-related data transfer.[147] This legal basis should take into account the case whether or not the purpose of collection of the data differs from the purpose of access. Several provisions of Council Decision 2008/633 allowing national law enforcement authorities and Europol to access the VIS data[148] have an exemplary function and might serve as an example on how such an instrument would look like. A harmonised AFSJ instrument could replace the different solutions chosen so far. Its provisions might include rules on the access of domestic law enforcement authorities to European databases serving a different purpose than law enforcement, but can also be limited to EU internal AFSJ information exchange. When developing a single instrument harmonising the AFSJ data exchange, the following reflections not yet recognised in the security-related personal data exchange between AFSJ actors could be considered.

8.4.1 Specifying Unclear Legal Terms and Restricting the Purpose of Access and Transfer

Avoiding ambiguous terms is an essential requirement of an instrument regulating information exchange in the AFSJ.[149] For this purpose, the databases of the respective actors in which the transferred data could be possibly introduced as well as the databases allowed to be accessed, should be precisely defined. This definition should

---

[147] Examples of data exchange in absence of a legal basis was Eurojust's data transfer in JITs or Eurojust's access to the CIS.

[148] Article 5 Council Decision 2008/633, OJ 2008, L-218/129.

[149] To the requirement to define terms such as "serious crime" in a legal act, compare ECtHR case law *Kennedy v. the United Kingdom*, Application no. 26839/05, para 159 from 18 May 2010.

not only relate, for instance, to the general description of AFSJ actors' databases, but should include specifications referring to the exact databases (EIS, analysis work files) in which the data could be entered or from which the data could be retrieved (e.g. exact description of the SIS II databases).

Moreover, essential terms repeatedly used in AFSJs' legal bases and information exchange instruments, such as "terrorist offences", "serious criminal offences" and above all "prevention of crime", are to be explained and defined in a harmonised way in order to avoid legal uncertainty.[150]

Inextricably linked with clear definitions is the respect of the rule of law. Therefore, the legal basis should always lay down the conditions under which the respective European actor or Member States may obtain access for consultation of the relevant database. To prevent unclear processing purposes, the purpose of access to another database should be limited to the prevention, detection and investigation of terrorist offences and serious criminal offences subject to the mandate of the accessing actors. To avoid unilateral and possible far-reaching changes, eventual amendments to the mandate of the accessing actor after the adoption of the access decision should not be covered by the instrument.

8.4.2 Designating the Accessing Actors and Authorities

To guarantee transparency in the AFSJ data exchange and to comply with ECtHR requirements demanding "explicit and detailed provisions" relating to the information, which may be handed out and to "the authorities to which information may be communicated"[151], the authorities, which are authorised to access the data of the respective database must be precisely defined. Member States as well as the European AFSJ actors should keep a list of the designated authorities or units and should notify in a declaration to the European Parliament, the Commission and the General Secretariat of the Council their designated authorities or units.[152] To improve transparency, the list and the declarations, including possible amendments to it, could be published by the Commission in the Official Journal of the European Union. At the national level, each Member State should be obliged to keep a list of the (operating) units within the designated authorities that are authorised to access the respective

---

[150] The definition of the terms "terrorist and serious criminal offences" could correspond to the offences under national law, which correspond or are equivalent to the offences in Articles 1–4 of Council Framework Decision 2002/475/JHA of 13 June 2002 on combating terrorism, OJ 2002, L-164/3 and to the forms of crime, which correspond or are equivalent to those referred to in Article 2 (2) of Framework Decision 2002/584/JHA on the European Arrest Warrant, OJ 2001, L-190/1. The not yet defined term "prevention of crime" needs specification and could, for instance, describe a situation in which criteria based on a verifiable prognosis, open to scrutiny by an external supervisor, suggest that somebody plans to commit a crime. Factual indications, which exclude individual assumptions or pure hypothetical reflections, should underpin this estimation.

[151] *Leander v. Sweden*, Application no. 9248/81, para 55 from 26 March 1987.

[152] Similar to Article 3 (2) Council Decision 2008/633, OJ 2008, L-218/129.

database. To further strengthen the internal handling and security of the data and to guarantee that only persons authorised to consult the files[153] access the personal data, only duly empowered staff of a special unit, which received special training in the handling of personal data of the accessing actor as well as the respective database should be authorised to access the respective database.

8.4.3 Harmonising the Access Procedure

Harmonising the access procedure with regard to data entailed in another database could be a further important development towards a coordinated approach to AFSJ data exchange.

Prior to accessing a database, a reasoned written or electronic request to the respective database should be submitted by the aforementioned special units of the AFSJ actor. Upon receipt of a request for access, duly empowered staff of the special unit within the respective database should verify whether the conditions for access are fulfilled. If all conditions for access are fulfilled, transmission of the requested data to the accessing actor should be carried out by the special unit of the database in such a way as not to compromise the security of the data.[154]

8.4.4 Coordinating the Access Conditions

Access for consultation of the respective database by the designated authorities and the respective EU actors should only take place within the scope and the limits of their powers and only if certain conditions applying in every AFSJ data exchange and respecting the rights of individuals are met.

In view of the increasing data exchange, the access for mutual consultation between the AFSJ actors should be always restricted to the necessity of the access in a specific case for the purpose of the prevention, detection or investigation of terrorist offences or serious criminal offences clearly defined in the access decision. Reasonable grounds to consider that the consultation of the data will substantially contribute to the prevention, detection or investigation of any of the criminal offences in question should be an additional access condition. Furthermore, to assure that interferences with the purpose limitation principles remain exceptional, if the

---

[153] *Rotaru v. Romania*, Application no. 28341/954, para 57 from 4 May 2000.

[154] Similar to Article 4 Council Decision 2008/633, OJ 2008, L-218/129. Alternatively, in exceptional cases of urgency, the special unit within the respective database may receive written, electronic or oral requests. In such cases, it shall process the request immediately and only verify ex post whether all access conditions are fulfilled, including whether an exceptional case of urgency existed. Such an exceptional case should be immediately reported to the supervisory authority of the respective database. The ex post verification shall take place without undue delay after the processing of the request.

grounds for access differ from the purpose of the collection of the requested data, a reasoned written or electronic request to the respective database justifying the reasons for access, should be required. In that case, upon receipt of a request for such processing, duly empowered staff of the special unit within the respective database should verify whether the conditions for processing for purposes different from the purpose of collection are fulfilled.[155]

Similar to the conditions of VIS access Decision 2008/633 allowing national law enforcement authorities and Europol to access the VIS data[156], consultation of the respective database should undergo a two-step process: in a first step, access could be limited to searching with a limited amount of data in the particular file depending on the respective database and including only a selection of the data actually stored in the relevant database, such as, for instance: surname, surname at birth (former surname(s)), sex, date, place and country of birth, residence, fingerprints, etc. Only in the event of a hit, consultation of the relevant database should give full access to all of the data entailed in the database (such as any other data taken from the respective file, photographs, etc.).

8.4.5 Data Protection and Data Security Rules

With regard to the level of data protection and in the absence of an overall approach to law enforcement and judicial data protection rules, the processing of personal data consulted should be at least equivalent to the level of protection resulting from the Council of Europe Convention of 28 January 1981 for the Protection of Individuals with regard to Automatic Processing of Personal Data as well as to the level of protection offered by the Recommendation R (87) 15 of 17 September 1987 of the Committee of Ministers of the Council of Europe Regulating the Use of Personal Data in the Police Sector, and for those Member States, which have ratified it, to the Additional Protocol of 8 November 2001 to that Convention. The provisions of Council Framework Decision 2008/977/JHA of 27 November 2008 on the protection of personal data processed in the framework of police and judicial cooperation in criminal matters should additionally be applicable.

---

[155] To assure transparency and to specify the conditions for Europol, some specifications could additionally apply; Europol's access could be, for instance, necessary for the purpose of a specific analysis in a specific case referred to in Article 14 Europol Decision or for an analysis of a general nature and of a strategic type, as referred to in Article 14 (4) of the Europol Decision, provided that the data is rendered anonymous by Europol prior to such processing and retained in a form in which identification of the data subjects is no longer possible; data obtained by Europol could be further prevented from being introduced in Europol's Information System, exemptions to this rule should require the consent of Europol's supervisory body; possible additional conditions for Eurojust could also relate to the restriction not to introduce data obtained in Eurojust's Case Management System whereby exemptions to this rule should require the consent of Eurojust's supervisory body.

[156] Council Decision 2008/633, OJ 2008, L-218/129.

The processing of personal data by the accessing actor should be in any case in accordance with the legal basis of the accessing actor and the rules adopted in implementation thereof and supervised by the supervisory body of the accessing actor. In the absence of one single AFSJ supervisory system and to guarantee effective supervision, personal data originally underlying the supervision of another authority must at any stage of the processing be accessible to this authority.

Special attention needs to be paid to the current violation of the purpose limitation principle in cases in which data collected for purposes outside of crime prevention are later used for law enforcement purposes. Enforcing and strictly applying the purpose limitation principle by introducing a general rule applicable to each AFSJ data exchange whereupon personal data obtained from the respective database shall only be processed for the specific purpose of the collection would counteract this worrying development. If, in exceptional cases, the purpose of collection differs from the purpose of the transfer, this purpose has to be evaluated by the duly empowered staff of the special unit within the respective database mentioned above. Particular attention thereby has to be paid to the question whether the change in the purpose is justified by evidence that indicates that the data in question substantially contribute to the prevention, detection or investigation of the criminal offences in question and that the change in the purpose is proportional in its means.

To limit data storing in time[157], any extension to the time limit originally applicable to the obtained data by the accessing actor should be subject to the approval of the supervisory bodies of both, the accessing actor as well as of the accessed database.[158]

Finally, the list laying down the data security measures of Council Decision 2008/633 allowing national law enforcement authorities and Europol to access the VIS data[159] regulates in detail the necessary security requirements, which the Member States have to apply. This list could serve as an example for similar provisions in every AFSJ data exchange. To guarantee a harmonised standard and to prevent provisions, such as in the Europol Decision, which make the establishment of data security rules dependent the necessity of such rules[160], its provisions should in any case be extended to all AFSJ actors.

8.4.6 Follow-up of the Transferred Data

Harmonising the criteria for the transfer of data obtained from another database to third states would contribute to an increased legal certainty in a currently rather

---

[157] *S. and Marper v. the United Kingdom*, Application nos. 30562/04 and 30566/04 from 4 December 2008, para 119.

[158] In addition, before being authorised to process data stored in the database, the staff of the authorities having a right to access the database should receive appropriate training about data security and data protection rules including being informed of any relevant criminal offences and penalties.

[159] Article 9 (2) Council Decision 2008/633, OJ 2008, L-218/129.

[160] Compare Article 35 Europol Decision (footnote 118).

under-regulated area.[161] The transfer of such data could be subjected to the following conditions:

- If the purpose of collection of the data differed from the purpose of access, such personal data obtained from the database should not be transferred or made available to a third country or to an international organisation. Exceptions must be justified by evidence proving the importance of the exceptional situation.
- If the purpose of collection of the data corresponded to the purpose of access, such personal data obtained from the database could be transferred or made available to a third country or to an international organisation under the conditions of an agreement concluded with the third state assuring an adequate level of protection in the sense of Article 25 of Directive 95/46 for the intended data processing, exclusively for the purposes of the prevention and detection of terrorist offences and of serious criminal offences and under the access conditions set out above, subject to the consent of the Member State having entered the data into the database and in accordance with the national law of the Member State transferring the data or making them available. Ad hoc transmission to third states in absence of an exchange agreement should be limited to very exceptional cases and only with the sole aim of taking urgent measures to counter imminent serious danger threatening a person or public security. An undertaking obliging the recipient to use the data only for the agreed purpose of transmission should be concluded before the transfer. In any case, if ad hoc data transfer is carried out, the supervisory authority of the transferring actor needs to be informed about the transfer and has the right to prevent further transfers when it comes to the conclusion that the data protection requirements are repeatedly not complied with.
- In both cases the respective EU actor and, in accordance with national law, Member States should ensure that records are kept of such transfers and make them available to national data protection authorities upon request. In addition, rules restricting the onward transfer of the already transmitted data are equally important to limit the risks arising out of the extension of the circle of recipients. The conditions relating to onward transfer entailed in the implementing rules governing Europol's relations with partners[162], could thereby have exemplary function. Above all, the provisions, which oblige the recipient to give an undertaking (relating to an obligation to delete incorrect or outdated data, to delete data in case they are not anymore necessary for the purpose of the transfer, to asks the transferring actor for consent before further transferring received data, etc.) to guarantee certain basic data protection rights, should serve as an example in the whole area of AFSJ-related data exchange.

---

[161] Europol is the only body providing for certain basic rules in cases of third-party transfer, compare: Council Decision 2009/934/JHA of 30 November 2009 adopting the implementing rules governing Europol's relations with partners, including the exchange of personal and classified information, OJ 2009, L-325/6.

[162] Council Decision 2009/934/JHA of 30 November 2009 adopting the implementing rules governing Europol's relations with partners, including the exchange of personal and classified information, OJ 2009, L-325/6.

8.4.7 Cooperation Between Data Protection Authorities and Penalties in Case of Misuse

To ensure the practical enforcement of data protection rights, the national supervisory authorities, the supervisory authority of the database and the supervisory authority of the accessing actor, should closely cooperate in contributing to a coordinated supervision of the transfer from the database to the respective European actor.[163]

A provision for penalties in form of administrative and/or criminal fines that are effective, proportionate and dissuasive if the data are used contrary to the rules of the decision regulating the transfer, would considerably contribute to an effective enforcement of the data protection rules entailed in the decision.

8.4.8 Access Right, Correction, Deletion and Notification

To improve transparency, the reasons to deny access could be unified (e.g. access can be denied when the access may jeopardise the fulfilment of the AFSJ actors' tasks, a national investigation or the rights and freedoms of third parties[164]) and their application should in any case be open to external supervision. The internal Data Protection Officer should be informed about each access request and involved in the decision whether access is to be granted or not. If access is denied, appeal should be possible to the respective supervisory authority, which then should have the possibility to get access to the respective documents justifying the refusal. A time limit (of three months) to reply to an access request would support the practical enforcement of the access right.

Transparency and a clear definition of the circumstances and limits of the storing require that information about the transfer of the data to another database is to be provided to the person concerned by the accessing actor or the Member States entering the data at the time of the transfer or as soon as notification can be carried out without jeopardising the purpose of the transfer. The protection of data of persons, which were entered in the database due to the person's incidental link to the actual

[163] The cooperation between national and European DPAs should include the exchange of relevant information, the assistance of each other in carrying out audits and inspections or the examination of difficulties of interpretation or application of the decision regulating the data exchange. Studying problems with the exercise of independent supervision or with the exercise of the rights of data subjects and supporting each other in cases where individuals exercise their right of access, correction, deletion and notification or drawing up harmonised proposals for joint solutions to any problems including the promotion of awareness of data protection rights would complement the cooperation. For this purpose, regular meetings resulting in an annual joint report should take place. This joint activity report should be sent to the European Parliament, the Council, the Commission and the supervisory authority managing the database and include a chapter of each Member State prepared by the national supervisory authority of that Member State containing an assessment of the cases where individuals exercised their right of access, correction, deletion and notification.

[164] Article 19 (4) Eurojust Decision.

targeted person (e.g. victims, witnesses, person issuing an invitation and/or liable to pay the applicant's subsistence costs during the stay, etc.), could be improved when introducing a general notification duty in case their data are transferred. This duty could embrace additional information on the identity of the actor receiving the data together with its contact details, the purposes for which the data will be processed at the actor receiving the data, the categories of recipients of the data, including the possible third parties, information on changes concerning the data retention period as well as information on the necessity and the purpose of the transfer.[165]

To prevent that the incorrect data obtained from a database are again transferred to possible third parties, the AFSJ actor should, upon receiving such a request or if it has any other evidence to suggest that data processed in the database are inaccurate, immediately inform the authority of the Member State, which has entered the data in the database, which shall check the data concerned and, if necessary, correct or delete them immediately.[166]

A duty to explain in writing to the person concerned without delay why the AFSJ actor or the Member State responsible is not prepared to correct or delete data relating to him if it does not agree that data recorded in the database are inaccurate or have been recorded unlawfully, would additionally improve the practical implementation of the correction or deletion right. This information should contain an explanation of the steps, which the requesting person can take if he does not accept the explanation provided including information on how to bring an action or a complaint before the competent authorities or courts and on any assistance that is available. Moreover, a follow-up given to the exercise of the rights of correction and deletion should be carried out as soon as possible by the responsible supervisory body.

8.4.9 Keeping of Records

To facilitate the monitoring and evaluation tasks of the supervisory authorities, an ex post control of the admissibility of all data processing operations resulting from access to the database for consultation should be introduced. All access requests should be recorded for the purposes of checking whether the search was admissible or not, for the purpose of monitoring the lawfulness of data processing, for self-monitoring, ensuring the proper functioning of the system as well as for checking the data integrity and security.[167]

---

[165] In case a person concerned exercises its right to challenge the accuracy of its data, the AFSJ actor or the Member State responsible should be obliged to check the accuracy of the data and the lawfulness of their processing in the database within a limited period.

[166] Similar to Article 14 (5) VIS access Decision 2008/633 the Member State or the AFSJ actor responsible shall confirm in writing to the person concerned without delay that it has taken action to correct or delete data relating to it.

[167] Compare Article 16 VIS access Decision 2008/633, OJ 2008, L-218/129. Such records must be subject to the necessary security requirements and should be deleted after the retention period of

8.4.10 Implementing Effective Monitoring and Evaluation

Effective monitoring and evaluation mechanisms contribute to an improved control of the effectiveness and the necessity in terms of output, security and quality of service of the access to other databases.[168] Consequently, the respective supervisory authorities in cooperation with the respective AFSJ actor should carry out checks and submit a report to the European Parliament, the Council and the Commission on the technical functioning, the need and the use of the access possibilities of the respective database.[169] Exceptional cases of urgency should be documented and an overall "evaluation of the application and the results achieved against the objectives and an assessment of the continuing validity of the underlying rationale" behind the access as well as the impact on fundamental rights should be made.[170] This report should be made public to allow for discussion of its results.

8.5 Conclusion

Summarising, the currently under-regulated data exchange between the different AFSJ actors (inter-agency exchange and access of Europol and Eurojust to EU databases) can only be effectively countered by the introduction of common principles regulating the data exchange and the protection rights of individuals in this area. After the adoption of the Lisbon Treaty, the chances to introduce such principles are better than ever. The pillars are abolished, decision making has improved and the participation of the European Parliament in the legislative process in the AFSJ assures an increased respect of fundamental rights. Article 16 TFEU introduced a comprehensive legal basis for the protection of personal data applicable to almost all Union policies, including police and judicial cooperation (Commission communication 2010, p. 13, para 2.3). The Commission repeatedly emphasises the need to have a "comprehensive protection scheme and to strengthen the EU's stance in protecting the personal data of the individual in the context of all EU policies, including law enforcement and crime prevention". (Commission communication 2010) The Data Protection Directive 95/46 is in the review process and common data protection principles, covering the former first as well as the third pillar, are likely to be

the data has expired. Comparable to Article 16 (1) VIS access Decision 2008/633 allowing national law enforcement authorities and Europol to access the VIS data, those records could show: the exact purpose of the access for consultation referred to in Article 5 (1), including the form of terrorist offence or other serious criminal offence concerned, the respective file reference; the date and exact time of access; where applicable that use has been made of the urgent access procedure; the data used for consultation; the type of data consulted and according to the rules of the respective AFSJ actor or to national rules, the identifying mark of the official who carried out the search and of the official who ordered the search or supply.

[168] Compare Article 17 (1) VIS access Decision 2008/633, OJ 2008, L-218/129.

[169] Analogous to Article 17 VIS access Decision 2008/633, OJ 2008, L-218/129.

[170] Article 17 (4) VIS access Decision 2008/633, OJ 2008, L-218/129.

introduced in the new version (Commission communication 2010, p. 4, para 1). This essay aimed at contributing to the current discussion by presenting one of several solutions to develop a practical and comprehensive approach, including common data protection principles, in the area of EU internal AFSJ information exchange.

References

Commission communication. 2010. On A comprehensive strategy on data protection in the European Union, COM(2010) 609 final of 4 November 2010, p. 13, para 2.3.

De Buck, Bart. 2007. Joint investigation teams: The participation of Europol officials. *ERA Forum* 8:263.

De Hert, Paul, and Luc Vandamme. 2004. European police and judicial information-sharing, cooperation: Incorporation into the community, bypassing and extension of schengen. *ERA Forum* 5:425–434.

De Moor, Stefan. 2009. The difficulties of joint investigation teams and the possible role of OLAF. *Eucrim* 3:94–99, 97.

De Schutter, Olivier. 2008. The two Europes of human rights: The emerging division of tasks between the Council of Europe and the European Union in promoting human rights in Europe. *Columbia Journal of European Law* 14:509–560.

Garside, Alice. 2011. The political genesis and legal impact of proposals for the SIS II: What cost for data protection and security in the EU?, 16, Sussex Migration Working Paper no. 30, March 2006. http://www.sussex.ac.uk/migration/documents/mwp30.pdf. Accessed 12 July 2011.

Gusy, Christoph. 2008. Europäischer Datenschutz. In *Alternativentwirf Europol und europäischer Datenschutz*, ed. Jürgen Wolter et al., 265–280. Heidelberg: C.F. Müller Verlag.

Hijmans, Hielke, and Alfonso Scirocco. 2009. Shortcomings in EU data protection in the third and the second pillars. Can the Lisbon Treaty be expected to help? *Common Market Law Review* 46:1485–1525.

Holzenberger, Mark. 2006. Europols kleine Schwester—Die Europäische Grenzschutzagentur Frontex. *Bürgerrechte und Polizei/CILIP* 2:56–63.

Horvatis, Lisa, and Bart deBuck. 2007. The Europol and Eurojust project on joint investigation teams. *ERA Forum* 8:239–243.

Lopes da Mota, José Luis. 2009. Eurojust and its role in joint investigation teams. *Eucrim* 3:88–90.

Mitsilegas, Valsamis. 2009. *EU criminal law*. 223. Oxford: Hart.

Ralf, Riegel. 2009. Gemeinsame Ermittlungsgruppen, Herausforderungen und Lösungen. *Eucrim* 3:99–106.

Rijken, Conny, and Gert Vermeulen. 2006. *Joint investigation teams in the European Union, from theory to practice*. The Hague: T.M.C Asser Press.

Siemen, Birte. 2006. *Datenschutz als europäisches Grundrecht*. Berlin: Duncker & Humblot.

Vervaele, John A. E. 2008. The shaping and reshaping of Eurojust and OLAF. *Eucrim* 184:3–4.

Chapter 9
The Adequacy of an EU-US Partnership

Els De Busser

9.1 Transatlantic Hopes and Wishes

The EU and the US represent a total of almost 800 million people and have set up a considerable cooperation in criminal matters by exchanging personal data for the purpose of prevention, detection, investigation, or prosecution of criminal offences. This cooperation is characterized by bilateral agreements as well as by agreements between the EU (representing its Member States) and the US and agreements by the EU's bodies (responsible for judicial and law enforcement cooperation in criminal matters) and the US. This cooperation is also characterized by differences in legislation and attitude towards the protection of personal data, however, resulting in reports on illegal transfers of personal data[1] and the rejection of the Agreement between the EU and the US on the processing and transfer of Financial Messaging Data from the EU to the US for the purposes of the Terrorist Finance Tracking Program (the Interim Agreement).[2] These experiences have led to attempts to bring both parties to the table to negotiate a general agreement that can govern the transatlantic exchange of personal data for the purpose of prevention, detection, investigation, or prosecution of criminal offences. The following questions remain: Which course should the transatlantic exchange of personal data in criminal matters take? How can we make a compromise between the conditions the EU wants to see fulfilled and the wishes that the US authorities have or is a compromise simply impossible? Let us first look at what both sides would like to achieve with regard to transatlantic data exchange.

On the EU side, the European Commission recognized that the EU's legal framework on the protection of personal data is in need of review. In spite of the

[1] See, for example, Lichtblau and Risen (2006), and Modderkolk and Wester (2011).

[2] Agreement between the European Union and the United States of America on the processing and transfer of Financial Messaging Data from the European Union to the United States for the purposes of the Terrorist Finance Tracking Program, O.J. L 8, January 13, 2010, 11–16.

E. De Busser (✉)
Max Planck Institute for Foreign and International Criminal Law,
Günterstalstraße 73, 79100 Freiburg, Germany
e-mail: e.busser@mpicc.de

S. Gutwirth et al. (eds.), *European Data Protection: In Good Health?*,
DOI 10.1007/978-94-007-2903-2_9, © Springer Science+Business Media B.V. 2012

technological advancements in data processing, the basic data protection principles are still considered valid, even though their application needs to be clarified.[3] However, the entry into force of the Lisbon Treaty and the dissolving of the three pillars require a new general instrument on data protection. In addition, the Commission has recognized that the 2008 Framework Decision on Data Protection in Criminal Matters[4] is not an adequate instrument.[5] Finally, even when the principles governing data protection still prevail amongst the plethora of new techniques and concepts, such as data mining, cloud computing, and behavioral advertising, the concrete rules on data gathering and processing need to be revised and updated in view of these new circumstances.

What the EU wants is, first of all, a revision of the current data protection legal framework. For this purpose, the Commission presented a comprehensive approach to data protection in the EU.[6] Since the exchange of data with third states, especially the US, has intensified significantly since 2001, the Commission's communication also includes a chapter on the global dimension of data protection. Under this heading, two main objectives are listed. Firstly, the procedure for allowing data transfers to third states should be simplified and clarified. Third states need to be checked as to whether their legal framework on data protection is adequate within the framework of EU rules before they can receive any personal data transmitted from a Member State. This adequacy procedure needs to be improved. Secondly, the Commission aims to promote universal principles of data protection. This means that cooperation with organizations such as the UN, the Council of Europe, and the Organization for Economic Cooperation and Development (OECD) should be strengthened as far as data protection is concerned. The Commission's approach was not presented with the transatlantic cooperation in criminal matters in mind. Nevertheless, one cannot discuss the EU-US negotiations on new data transfer agreements without considering the Commission's plans.

Besides the review of its own data protection framework, the EU has been active in negotiating agreements with the US involving the transfer of personal data for the purpose of prevention, detection, investigation, or prosecution of criminal offences. The idea of introducing a general agreement on data protection in transatlantic cooperation in criminal matters took shape and negotiations were taken up in December

---

[3] European Commission, Comparative study on different approaches to new privacy challenges, in particular in the light of new technological developments, Final Report, 21 (2010).

[4] Framework Decision of November 27, 2008 on the protection of personal data processed in the framework of police and judicial cooperation in criminal matters, O.J. L 350, December 30, 2008, 60–71.

[5] Communication from the Commission to the European Parliament, the Council, the Economic and Social Committee and the Committee of the Regions, A comprehensive approach on personal data protection in the European Union, COM (2010)609 final, November 4, 2011, 13–15 (further: COM (2010) 609 final). See also European Data Protection Supervisor, Opinion on the Communication from the Commission to the European Parliament, the Council, the Economic and Social Committee and the Committee of the Regions—A comprehensive approach on personal data protection in the European Union, January 14, 2011, 26–28.

[6] COM (2010) 609 final.

2010. It is the intention of the European Commission to use this future agreement as an umbrella instrument, not only for future agreements with the US but also for application to the existing agreements.[7]

Thus, the objectives on the EU side are threefold a simple and clear procedure for allowing data transfers to third states, universally valid data protection principles, and for the existing and future transatlantic cooperation to be governed by standards equivalent to the European standards.

On the US side, the objectives are also clear: smooth delivery of personal data from the EU judicial and law enforcement authorities, EU air carriers (passenger name record data), and financial data controllers (the Society for Worldwide Interbank Financial Telecommunication or SWIFT). Existing agreements with the EU and with Europol and Eurojust do not hide the fact that there should not be too many restrictions on the transatlantic flow of data. Ultimately, it was the European Council that asked the Commission to prepare a recommendation for the "negotiation of a data protection and, where necessary, data sharing agreements for law enforcement purposes with the United States." According to the US, the existing agreements should nonetheless remain untouched. The planned retroactive application of the future umbrella instrument was thus not well received by the US delegation to the EU. For these reasons, we can state that the goals on the US side are transparent and straightforward, namely trouble-free data transfers.

Are the EU's and the US' aims for transatlantic cooperation in criminal matters compatible, and how should we go about forming them into an agreement? This is the central question I will attempt to answer in this contribution. In Sect. 9.2, the scope of this agreement is analyzed, including the meaning of the key concepts. Sect. 9.3 focuses on the prerequisite for personal data transfers from the EU to a third state, that is the decision—based on an assessment of the legal framework—that the requesting state has an adequate level of data protection. Such an assessment has not been made so far for the US. In addition, the procedure of assessing a state's level of data protection is under review. Therefore, it should first be clarified whether and how the adequacy procedure should be carried out. In this part of the contribution, the adequacy procedure is studied as to its theoretical framework and its practical implementation. The lack of consistency is highlighted followed by three significant remaining questions with regard to the adequacy procedure: which is the authority that should decide upon the adequate level of data protection of a state; what is the precise content of this assessment and when should this assessment take place?

In Sect. 9.4, the future of this adequacy procedure in the transatlantic cooperation is studied. The Commission is working on a new agreement with the US as to which adequacy requirement is applicable. At the same time, the Commission suggests having the US ratify the Council of Europe Convention for the Protection of

[7] Commission européenne, Proposition de recommandation du Conseil autorisant l'ouverture de négociations en vue d'un accord entre l'Union Européenne et les Etats Unis d'Amérique sur la protection des données personnelles lors de leur transfert et de leur traitement a des fins de prévention, d'investigation, de détection ou de poursuite d'actes criminels y compris le terrorisme, dans le cadre de la coopération policière et judiciaire en matière pénale, COM (2010) 252/2, Annex, May 12, 2010.

Individuals with regard to the Automatic Processing of Personal Data (further: Data Protection Convention)[8] and its Additional Protocol.[9] If this occurs, the adequacy procedure would no longer be needed in the transatlantic cooperation. To date, however, the US' data protection regime is based on other ideas than those of the Data Protection Convention.

It is important to note here that this contribution is written from the perspective of the EU and the EU legal framework and policy on data protection. The US legal framework and policy on data protection have only been included in the analysis when relevant for studying the transatlantic cooperation in criminal matters.[10]

9.2 Definition of Law Enforcement

It seems rather obvious when two parties are negotiating an agreement on exchanging information for the purpose of law enforcement that both have the same idea on what exactly law enforcement is. Nevertheless, it was—and still is—surprisingly difficult to define the term "law enforcement" or "law enforcement authority" in the context of transatlantic cooperation. In the 2006 Framework Decision on simplifying the exchange of information and intelligence between law enforcement authorities, a transparent definition of law enforcement is given for the EU Member States.[11] But the US has a complex landscape of state and federal authorities and of authorities involved in law enforcement and intelligence, often having double competences—such as the FBI and the CIA, which are both responsible for law enforcement and intelligence activities.[12]

The difficulties in joining the two approaches became clear during the negotiations on the 2002 Europol-US Supplemental Agreement.[13] Europol issued an informal explanatory note only representing Europol's opinion, in which the following statement was made: "From the start, the US side made it clear that it was impossible for them to indicate with any degree of accuracy, which authorities could be involved in using such information, given the fact that there are many different authorities, which would qualify as being involved in preventing, investigating and prosecuting criminal offences. This was especially true given the many different State and local authorities

[8] Convention for the Protection of Individuals with regard to the Automatic Processing of Personal Data, ETS no. 108.

[9] Additional Protocol to the Convention for the Protection of Individuals with regard to Automatic Processing of Personal Data, regarding Supervisory Authorities and Transborder Data Flows, ETS no. 181.

[10] For a more detailed analysis of the US legal framework and policy on data protection see De Busser (2009).

[11] Framework Decision on simplifying the exchange of information and intelligence between law enforcement authorities between the Member States of the EU, O.J. L 386, December 29, 2006, 91.

[12] See inter alia, Fijnaut (2004), Vervaele (2005), and Manget (2006).

[13] Supplemental Agreement between the Europol Police Office and the United States of America on the Exchange of Personal Data and Related Information, November 6, 2002.

responsible for such issues."[14] When talks on a general data exchange agreement with the US started, a High-Level Contact Group (HLCG) was established to prepare for this agreement inter alia by developing common definitions of data protection principles.[15] The HLCG, which included senior officials from the European Commission, the EU Presidency (supported by the Council Secretariat), and the US Departments of Justice, Homeland Security, and State, agreed on 12 common data protection principles, such as purpose specification/limitation and information security.

When defining the scope of the principles under consideration, the HLCG recognized that the EU and the US have different ways of describing "law enforcement purposes." In the EU, this covers the use of data for the prevention, detection, investigation, or prosecution of any criminal offense. In the US, this encompasses the prevention, detection, suppression, investigation, or prosecution of any criminal offense or violation of law related to border enforcement, public security, and national security as well as noncriminal judicial or administrative proceedings related directly to such offenses or violations. According to the HLCG, in practice, these different views on law enforcement may coincide to a large extent.[16]

To base a new agreement on the possibility that both parties may consider the scope of the agreement to correspond is not a secure basis.[17] The gap that inevitably occurs where two interpretations of scope do not coincide could lead to practical and legal difficulties in deciding whether a transfer of data is governed by the agreement or not. In fact, violations of data protection principles could be caused if data falling within this gap are transferred and considered to be transferred under the terms of the agreement by one party but not by the other. For example, the US would consider intelligence data to be exchanged for the purpose of national security, that is, for law enforcement purposes, but this transfer would not fall within the scope of the agreement from the EU's point of view. This means additional work is still needed to clearly define the scope of this future agreement on data transfers for law enforcement purposes.

9.3 Adequacy Procedure

Transferring personal data from an authority within the EU—including Europol or Eurojust as the EU's bodies for law enforcement and judicial cooperation in criminal matters—to an authority in a third state still means that the EU standards on data

[14] Council, 13696/1/02, Informal explanatory note regarding the draft supplemental agreement between the United States of America and the European Police Office on the exchange of personal data and related information, November 28, 2002, 11.

[15] Council, 9831/08, EU US Summit, June 12, 2008—Final Report by EU-US High-Level Contact Group on information sharing and privacy and personal data protection, May 28, 2008, 2.

[16] Ibid., 3–4.

[17] See also European Data Protection Supervisor, Press Release November 11, 2008, Opinion on transatlantic information sharing for law enforcement purposes: Progress is welcomed, but additional work is needed, 13.

protection need to be respected. This can take place either by making the transfer to a state that has ratified the Council of Europe's Data Protection Convention or by ensuring that the third state has an adequate level of data protection. Data transfers to a third state that is not a party to the Data Protection Convention is thus still possible if this state has a data protection regime that offers safeguards for the requested data that are appropriate from the EU's point of view. This does not mean that the data protection regime of the requesting state must be identical, but an assessment needs to take place as to whether it is adequate. It is not the Data Protection Convention but the above mentioned Additional Protocol that lays down the adequacy procedure in Article 2 on transborder flows of personal data to a recipient, which is not subject to the jurisdiction of a Party to the Convention.

Article 2 of the Additional Protocol allows for derogations from the adequacy requirement that should be interpreted restrictively. Similar to the derogations from the provisions on human rights in the European Convention for Human Rights and Fundamental Freedoms (ECHR), they should at least be laid down by (national) law and be necessary for the protection of legitimate prevailing interests. The explanatory report to the Additional Protocol also refers to the same interests, based on which the right to privacy and data quality principles can be lawfully derogated from. This means derogations from the adequacy procedure are allowed to protect an important public interest, the exercise or defense of a legal claim, or the extraction of data from a public register. Derogations can also be made for the specific interest of the person whose data are transferred for the fulfilment of a contract with this person or in his interest, to protect his vital interests or if he has given his informed consent.

In case an adequate level of data protection cannot be assured by the requesting third state, another possibility for exchange still exists if the receiving state provides sufficient safeguards that are deemed adequate by the providing state. The safeguards can be limited, however, to include only the relevant elements of data protection and are only applicable to a specific transfer of data.

9.3.1 Theory and Practice

The adequacy procedure and the assessment that is part of it are thus significant elements of data transfers to third states and aims to protect the data protection standards that the EU Member States ensure. Unfortunately, this is not the case for all data transfers, as the requirement of making an adequacy assessment is not a uniform requirement. It is not even a uniform requirement in the field of law enforcement and judicial cooperation in criminal matters, which is—due to the sensitive nature of the data—a field that would surely benefit from consistently protecting EU standards in matters of transfer to a third state. On the contrary, the Framework Decision on Data Protection in Criminal Matters includes the adequacy requirement but is only applicable to data that the providing Member State receives from another Member State. This means that data gathered by a Member State itself can be sent to a third state without having to check the third state's level of data protection. Obviously, if the providing Member State has laid down the adequacy procedure in its own national law, it would still be obliged to check the requesting third state's level of data protection.[18]

The only type of data transfer for which an adequacy assessment should be made in every case concerns the transfer of data for activities falling within the scope of the former first pillar, that is, Community law. However, research has proven that for these transfers there is no consistency in compliance with the provisions of Directive 95/46/EC.[19]

In the field of law enforcement and judicial cooperation in criminal matters, Europol and Eurojust should not be overlooked. These two bodies each have binding adequacy procedures in their own respective data protection rules that are independent from the Framework Decision on Data Protection in Criminal Matters. The differences between the procedures that Europol[20] and Eurojust[21] have laid down for themselves are significant, and the mandatory nature of the adequacy assessment as a prerequisite for data transfers to third states is clear. Nonetheless, compliance with this procedure is also problematic here, especially as regards cooperation with the US.

Europol has declared that the US ensures an adequate level of data protection, but no complete assessment has been made. Still, personal data transfers are made under the terms of the 2002 Europol-US Supplemental Agreement. Eurojust has laid down in its agreement with the US that no general—"generic"—restrictions for processing of data with respect to the legal standards of the receiving party may be imposed as a condition for delivering information.[22] Obviously, this should be read as a denial of any adequacy procedure whatsoever, since the assessment that is part of the procedure is exactly that: it is a condition without which information should not be transmitted; it is a restriction with respect to the third state's legal standards on processing the received data, and it is a restriction of a generic nature. This means that it is not applicable to only a specific group of data but is binding for all personal data transfers falling within the scope of the agreement.

When considering the agreements the EU has made to represent its Member States, two cooperation agreements with the US are relevant: the 2003 EU-US Agreement

[18] A recent study ordered by the European Commission, revealed that the national laws of the member states do not fully comply with Article 26 of Directive 95/46/EC that provides in the adequacy assessment for transborder data flows. Inconsistencies lie in the explicit or implicit nature of legal provisions concerning adequacy, the authority deciding upon adequacy (Commission or member state authority) and divergences in the application of the special conditions under which data may be sent to third countries without adequate data protection. See for a full report: Korff (2010).

[19] Ibid., 91–94.

[20] Council of the European Union, Act March12, 1999 adopting the rules on the transmission of personal data by Europol to third states and third bodies, O.J. C 88, March 30, 1999, 1.

[21] Council Decision of February 28, 2002 setting up Eurojust with a view to reinforcing the fight against serious crime, O.J. L 63, June 6, 2002, Article 27, § 4 and Council, Rules of procedure on the processing and protection of personal data at Eurojust, O.J. C 68, March 19, 2005, Article 28, §§ 2 and 3.

[22] Article 10, Agreement between Eurojust and the United States of America, November 6, 2006.

on Mutual Legal Assistance in Criminal Matters (2003 MLA Agreement)[23] and the 2010 EU-US Agreement on the processing and transfer of Financial Messaging Data from the EU to the US for the purposes of the Terrorist Finance Tracking Program (2010 TFTP Agreement).[24] The 2003 MLA Agreement included exactly the same provision prohibiting all generic restrictions in order to facilitate the data flow as was the case in the above-mentioned Eurojust-US Agreement.

The 2010 TFTP Agreement was not only unique before it was enacted, due to the historic rejection of the first text—the Interim Agreement—by the European Parliament in February 2010.[25] It also demonstrated another creative take on avoiding the adequacy procedure by laying down the following provision: "subject to ongoing compliance with the commitments on privacy and protection of personal data set out in this Agreement, the U.S. Treasury Department is deemed to ensure an adequate level of data protection for the processing of financial payment messaging and related data transferred from the European Union to the United States for the purposes of this Agreement."[26] This provision was identical in the Interim Agreement and in the adopted 2010 Agreement. The Agreement was thus based on the assumption of an adequate level of data protection rather than on a genuine assessment. Needless to say, the Article 29 Working Party—the independent EU Advisory Body on Data Protection and Privacy—did not like this provision. In fact, when the European Parliament's Committee on Civil Liberties, Justice, and Home Affairs asked the Article 29 Working Party and the Working Party on Police and Justice—a specific working group of the Conference of Data Protection Authorities—to evaluate the Interim Agreement, the chairmen of both Working Parties expressed their concerns in a letter to the Committee. Their statement is clearly one of dissatisfaction: "Furthermore, the wording of Article 6 of the Interim Agreement, according to which the 'U.S. Treasury Department is deemed to ensure an adequate level of data protection', has brought about a certain degree of perplexity amongst the Working Parties' members."[27] Both chairmen stress the fact that no independent assessment of the level of data protection by the US Department of the Treasury (UST) was made before concluding the Agreement and wonder whether the joint review that should be conducted by the parties (at the latest 6 months after entry into force) could take the form of an adequacy check. This could be the case; however, it would still not replace an assessment made before deciding upon an agreement, as it would be post factum and many data have been transferred already under the terms of a legal instrument that

[23] Agreement June 25, 2003 on mutual legal assistance between the European Union and the United States of America, O.J. L 181, July 19, 2003, 34–42.

[24] Agreement between the European Union and the United States of America on the processing and transfer of Financial Messaging Data from the European Union to the United States for the purposes of the Terrorist Finance Tracking Program, O.J. L 195, July 27, 2010, 5–14.

[25] See European Parliament Recommendation, A7–0013/2010, February 5, 2010 and Press Release, SWIFT: European Parliament votes down agreement with the US, February 11, 2010.

[26] Article 6, 2010 TFTP Agreement.

[27] Article 29 Data Protection Working Party and Working Party on Police and Justice, Letter to Juan Fernando López Aguilar, Chairman of the Committee on Civil Liberties, Justice, and Home Affairs, January 22, 2010.

may be considered to be inappropriate when it comes to the level of data protection of the UST. Furthermore, the first joint review of the 2010 TFTP Agreement that has been carried out on 17 and 18 February 2011 clearly states that it is a valuable tool for the assessment of the level of implementation of the Agreement and the UST's compliance with the safeguards included therein. This does not include the assessment of the adequacy of the UST's level of data protection.[28]

It would be unrealistic to state that consistency would be ensured if all Member States ratified the Additional Protocol to the Data Protection Convention,[29] since this neither guarantees practical compliance nor does it guarantee proper assessments that are not guided by political or economic objectives. Still, ratification by all Member States would be a first step towards introducing some uniformity into the adequacy procedure. The second step should be made by the Commission, which promised to clarify and simplify the establishment of an adequacy assessment.[30]

Establishing an assessment as to whether a data protection legal framework is adequate or not is a procedure that has raised many questions, which the European Commission is determined to solve. The most important questions relate to the authority that makes the adequacy assessment on the one hand and the content of such assessment on the other. Raab and Bennett already discussed five interrelated concerns also identified by other scholars at the time Directive 95/46/EC was adopted.[31] These concerns will form the basis of the following analysis, which will focus more on judicial and law enforcement cooperation in criminal matters than on Directive 95/46/EC.

9.3.2 Equal Rules

The first main concern is the emergence of "data havens" or "data shopping," which is a potential consequence of having different rules on data protection in different Member States. It cannot be prevented that Member States exchange personal data amongst themselves in accordance with the legislation on information exchange in criminal matters. This can result in third states relying on the Member State that seems to be the "easiest" to deal with in order to obtain the data they want. A Member State that has not ratified the Additional Protocol to the Data Protection Convention and is also not bound by an adequacy procedure in another way, for

---

[28] Commission report on the joint review of the implementation of the Agreement between the European Union and the United States of America on the processing and transfer of Financial Messaging Data from the European Union to the United States for the purposes of the Terrorist Finance Tracking Program, March 16, 2011, 15.

[29] The protocol entered into force for Austria, Bulgaria, Cyprus, Czech Republic, Estonia, Ireland, Germany, Hungary, Latvia, Lithuania, Luxembourg, Netherlands, Poland, Portugal, Romania, Spain, Slovakia, and Sweden.

[30] COM (2010) 609 final, 15–16.

[31] Bennet and Raab (1997).

example, by national law, would be an easy target.[32]A similar situation could occur when a Member State that is bound by an adequacy procedure and has issued a positive adequacy assessment for a particular third state transfers personal data to this third state from other Member States or from a database set up among the Member States. Transmitting personal data received from other Member States for the purpose of prevention, detection, investigation, and prosecution of criminal offences corresponds to the scope of the Framework Decision on Data Protection in Criminal Matters. Article 13 of this Framework Decision makes the Member States—and not the Commission—responsible for making adequacy assessments of third states. Therefore, the situation described above is not unimaginable in the case of personal data related to a criminal case.

A solution would be to ensure uniform rules in all Member States. If all Member States would ratify the Additional Protocol to the Data Protection Convention, the adequacy procedure would at least be mandatory for all. Nevertheless, this does not ensure its equal application, that is, equal adequacy assessments, by the national authorities regarding the same requesting third state. As mentioned above, putting Member State authorities in charge of adequacy assessments for data exchange in criminal matters entails the risk of data shopping.

9.3.3 Which Authority?

The question of which authority makes the assessment has not yet been answered, because EU legal instruments on data protection allow adequacy assessments made by the Member States as well as by the European Commission. Directive 95/46/EC mentions both options, while the Framework Decision on Data Protection in Criminal Matters only mentions the Member States. Both options have advantages and disadvantages. The advantage of the Commission making the assessment is that one uniform decision on a third states' level of data protection is introduced on which all Member States can rely. The risk, however, is that evaluations made by the Commission are directed by wider political and economic concerns in relation to the third state concerned.[33] Furthermore, this may not just be a concern in the case of Commission assessments, as the EU-US agreements negotiated by the EU Member States—represented by the EU presidency—also skipped a full evaluation of the US level of data protection (cf. supra).

The main drawback of Member State assessments is that different conclusions in different states can create confusion. Due to the lack of uniform rules on how the evaluations are performed, Member States can apply diverging methods or include different elements of data protection legislation. For example, one Member State may also include the practical application of data protection legislation in the assessment, while another may only rely on "law in the books." The concern regarding divergent

[32] See also Korff (2010).

[33] Bennet and Raab (1997).

implementation laws in the Member States has been confirmed by the European Commission in its first report on the implementation of Directive 95/46/EC that provides for the adequacy requirement in its Articles 25 and 26.[34] This concern was recently also confirmed by a study of the national legislations.[35]

Furthermore, national authorities making the adequacy assessment tend to evaluate a third states' data protection regime from the point of view of their own legislation. Even when the national laws are implementations of EU legal instruments on data protection, they still differ considerably.[36]

There are significant advantages to introducing Article 29 Working Party (hereinafter 29 WP) as the central authority deciding upon the adequacy of the level of data protection in third states for all Member States. Firstly, the above-mentioned disadvantages, which are generated by a Member State's authority or the Commission making the assessment, are in principle avoided. The 29 WP consists of representatives of the data protection authorities of the 27 Member States, a representative of the European Data Protection Supervisor, and a representative of the Commission. In accordance with Directive 95/46/EC, the 29 WP members act independently and decide by majority voting. Thus, one could expect there to be fewer chances of economic or political interests prevailing over data protection interests. Obviously, chances of this happening can never fully be excluded. Secondly, the 29 WP as the central authority helps avert the fact that national data protection legislations differ, which causes national assessments of a third state's adequacy level to differ. Thirdly, it is already the task of the 29 WP to advise the Commission on the adequate level of data protection in third states.[37] Making these evaluations binding decisions for all Member States would thus not require a change in its working procedure, although it would naturally increase the workload of the members. In addition, this new competence would require an amendment to Directive 95/46/EC. In view of the disappearance of the former three pillars and the current review of the legal framework on data protection, it would be appropriate to amend the tasks of the 29 WP and widen its function to also include criminal matters.

It is not necessary to set up a new authority. Utilizing the expertise and working procedure of the 29 WP would promote clarity regarding the deciding authority as well as uniformity regarding data transfers from Member States to third states.

---

[34] European Commission, First report on the implementation of the Data Protection Directive (95/46/EC), COM (2003) 265, May 15, 2003, 18–19.

[35] European Commission, Comparative study on different approaches to new privacy challenges, in particular in the light of new technological developments, Final Report, 2010, 92–93.

[36] Bennet and Raab (1997).

[37] See in the same sense: Working Party on the Protection of Individuals with regard to the Processing of Personal Data, XV D/5020/97-EN final, WP 4, First Orientations on Transfers of Personal Data to Third Countries—Possible Ways Forward in Assessing Adequacy, Discussion Document adopted by the Working Party on June 26, 1997, 3.

9.3.4 Content of Adequacy

The type of data transfer determines the content of the adequacy assessment. This is implied by the Explanatory Report to the Additional Protocol to the Data Protection Convention, which states that the provisions of Chapter II (basic principles of data protection) of the Convention should be taken into account when assessing the adequacy of the third state's legal framework. Nonetheless, this clarification is only valid as far as these principles are relevant for the specific case of transfer. Thus, the basic principles of data protection do not necessarily have to be taken into account for every data transfer.

The 29 WP already examined the content of an adequacy assessment in 1997 and published a discussion document on the central question of adequacy in the context of Directive 95/46/EC.[38] Even though it is not applicable to the field of criminal matters, the document provides solid guidelines on what an adequacy assessment should include. In this discussion document, the 29 WP identified three types of data transfers within the scope of Directive 95/46/EC: a transfer between an EU-based data controller and a data controller based in a third state; a transfer between an EU-based data controller and a data processor based in a third state who processes the data on behalf of the data controller, and a transfer between an EU-based data subject and a data controller based in a third state. In the field of information exchange in criminal matters, the first type of transfer is the most common one, as these exchanges are organized between law enforcement and prosecution authorities of different states. This means that the data are transferred from an authority that determines the purpose and means of processing the data to an authority that also has that competence, yet within the framework of different data protection legislation.[39]

Besides the 29 WP, the Europol Decision has incorporated a list of items to consider when evaluating a third state's level of data protection.[40] Unlike the 29 WP, Article 23 of the Europol Decision focuses on data exchange in criminal matters rather than on data exchange in commercial matters—which is focused on by Directive 95/46/EC—and includes elements of data processing rather than the principles governing data processing. The list contains: the nature of the data, the purpose for which the data is intended, the duration of the intended processing, the general or specific data protection provisions applying to the requesting authority, and whether or not the entity has agreed to specific conditions required by Europol concerning the data.

---

[38] Working Party on the Protection of Individuals with regard to the Processing of Personal Data, XV D/5020/97-EN final, WP 4, First Orientations on Transfers of Personal Data to Third Countries—Possible Ways Forward in Assessing Adequacy, Discussion Document adopted by the Working Party on June 26, 1997.

[39] Ibid., 9.

[40] Council Decision of 6 April 2009 establishing the European Police Office (EUROPOL), O.J. L 121, May 15, 2009, 49.

The 29 WP defined a list of content principles to consider when assessing adequacy and heads off this list with the purpose limitation principle.[41] The Europol list also starts with the purpose for which the data is intended. When exchanging personal data for the purpose of prevention, detection, investigation, or prosecution of criminal offences, one main concern is data regarding a person to whom the presumption of innocence (Article 6 ECHR) still applies. It is thus highly important that these data are only processed for this specific purpose or a purpose that is compatible therewith. For this reason, the purpose limitation principle should be a minimum requirement to be fulfilled when deciding upon the adequacy of a third state's data protection framework. Careless handling of the data and improper safeguarding of the proportionality principle can have crucial repercussions for an individual involved in a criminal investigation either as suspect, witness, or victim. Measures protecting the quality of the data and their proportionality in relation to the purposes for which they are processed, should therefore also be laid down in the legal framework of the third state concerned. The onward transfer of data to other third states should be restricted in the case of criminal matters. In investigations or prosecutions of criminal offences that have links to several states, however, an onward transfer could become necessary. Nevertheless, an adequate level of data protection should also be provided by the receiving third state. Finally, technical and organizational security measures should be in place in order to prevent tampering or loss of data. These measures may not be laid down in national law, yet the data controller in the third state should provide for a level of data security that is sufficient and appropriate for the risks that the processing of data presents.

Two other principles were identified by the 29 WP in the aforementioned discussion document: the transparency principle and rights of access, rectification and opposition. In information exchanges for the purpose of prevention, detection, investigation, or prosecution, these principles cannot be guaranteed in every case—in the interest of the criminal investigation. For this reason, they cannot be part of the minimum data protection rules included in an adequacy assessment.

In addition to content principles, enforcement and supervision mechanisms should be installed in a third state in order to provide for adequate protection of data transferred from an EU authority. The 29 WP rightfully stated that it is more efficient to define the objectives to be achieved by these mechanisms rather than requiring their mere presence.[42] This means that the assessment of a state's data protection system should go beyond the "law in the books" and evaluate whether the system provides for support to data subjects and appropriate redress as well as a substantial level of compliance with the data protection rules. The independence of the authorities involved is a prerequisite.

---

[41] Working Party on the Protection of Individuals with regard to the Processing of Personal Data, XV D/5020/97-EN final, WP 4, First Orientations on Transfers of Personal Data to Third Countries—Possible Ways Forward in Assessing Adequacy, Discussion Document adopted by the Working Party on June 26, 1997, 6.

[42] Ibid., 7.

9.3.5 When to Assess or Reassess?

An additional question, especially if the 29 WP is to be introduced as the authority deciding upon the adequacy of third states' legislation, is the moment at which the assessment should be made. Since fast information exchange is of utmost importance in most international criminal investigations, the duration of an adequacy procedure should be considered.

A case-by-case approach as foreseen in Directive 95/46/EC can be quick but is unrealistic due to the high amount of data transfers, particularly in the case of criminal matters. Nonetheless, in the case of criminal investigations, an urgency procedure could be introduced by which a decision on the adequate level of data protection is made for one specific data transfer.

The 29 WP itself put forward the idea of white lists of third states with an adequate level of data protection. Even partial listing of a particular type of data transfers is suggested by the 29 WP.[43] An a priori list consisting of all third states with which Member States could safely exchange personal data would take the 29 WP a long time to compile, principally blocking data transfers in the meantime. However, it still seems to be the best option. The list would have to be reviewed regularly in order to keep up with amendments to legislation in the third states. Obliging third states to inform the 29 WP whenever data protection legislation is modified would be another option.

9.4 Future of Adequacy: Negotiating a New EU-US Agreement

The planned review of EU legislation on data protection happens to be in progress at the same time as negotiations for a general data protection agreement between the EU and the US. The background of this development is the transatlantic cooperation that has been intensified since the terrorist attacks of 2001 in the US. The European Council asked the Commission to propose a general agreement with the US on data protection and, where necessary, on data sharing for law enforcement purposes (future EU-US Agreement on data protection).

One of the main questions to be answered in this respect is whether the principles laid down in such an agreement would apply not only to future agreements covering data exchange but also to existing ones. The negotiating directives include the statement that the future agreement shall also apply to "*existing EU or Member States personal data transfer and processing agreements with the US for the purpose of preventing, investigating, detecting or prosecuting, criminal offences, including terrorism, in the framework of police cooperation and judicial cooperation in criminal*

[43] Working Party on the Protection of Individuals with regard to the Processing of Personal Data, XV D/5020/97-EN final, WP 4, First Orientations on Transfers of Personal Data to Third Countries—Possible Ways Forward in Assessing Adequacy, Discussion Document adopted by the Working Party on June 26, 1997, 3–4.

matters after a transitional period of three years."[44] The effect that this statement could have is considerable, especially with regard to the adequacy procedure.

Even when the focus is not on the bilateral agreements but just on the data transfers between the EU and the US, the future agreement would change the terms of four existing agreements mentioned above. They not only include the 2003 MLA Agreement and the 2010 TFTP Agreement, but also the 2002 Europol-US Supplemental Agreement and the 2006 Eurojust-US Agreement, as the latter two equally provide for personal data transfers in the framework of police and judicial cooperation in criminal matters. As explained in the previous section, it is precisely in preparing these four agreements that the EU, Europol, and Eurojust have not complied with the requirement of evaluating the US level of data protection.[45] Due to adequacy requirements laid down in the Framework Decision on Data Protection in Criminal Matters, the future EU-US Agreement on data protection should not be concluded without making an adequacy assessment.

Considering the bad experience that the European Parliament had with the 2010 TFTP Agreement that was recently reviewed and revealed the existence of oral instructions from the UST to Europol concerning the data transfers,[46] it can be expected that the parliamentarians will push strongly for a genuine evaluation of the US data protection system.[47]

If a genuine assessment of the US level of data protection were made, it would have significant effects on the content of existing legal instruments. Regarding the Europol-US Supplemental Agreement and the EU-US TFTP Agreement, this would mean that the assumption of an adequate level of data protection would finally be backed up by an assessment of the US data protection framework, followed by a decision on its adequacy regarding data transfers with the EU. With regard to the EU-US MLA Agreement and the Eurojust-US Agreement, it would follow that the prohibition of generic restrictions should be amended or deleted. Nevertheless, on October 26, 2010, the Ambassador of the US Mission to the EU, William E. Kennard, declared during a hearing on the future EU-US Agreement on data protection in the European Parliament that the US does not wish to renegotiate the existing agreements.

The negotiating directives themselves could be the answer to this problem, as the Commission considers it a desirable step for the US to ratify the Council of Europe

---

[44] Commission européenne, Proposition de recommandation du Conseil autorisant l'ouverture de négociations en vue d'un accord entre l'Union Européenne et les Etats Unis d'Amérique sur la protection des données personnelles lors de leur transfert et de leur traitement a des fins de prévention, d'investigation, de détection ou de poursuite d'actes criminels y compris le terrorisme, dans le cadre de la coopération policière et judiciaire en matière pénale, COM (2010) 252/2, Annex, Negotiating Directives, § 4, May 12, 2010.

[45] See also De Busser (2010).

[46] See the report by the Europol Joint Supervisory Body, Europol Inspection Report 11-07, Report on the Inspection of Europol's Implementation of the TFTP Agreement, conducted in November 2010 by the Europol Joint Supervisory Body, accessed on April 1, 2011.

[47] European Parliament, "SWIFT implementation report: MEPs raise serious data protection concerns," Press Release, March 14, 2011.

Data Protection Convention and its Additional Protocol.[48] As mentioned above, the Data Protection Convention lays down the basic principles of data protection that have been implemented in the EU, and its Additional Protocol is the general legal basis for the adequacy procedure. Thus, if the US would agree to accede to these two legal instruments, there would be no need for the entire discussion surrounding the adequacy procedure, as the US would have to implement the same data protection principles in its system. This idea is neither desirable nor realistic.

The US accession to the Data Protection Convention and its Additional Protocol is not desirable due to the significant differences between the US system of data protection and that of the EU. These differences already led to the creation of the Safe Harbor principles,[49] the so-called "undertakings" attached to the Commission's adequacy assessment concerning the transfer of passenger name records[50] and the rejection of the first TFTP Agreement in February 2010.[51] Research has proven that data protection legislation in the US and the EU is divergent rather than similar.[52]

The US accession to both legal instruments is unrealistic for two reasons. Firstly, it is questionable whether it is a realistic option to ask a state with a legal history that has—in comparison to the elaborate EU rules—not been characterized by detailed data protection rules to change its attitude as well as its legislation and adhere to a set of formerly unknown principles that would have to be implemented in national law. Secondly, the (recent) history of EU-US cooperation in criminal matters has demonstrated that it is also not reasonable to expect the US to embrace our umbrella data protection system. As explained above, the prohibition of generic restrictions indicates that a smooth and trouble-free data exchange should be the goal and not a complete transformation of the US data protection regime.

---

[48] Commission européenne, Proposition de recommandation du Conseil autorisant l'ouverture de négociations en vue d'un accord entre l'Union Européenne et les Etats Unis d'Amérique sur la protection des données personnelles lors de leur transfert et de leur traitement a des fins de prévention, d'investigation, de détection ou de poursuite d'actes criminels y compris le terrorisme, dans le cadre de la coopération policière et judiciaire en matière pénale, COM(2010) 252/2, Annex, Negotiating Directives, § 17, May 12, 2010.

[49] Commission Decision of July 26, 2000 pursuant to Directive 95/46/EC of the European Parliament and of the Council on the adequacy of the protection provided by the safe harbor privacy principles and related frequently asked questions issued by the US Department of Commerce, O.J. L 215, August 25, 2000, 7–47.

[50] Commission Decision of May 14, 2004 on the adequate protection of personal data contained in the Passenger Name Record of air passengers transferred to the United States' Bureau of Customs and Border Protection, O.J. L 235, July 6, 2004, 11–14; Annex Undertakings of the Department of Homeland Security Bureau of Customs and Border Protection (CBP), O.J. L 235, July 6, 2004, 15–22.

[51] European Parliament Recommendation, A7-0013/2010, February 5, 2010 and Press Release, SWIFT: European Parliament votes down agreement with the US, February 11, 2010.

[52] De Busser (2009).

9.5 Conclusion

Returning to the hopes and wishes of both the EU and the US, the following remarks can be made. The listed goals that the EU set out for itself are exactly the goals for the future. All three of these goals are part of an approach that has not been realized yet while the US' desire of an effortless data transfer seems to be in full progress.

The objectives on the EU side were to have a simple and clear procedure for allowing data transfers to third states, universally valid data protection principles, and for the existing and future transatlantic cooperation to be governed by standards equivalent to the European standards. However, simplifying the adequacy procedure is an exercise that is more complicated than it looks at first sight. Only a portion of the questions that it raises have been touched upon in this contribution. Introducing the 29 WP as the central authority deciding upon the adequacy of third states' level of data protection is in principle a good idea and solves several of the questions mentioned above. Nevertheless, it should be stressed that the 29 WP is not yet equipped for this challenging task.

The adequacy procedure remains to be a thorny issue in the transatlantic cooperation in criminal matters. By drafting the future agreement on data protection both parties could attempt to solve this, however, due to the differences between the data protection framework of the EU and the US, additional safeguards will always have to be guaranteed as was done in the past.

Universally valid data protection principles and the equivalence of the data protection standards in the EU-US cooperation to the EU standards are both objectives that are unrealistic. The transatlantic cooperation in criminal matters in the past decade is a good example thereof. As agreements that have been concluded to exchange personal data for the purpose of prevention, detection, investigation, or prosecution of criminal offences in the transatlantic cooperation have all been drafted in order to facilitate the flow of personal data rather than to safeguard EU data protection standards, the US seems to have realized more of its hopes and wishes than the EU.

Especially when considering the recent inspection by the Europol Joint Supervisory Board of the implementation of the 2010 TFTP Agreement by Europol, including the lack of time for Europol to prepare for its new role and the receiving of oral instructions by the UST regarding the data transfers, it seems that the transatlantic flow is dictated more by the US' wishes than by the EU's.

References

Bennet, Colin J., and Charles D. Raab 1997. The adequacy of privacy: The European Union data protection directive and the North American response. *The Information Society* 13:245–263.
De Busser, Els. 2009. *Data protection in EU-US criminal cooperation.* Antwerp-Apeldoorn: Maklu.
De Busser, Els. 2010. Transatlantic adequacy and a certain degree of perplexity. *Eucrim* 1:30–36.
Fijnaut, Cyrille. 2004. Inlichtingendiensten in Europa en Amerika: de heroriëntatie sinds de val van de Muur en 11 September 2001. *Justitiële Verkenningen* 3:10–42.

Korff, Douwe. 2010. Comparative study on different approaches to new privacy challenges, in particular in the light of technological developments. Working Paper no. 2: Data protection laws in the EU: The difficulties in meeting the challenges posed by global social and technical developments, *European Commission DG Justice, Freedom and Security Report*.

Lichtblau, Eric, and Risen James. 2006. Bank data is sifted by U.S. in secret to block terror. *The New York Times,* 23. June.

Manget, Fred F. 2006. Intelligence and the criminal law system. *Stan. L. & Pol'y Rev.* 17:415–435.

Modderkolk, Huib, and Wester Jeroen. 2011. Via Zoeterwoude kon CIA 'iedere' euro zien. *NRC Handelsblad,* March 19–20:3.

Vervaele, John A. E. 2005. Gegevensuitwisseling en terrorismebestrijding in de VS en Nederland: Emergency criminal law? *Panopticon* 2005:27–52.

Chapter 10
Law Enforcement in the Clouds: Is the EU Data Protection Legal Framework up to the Task?

Maria Grazia Porcedda

10.1 Introduction

The Internet was born as an ensemble of connections between computers and networks. Now, it is turning into a place where we store and process data. Cloud computing is a new way of exploiting existing computing techniques, whose potential has been rapidly realized by businesses, citizens and institutions; it takes place against the background of, and further contributes to, the legal complexity of the Internet (i.e. applicable law, law enforcement and security issues).

This situation must be addressed if all the potential benefits of cloud computing are to be reaped (ENISA 2009; European Commission 2010c, 0245; Gellman 2009). This concerns not only the legal-technical obstacles hindering its development, but also the fact that cloud computing has become both a source and a target for crime, specifically cybercrime. As a consequence, the use of cloud computing naturally draws the attention of law enforcement agencies (LEAs) while affecting fundamental rights, such as privacy and data protection (The Stockholm Program 2010).[1] In fact, in the European Union (EU) these partially overlapping—but nonetheless different—concepts, are intended as individual rights (Rodotà 2005, 2009),[2] enshrined in

[1] The political priority in the EU is to "ensure respect for fundamental freedoms and integrity while guaranteeing security". This should translate into a high level of data protection and privacy, which is overarching in justice, liberty and security and should be adequately protected (ibid.).

[2] In particular, privacy "consists in preventing others from interfering with one's private family and life. In other words, it is a static, negative kind of protection. Conversely data protection sets out the rules on the mechanisms to process data and empowers one to take steps—i.e., it is a dynamic kind of protection, which follows a data in all its movements. [...] data protection contributes to the 'constitutionalisation of the person' [...] can be seen to sum up a bundle of rights that make up citizenship in the new millenium". As such, LEAs access to the data will have a different impact on the two rights. Because of space constraints, and because of the dynamic nature of cloud computing and the subject analysed, the chapter focuses on data protection only. For an account of the evolution and separation of privacy and data protection, see, inter alia (Rodotà 2009).

M. G. Porcedda (✉)
European University Institute, Florence, Italy
e-mail: maria.porcedda@eui.eu

S. Gutwirth et al. (eds.), *European Data Protection: In Good Health?*,
DOI 10.1007/978-94-007-2903-2_10,

Article 8 of the European Convention on Human Rights (hereafter the ECHR; Council of Europe 1950) and Articles 7 and 8 of the Charter of Fundamental Rights of the European Union (2000; hereafter the Charter).[3]

As for the reasons why LEAs may want to access data stored in the cloud, there are at least three possibilities:[4]

1. The data could be relevant for an on-going investigation on crimes unrelated to cloud computing;
2. LEAs are investigating a criminal attack (threat) to computer infrastructure ("operations to disrupt, deny, degrade, or destroy information resident in computers and computer networks, or the computers and networks themselves").
3. LEAs are investigating a computer-assisted crime (threat): "malicious activities (i.e. fraud, drug trafficking, money laundering, infringement to intellectual property rights, child pornography, hoaxes, gathering of information and illegal copying of data) which are facilitated by a computer." Here, the computer is the means to threaten or offend (European Commission 2002).

The purpose of this chapter[5] is to address data protection[6] issues concerning LEAs' access, processing, and use of data stored in cloud computing services with reference to the above mentioned likely scenarios. In particular, I shall focus on the legal framework of the EU and consider its adequacy to the challenge of (LEA's access and use of data in) cloud computing.

I argue that, on the one hand, these activities may seriously interfere with data protection; yet, the analysis will hopefully demonstrate that the issues raised by the cloud are ultimately not new, and that the cloud simply deepens them, urging immediate action. In fact and on the other hand, I argue that, in order to provide adequate protection for individuals, various aspects of current data protection laws need to be updated (thus avoiding a mismatch between different legal instruments).

Before discussing how these arguments can be supported, a short presentation on cloud computing can be helpful.

[3] However, in other countries, notably the US, a right to data protection is still questioned and the notion of privacy is more open-ended (Solove 2007).

[4] It is to further the scholarship to research the impact on data protection of LEAs' requests different from the ones listed.

[5] This paper is a condensed version of Sects. 1, 3 and 4 of "Data Protection in the Clouds: Regulatory Challenges," (Working paper for the Conference (Porcedda and Walden 2011) "Law Enforcement in the Clouds: Regulatory Challenges" Brussels, Belgium, February 24, 2011, available at: http://www.crid.be/cloudcomputing/paper.html), which dealt with European Union-related data protection issues of LEAs' access to data stored in the cloud. For an excellent account of the "European Union and international legal rules, particularly the Council of Europe Cybercrime Convention (2001), governing the obtaining of data for investigative and subsequent prosecutorial purposes, and how such rules may, and do, interact and potentially conflict with foreign laws and rules [as well as...] some of the forensic challenges addressed all relevant law enforcement issues", see the Sect. 2 of the same, written by Ian Walden (see also at: http://papers.ssrn.com/sol3/papers.cfm?abstract_id=1781067).

[6] Privacy concerns raised by cloud computing are not the object of the analysis here (see fn. 2).

10.1.1 What is Cloud Computing?

"Cloud computing" refers to an established technique of computing used for a variety of services, ranging from those offered for the benefit of individuals (such as services offered by social networks) to those proposed for the benefit of companies, either through sharing common software (cloud service providers) or by using shared information infrastructures (cloud infrastructure providers). The term itself is contested: it may well be said that it is a term of hype and,[7] as a result, several varying definitions highlighting different characteristics exist, to the detriment of possible regulatory actions (Porcedda and Walden 2011). In order to sketch the basic characteristics of cloud computing for this discussion, the NIST definition will be followed:

> Cloud computing is a model for enabling convenient, on-demand network access to a shared pool of configurable computing resources (e.g., networks, servers, storage, applications, and services) that can be rapidly provisioned and released with minimal management effort or service provider interaction (Grance and Mell 2009).

This is by no means an attempt to elevate one definition above another. Rather, this brief paragraph aims: (i) to demonstrate what is at stake is a complex system requiring careful understanding, especially in the field of law enforcement; (ii) to establish a typology of cloud computing services for regulatory purposes is quite important.[8]

Recalling some characteristics that will be further analysed in this contribution can be helpful. According to the NIST paper, clouds can follow four deployment models: *private* ("solely for an organization"), *community* ("shared by several organizations and supporting a specific community that has shared concerns"), *public* ("available to the general public and owned by the cloud provider") and *hybrid* ("a composition of two or more clouds [...] bound together by standardized or proprietary technology"; Grance and Mell 2009). Furthermore, clouds can provide three types of services (Armbrust et al. 2009), namely "software as a service" (*SaaS*: "users access applications on the Web; the cloud provider is technically responsible for the application services and for the data of the users"), "platform as a service" (*PaaS*: "an operating system where users can install their own applications, but data are stored according to the "application's decisions") and "infrastructure as a service" (*IaaS*: "a 'logical hardware' infrastructure organized by the users"; Grance and Mell 2009). "At a high level, these three services are carried out through two kinds of elements in the cloud: datacentres and clusters. The datacentres are specialized hardware where data are stored. They generally provide security for access and recovery services. The clusters offer the execution of programs with a high level of speed" (Gayrel et al. 2010; Porcedda and Walden 2011).

---

[7] "Cloud computing has been talked about, blogged about, written about [...] Nevertheless, confusion remains about exactly what it is and when it is useful..." (Armbrust et al. 2009, 3).

[8] Especially since the legal problems raised by each kind of computer service might to a certain extent differ.

Fig. 10.1 The "onion"

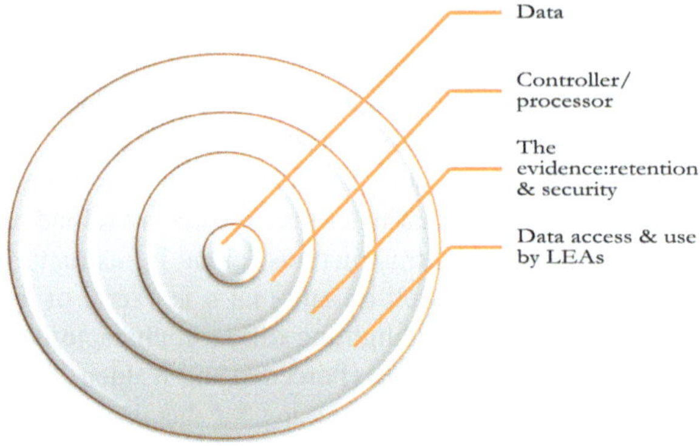

10.1.2 The Interplay Between Clouds, LEAs and Data Protection: Problems and Hypotheses

I shall illustrate the several issues raised by the interplay between cloud computing, LEAs' action and the current privacy/data protection legal framework, and the order of the sections, by means of an imaginary "onion-like" multilayered structure (see Fig. 10.1). Layer 4, the outermost, and most evident for the purpose of this paper, is constituted by the issues relating to the data accessed by LEAs. Technically, though, data are evidence, which are affected by specific problems of retention and security (layer 3). Behind the collection of data as evidence, there exist a constellation of private entities acting as data controllers and processors (layer 2). Finally, layer 1 addresses the issues affecting the data, the core element of the matter. I shall start with layer 1, preceding it with some preliminary remarks on the legal framework.

In greater detail, Sect. 10.2 provides a general introduction of the data protection (and privacy) legal framework after the entry into force of the Lisbon Treaty (TEU and TFEU 2010), with two purposes. First, to establish a set of principles on data protection against which to compare the quality of data protection rules in the area of justice, freedom and security (in the case of cloud computing).

Yet, the analysis shall not only focus on an investigation. In fact, although LEAs' access to the data implies that one of the three reasons (at least) why LEAs may want to access the data has occurred, the conditions prior to the investigation/crime are important in defining what data can be protected, by whom and, very importantly, what is the applicable law and who is liable for the absence of security measures (where this applies). The conditions prior to investigation/crime will also lead to establish liability in case of attacks against information systems. As a result, it is relevant to analyse issues (and legal instruments) that, at a first glance, seem to concern only the business relationship around cloud computing.

Secondly, Sect. 10.2 aims to explain that, pending new legislation pursuant to Article 16 of the Lisbon Treaty (TFEU), layers 1–3 are still regulated under the former "first pillar", while layer 4 is properly regulated by police and judicial cooperation rules, but still within the former "third pillar". Therefore, the analysis of the issues raised by cloud computing, and the demonstration of the argument that data protection laws are not up to the task, will simultaneously move, on the one hand, from the (former) first to the third pillar, and on the other, from the situation prior to the investigation, to the moment of the investigation itself, as far as data protection is concerned.

Section 10.3 addresses the definition of personal data (layer 1, the core of data protection) and the provisions on data controller-processor (layer 2), as defined in Directive 95/46/EC (hereafter the Data Protection Directive, 1995). These do not seem well designed to face the challenges of cloud computing, and call for new solutions, as with other challenged tenets of the Data Protection Directive, i.e. the rules on applicable law and data transfers.

Section 10.4 analyses other challenged rules of data protection, namely those concerning security, which is particularly relevant both ex ante the investigation/crime and ex post, when it comes to data as evidence (layer 3), whose volatility requires some form of retention.

Finally, Sect. 10.5 addresses layer 4, which is represented by LEAs' access to and use of data stored in cloud computing services, and in particular the adequacy of LEAs' rules on data protection as explained above. Section. 10.6 summarises the issues raised and concludes that, while they are not new, action should be taken to address them before it is too late.

10.2 The Data Protection (and Privacy) Legal Framework after Lisbon

The data protection (and privacy) landscape in the EU is likely to undergo major transformation due to the innovations introduced by the Lisbon Treaty. Before discussing the problems relating to the implementation of these innovations, let us first examine their nature. In particular, four are relevant to this discussion.

First of all, the Lisbon Treaty requires that the EU accede to the European Convention on Human Rights. This will provide a direct connection between the Court of Justice of the European Union (hereafter the ECJ) and the European Court of Human Rights (hereafter the ECtHR) case law.

Second, it makes the Charter legally binding, granting it the same force as the Treaties (a constitution-like force, since the Charter's status is equal to EU primary law). Although the Charter does not create new rights (House of Lords 2008), it offers a comprehensive collection of the fundamental rights protected under EU law.

In particular, it is indeed crucial for data protection because it distinguishes between the right to private life (Article 7) and to data protection (Article 8), and provides a refined definition of the latter. In fact, Articles 7 and 8 represent the latest definition of the right to respect for private and family life and data protection offered by previous instruments, namely Article 8 of the ECHR, the Council of Europe Convention 108 (1981; both of them binding on all Member States) and Directive 95/46/EC.

Since these instruments follow a progression, they should be read and interpreted together. First, Convention 108 clearly refers to Article 8 ECHR, both in the explanatory report to the Convention and in its preamble.[9] The ECtHR has recalled this in several judgements.[10] It should also be pointed out that by means of this reference that the Convention acquires a more ample purview than simply data protection (ECtHR 2002).[11] As a result, a strong link is created between the right to private and family life and the right to data protection.

Next, recitals (10)[12] and (11)[13] of the Data Protection Directive also establish a strong relation vis-à-vis Article 8 ECHR and Convention 108, respectively, (thus confirming the strong link between the two rights).

Finally, since the Charter is the latest in a line, the same logic also applies to it: Article 52.3 of the Charter reads "In so far as this Charter contains rights *which correspond to rights guaranteed by the Convention for the Protection of Human Rights and Fundamental Freedoms, the meaning and scope of those rights shall be the same as those laid down by the said Convention.* This provision shall not prevent Union law providing more extensive protection". Articles 7 and 8 not only encompass previous definitions, but they also improve them and are therefore the most appropriate benchmark for this reflection.

The definition of Article 7 reads:

> Everyone has the right to respect for his or her private and family life, home and communications.

while the definition of Article 8 reads:

1. Everyone has the right to the protection of personal data concerning him or her.
2. Such data must be processed fairly for specified purposes and on the basis of the consent of the person concerned or some other legitimate basis laid down by law. Everyone has the right of access to data which has been collected concerning him or her, and the right to have it rectified.
3. Compliance with these rules shall be subject to control by an independent authority.

[9] "'Information power' (i.e. further growth of automatic data processing) brings with it a corresponding social responsibility of the data users in the private and public sector." "It is desirable to extend the safeguards for [. . .] the right to the respect for privacy [. . .]" and "it is necessary to reconcile [. . .] the respect for privacy and the free flow of information between peoples [. . .]".

[10] For a detailed analysis of the ECtHR's case law, see De Hert and Gutwirth (2009).

[11] Although I stress again that only the latter is addressed here. "The very essence of the Convention is respect for human dignity and human freedom. Under Article 8 of the Convention in particular, where the notion of personal autonomy is an important principle underlying the interpretation of its guarantees, protection is given to the personal sphere of each individual, including the right to establish details of their identity as individual human beings" (ECtHR, Goodwin v. UK (28957/95), judg. 11.07.2002, par. 90).

[12] "[. . .] The object of the national laws on the processing of personal data is to protect fundamental rights and freedoms, notably the right to privacy, which is recognized both in Article 8 of the European Convention for the Protection of Human Rights and Fundamental Freedoms and in the general principles of Community law; [. . .] for that reason, the approximation of those laws must not result in any lessening of the protection they afford but must, on the contrary, seek to ensure a high level of protection in the Community."

[13] " [. . .] The principles of the protection of the rights and freedoms of individuals, notably the right to privacy, which are contained in this Directive, give substance to and amplify those contained in the Council of Europe."

It is worth noting that data protection and the right to respect for one's private life, home and communications should be ensured as a general rule.

With regards to data protection in particular, an analysis of the article includes the following:

a. **Substantive principles on processing** (which correspond to the substantive principles listed in Article 6 of the Data Protection Directive):

   1. *Fairness*: to be fair, the processing must be (1) done for a legitimate purpose (legitimacy), which is defined either by the consent of the person (in the terms of Article 6 of the Data Protection Directive), or by law (i.e. Article 7 of the Data Protection Directive); (2) transparent, i.e. the data subject must be adequately informed (compare Articles 10 and 11 of the Directive);
   2. *Legality*: all phases of the processing operations (including collection) must be carried out in accordance with the law, which must be clear, i.e. leaving no room for ambiguous interpretations, and foreseeable, i.e. the consequences of each provision must be known ex ante (lawfulness).
   3. *Purpose limitation*: each processing operation must be tied to a specific, limited purpose (necessity and proportionality). The use of the same set of data for different purposes constitutes a new processing, subject to the conditions listed. The respect of purpose limitation is therefore crucial to an effective data protection regime.

b. **Procedural principles on processing**:

   4. *Substantive rights*: the data subject has the right of access to data concerning him or her, and to rectify them if they are not correct (compare with Article 12 of the Data Protection Directive);
   5. *Control by an independent authority*: no right is effective if it is not implemented and only the oversight of an independent authority can ensure compliance with these rules.

Article 8 must be further read in conjunction with Articles 51 and 52 of the Charter. The former limits the application of the Charter to Union law, while the latter recognises that "Subject to the principle of proportionality, limitations may be made only if they are necessary and genuinely meet objectives of general interest recognised by the Union or the need to protect the rights and freedoms of others". Such interests are those listed in Article 8.2 ECHR, or in Article 9 of the Convention 108 or Article 13 of the Data Protection Directive, which includes, among others, *(d) the prevention, investigation, detection and prosecution of criminal offences, or of breaches of ethics for regulated professions; [. . .] (f) a monitoring, inspection or regulatory function connected, even occasionally, with the exercise of official authority in cases referred to in (c), (d) and (e); (g) the protection of the data subject or of the rights and freedoms of others*.

As a consequence, data processing for police and judicial cooperation falls under the scope of the exceptions.[14] Nonetheless, the derogations listed must be provided

[14] Although the scope of the exceptions is not going to be discussed here, it is worth noting that it has already been questioned long ago (Rodotà 1973).

for by legislative measures, and therefore have to respect the parameters established by the existing instruments.

Furthermore, Article 52.1 of the Charter reads as follows "Any limitation on the exercise of the rights and freedoms recognised by this Charter must be provided for by law and respect the essence of those rights and freedoms". The ECJ has made clear in several judgements that exceptions must be interpreted restrictively—as any exception; therefore, (necessary and proportional pursuant to Article 52 of the Charter) exceptions have to fulfil the essence, or the core (Scheinin 2009), of data protection,[15] as defined by Article 8 of the Charter: legality, preciseness and foreseeability (lawfulness); fairness, legitimacy (consent, but not only) and transparency; purpose limitation (proportionality and necessity); recognition of subjective rights; and independent supervision. Consequently, LEAs' practices should respect the substance of these principles[16] without jeopardising investigations.[17] These principles are used in Sect. 10.5 as a minimum standard to evaluate the quality of data protection in the area of justice, freedom and security (in the case of cloud computing).

Third, the so called "three pillars structure" has been abolished. Data protection was deeply affected by the pillars structure, in that the main rules regarding data protection, i.e. the Data Protection Directive, Regulation 45/2001/EC (2001) and Directive 58/2002/EC (2002, hereafter the e-Privacy Directive, revised in 2009) applied only to the first pillar, or community law, whereas data protection in police and judicial cooperation, the so-called Third Pillar, was regulated by different rules.

To conclude, the Lisbon Treaty contains a new, specific legal basis for data protection: Article 16 TFEU[18] (and Article 6 TEU).[19] Most importantly, Article 16 applies to both former first and third pillars and obliges the legislator to adopt common data protection rules (Hustinx 2009; Hijmans and Scirocco 2009) applying to both areas,

[15] It must be pointed out that the question of what constitutes the "core" of data protection has not been closed by Article 8: further principles specify the right to data protection, while others are being questioned, such as consent, which has long been considered a "rubber-stamp" principle (see Rodotà 1973). The discussion on the core of data protection is too wide to be developed here. For an excellent account of principles deriving from Convention 108, see De Busser (2009). For a specific analysis of the principles of data protection affected by the use of cloud computing, in the context of Council of Europe's Convention 108, see Gayrel et al. (2010). For a more detailed analysis of data protection in the area of Freedom, Security and Justice, see, inter alia, Gutwirth et al. 2009 Reinventing Data Protection? (De Hert et al. 2008; De Busser 2009; Dumortier et al. 2010; Hijmans and Scirocco 2009; Rodotà 1973; Gayrel et al. 2010).

[16] Or any principle constituting the core of data protection, consistent with what is discussed above.

[17] In other words, the application of these principles should be accommodated to the needs of investigations taking into account the specificity of the situation. For instance, giving information to the data subject beforehand is not conceivable without ruining the investigation. In this case, the individual should be informed after the fact. Access to data as normally intended could also disrupt investigations; an indirect form, such as access by the supervisory authority, can therefore be the alternative. Supervisory authority would be, in this case, the liaison between LEAs and the data subject.

[18] TFEU, at Art. 16.

[19] TEU, at Art. 6.

whereas the Common Foreign and Security Policy (CFSP) should be dominated by special rules, pursuant to Article 39. In fact, the Commission is currently studying the existing framework to propose a new one (European Commission 2010a).

Nevertheless, it should be recalled that Article 16 needs to be read in conjunction with Declaration n°21 on the protection of personal data in the fields of judicial cooperation in criminal matters and police cooperation, and the Article 39 TEU. Therefore, until legislation pursuant to Article 16 is enforced, not only will the existing rules adopted within the former third pillar still be valid (unless repealed/amended), but also, the whole existing legal architecture will still be based on the pillars structure, and on a—rather—different technological reality, as the following sections will demonstrate.

After having clarified the state of the art of the legal framework and established the parameters to judge data protection rules for LEAs' purposes, the specific issues raised by cloud computing can now be analysed, starting with the (former) first pillar and the circumstances prior to an investigation.

10.3 The (Former) First Pillar—Layers 1 and 2 or Personal Data, Scope of Application and Provisions on Data Controller and Data Processor

Cloud computing defies several tenets of the Data Protection Directive. We start by analysing the definition of personal data and the provisions on controller-processor, which are essential for data protection. While these specifically concern the business (user-provider) relationship, they are relevant also in the case of LEA's access to data in the cloud.

First, the definition of personal data laid down by Article 2(a) of Directive 95/46/EC[20] marks the division between data deserving protection or not. As unpacked by the Article 29 Data Protection Working Party (2007; hereafter the Article 29 Working Party), the definition excludes, inter alia, data relating to legal persons[21] and know-how. Moreover, unstructured data and data processed for domestic purposes fall outside of the scope of application of the Directive.[22]

Second, the definition of data controller and data processor, i.e. the "inner circle of data processing", allocates responsibilities for effective application of and compliance with data protection rules (Hustinx 2010; Gayrel et al. 2010). Inter alia, the

[20] "Any information relating to an identified or identifiable natural person ('data subject'); an identifiable person is one who can be identified, directly or indirectly, in particular by reference to an identification number or to one or more factors specific to his physical, physiological, mental, economic, cultural or social identity."

[21] In the case Lindqvist, the ECJ clarified that member states can extend the protection to legal persons; however, only Italy, Austria, Luxembourg have indeed extended some of the provisions to legal persons.

[22] While this may not be immediately relevant from LEA's point of view, it implies that data processed for domestic purposes may enjoy reduced protection from the very outset, and it may be difficult to establish responsibility if problems arise.

controller offers an essential criterion when choosing what is the applicable law (Article 4) and he/she ensures the enforceability of data protection rights, both proactively (ensuring implementation) and reactively (ensuring compensation). The identification of the data processor is highly relevant, too, in order to ensure the confidentiality and security of processing (Articles 16–17), and the applicable law for the security of processing, which depends on whether or not the processor is established in the EU (The Article 29 Working Party 2010c).

These different responsibilities are crucial for the purposes of this discussion, such as notification of access to data by LEAs, notification of security breaches, and responsibility for security and liability. The controller may also choose to take data protection even beyond the mandatory requirements of the law. For instance, he/she can also specify that data protection rules extend to legal persons, for example to avoid applying two standards to the same processing.

10.3.1 Personal Data, Controller-Processor and Cloud Computing

The aforementioned features are especially problematic in the case of cloud computing services (and their access by whomsoever). In fact, "Internet users act as data controllers ex Article 2d of the Directive for the data that they upload. However, in most cases [social networking] processing falls within the household exception ex Article 3.2 of the Directive. At the same time, special networking services are considered data controllers insofar as they provide the means for processing user data and provide all the basic services related to user management (e.g. registration and deletion of accounts)" (EDPS 2010).

Indeed, first, many of the services which qualify as software as a service (such as e-mail services, Google docs or social networks), would fall under the definition of "domestic" or "house holding" processing and therefore outside the scope of Directives 95/46/EC and 2002/58/EC.

Platform and infrastructure as a service include many services addressed to enterprises (but not only). In this case, the data implied could relate to legal persons, including know-how or intellectual property. These categories are, indeed, excluded by the existing definition of data protection, unless falling under the restricted cases of the e-Privacy Directive (which explicitly mentions the legitimate interest of the subscribers who are legal persons, as regards Articles 12 and 13 on unsolicited communications).

Second, the distinction between data controllers and data processors is blurred in cloud computing services, as recognised by the Article 29 Working Party (2010c, 169). According to both the Directive and many privacy policies, the controller would be the user, who in many cases lacks the technical competence and knowledge (the control of the means and purposes) to act as such. In fact, forms of co-control may exist *de facto*. This complicates the attribution of responsibility for the implementation of security measures, which is a controller's duty pursuant to the Data Protection Directive.

Such a legal uncertainty should be kept in mind—and swiftly addressed by means of regulation—whenever considering the potential malicious actions against the cloud environment and access to the data stored therein by law enforcement authorities.[23] Such a legal uncertainty also affects the principle of confidentiality (Directive 95/46/EC).[24] Indeed, especially because of this grey zone, cloud computing providers may be using users' data for profitable *meta*-processing activities.[25]

The Article 29 Working Party has argued that the idea of a legislative grey zone would be avoided by either the applicability of national data protection laws or the protection afforded by Article 8 ECHR (and now, also Article 8 of the Charter). In addition, other sets of rules, such as those on intellectual property, may indirectly protect the data subjects (The Article 29 Working Party 2007, 136).

However, it is debatable how effective this "indirect" protection would be. In any case, there is a clear imbalance between individuals and companies, and between companies themselves. This in turn may hinder the effectiveness (compliance) of data protection rules. The Article 29 Working Party has suggested that the lack of compliance with data protection rules may be countered by introducing the principle of accountability, possibly coupled with the use of sanctions (The Article 29 Working Party 2010a, 173).[26]

10.3.2 The Data Protection Directive and Applicable Law: "Mono-" and "Trans-Jurisdictional" Clouds

Due to some of the features of cloud computing, when analysing the interplay between data protection laws and law enforcement, the EU legal framework is necessary, but not sufficient: this leads to the principle of adequacy and the rules on applicable law.

[23] If, for instance, the definition of personal data covered know-how (which is different from, and not covered by, legislation on patents), the latter would (should) be technically and procedurally protected against security breaches like any personal data and, in case of a breach of security, clear liability for the consequent losses could be established. Conversely, certain categories of information would not be adequately protected, despite their importance for legal persons. On the other hand, information on employees would undoubtedly be considered personal data.

[24] Article 16 of the Data Protection Directive reads as follows "Any person acting under the authority of the controller or of the processor, including the processor himself, who has access to personal data, must not process them except on instructions from the controller, unless he is required to do so by law".

[25] Deciding the means and purposes of processing automatically qualifies somebody as a data controller, as we have seen. But this would not be possible under the current legal framework, given that the processing does not relate to personal data. This lack of protection may, in the long run, act as a boomerang effect for companies, too: cloud computing services could become unappealing to individuals and companies.

[26] Accountability would both mean to take the appropriate measures or follow the procedures to implement data protection principles (also when transferring data abroad), and to be able to demonstrate that appropriate and effective measures have been taken (evidence). This could be done by means of monitoring or conducting internal/external audits. It follows that transparency is an integral element of accountability.

A small digression is needed to clarify this point. As explained in the introduction, the term "cloud computing" encompasses different types of services. Providers may be offering their services to users located in one or several countries; also, the first provider may outsource a portion of the processing to another cloud provider or may in turn be renting its cloud infrastructure from a bigger cloud provider. All of these providers may be located in different countries or under different jurisdictions. Furthermore, clouds established under different jurisdictions may merge, thus entailing a migration of data from one legal system to another. Also, there may be no provider at all, that is an enterprise may simply be implementing its own system of cloud computing. In addition, the provider may forward (sell) the data in her/his possession for its own purposes (i.e. behavioural advertising). Finally, cloud computing often implies the use of data centres dispersed in several countries (from which comes the "feeling of location independence") and access to different markets.

As a result, cloud computing services can be roughly divided into two categories: domestic, or mono-jurisdictional, and trans-national, or trans-jurisdictional, clouds. If a cloud is physically located under the same jurisdiction, it is a domestic, or mono-jurisdictional, cloud. Conversely, we can talk about trans-national, or trans-jurisdictional, clouds (Clarke and Stavensson 2010). This is why cloud computing in general, and trans-jurisdictional (trans-national) clouds in particular, are so complicated from a legal standpoint.

In general legal terms, trans-national clouds complicate the conduct of investigations and the collection of evidence (not to mention the difficulty in obtaining redress for citizens located under several jurisdictions). From a data protection standpoint, this trans-national character (data-centres, outsourcing etc.) means that in many circumstances the use of cloud computing services will entail international data transfers (and therefore, accessing data across jurisdictions). This calls into question the validity of the concept of adequacy, the helpfulness of the existing rules on data transfers and on applicable law, as well as the enforceability of data protection and privacy rights. A commonly proposed solution is to build domestic clouds only.

10.3.2.1 Mono-Jurisdictional (Domestic) Clouds

In the EU, a cloud is "domestic" or "mono-jurisdictional" if the conditions laid down by Article 4 of the Data Protection Directive are satisfied: either the controller is located within the EU or, it uses equipment located in the EU for purposes other than those of transit. There are, however, two main problems.

First, the reference to the data controller is likely to be controversial in cloud computing, as the attribution of the role of data controller and data processor, to the user and cloud provider respectively, is not clear for the reasons stated above. The "equipment" criterion is also likely to raise important issues because of the characteristics of cloud computing (Leenes 2010).[27] The question of the applicable

[27] Not only laptops, but cookies can be considered equipment. Provided that the user has not blocked the latter, therefore, the Directive 95/46/EC would apply to most cloud computing services. On the matter, see Leenes (2010).

law is being revised by the Article 29 Working Party (2010d); the Future of Privacy Working Paper also refers to it as one of the items in need of revision (The Article 29 Working Party 2009, 168).

Second, even if the jurisdiction is the same, member states' e-law varies, i.e. the e-market is fragmented. This creates uncertainty about which rights and legal protection apply, which in turn undermines users' confidence and damages businesses (European Commission 2010c): the latter, by complying with one country's law, can risk breaching another's (European Commission 2010c; ENISA 2009).

Currently, when the EU legal framework applies, a controller may only transfer data outside the EU under the conditions established by either Articles 25 or 26 of the Data Protection Directive: either the recipient offers an adequate level of protection (i.e. as offered by the Safe Harbour scheme), or it falls under one of the exceptions to the rule, to be interpreted restrictively, listed in Article 26, among which are the use of contractual clauses and binding corporate rules (The Article 29 Working Party 1998, 12).

Data sharing with third countries that do not offer an adequate level of protection, though, is becoming more common (including those related to law enforcement; Council Decision 2009b),[28] and this is especially true, although implicit, in the case of cloud computing. This increases the risks of threats, which acquire a stronger international character.[29] Therefore, in order to both prevent threats and protect its citizens, the EU tries to export its principles, specifically by means of agreements and by leading negotiations on international standards (European Council 2010).[30]

An example is offered by the several agreements signed with the US (Bellanova and De Hert 2008),[31] at the EU but also at the member states' level (Bellanova 2010), sometimes with mixed results.[32]

Thought of in this way, the notion of adequacy entails an extraterritorial character (Hijmans 2010). In fact, the attitude of attempting to apply one's rules extraterritorially is diffused, although this clashes with the feasibility of enforcing them, with strong negative repercussions on redress mechanisms.

---

[28] See, inter alia, the implementing rules of the Council Decision (2009b) at: <http://www.europol.europa.eu/index.asp?page=legal_other>.

[29] As the Stockholm Program has explicitly recognized "the internal security is interlinked with the external dimension of the threats", (European Council 2010, 36).

[30] Ibid.

[31] For an account of the initial steps of the several EU-US information exchange agreements, see Rocco Bellanova and Paul De Hert (2008).

[32] As in the case of the Agreement on Mutual Legal Assistance of 25 June 2003 (EU-US Agreement on Extradition and Multilateral Legal Assistance 2003). In fact, data protection can be invoked as a ground for refusal only in exceptional cases. Furthermore, the case-by-case structure of the MLAT would make it possible to "bypass most of the sensitive issues of data protection" (Tanaka et al. 2010).

10.3.2.2 Trans-Jurisdictional (Trans-Border) Clouds

Although domestic clouds are desirable from a policy perspective (provided their weaknesses are addressed), they are accompanied by the existence of trans-national, or trans-jurisdictional, clouds, which encompass both proper multinational actors (i.e. Amazon, Google, Microsoft etc.), which are therefore faced with the legislation of several jurisdictions, and clouds based under one jurisdiction only but operating through several data centres in the world (i.e. Facebook). The unknown location of the data, coupled with the presence of multinational actors faced with different rules for different markets (Clarke and Stavensson 2010) is typical of trans-national clouds as they are built now. While the place of the stored data should not matter from the standpoint of Article 4 of the Data Protection Directive (Leenes 2010), it may further complicate the problem of the applicable law in practice.

Again, trans-national clouds entail continuous data transfers abroad and demonstrate how provisions on data transfers are outdated. Besides the points stated above, these rules reflect an earlier stage of development of technology, i.e. they interpret data transfers in terms of a "point-to-point transfer", taking place under a contract and involving notification (Hustinx 2010). The privacy community agrees on the need to revise these rules (The Article 29 Working Party 2009); cloud computing makes this necessity even more urgent.

The concepts of, and provisions on, personal data, controller, processor, adequacy and applicable law do not exhaust the list of challenged tenets. Security and retention must still be discussed, which is the purpose of the next section.

10.4 The (Former) First Pillar—Layer 3, or Data as Evidence: Retention and Security

The data accessed are evidence and, in the case of cloud computing, two problems coexist: that of volatility and that of integrity of the data. This makes it necessary to discuss the data protection issues concerning data retention and security in the cloud.

10.4.1 The Problem of Evidence in Cloud Computing Systems

The evidence-related problems in a cloud computing environment "concern access to the data prior to it being seized, and the preservation of the data being done correctly, since due to the dynamic nature of the operation of a cloud computer system, it would not be possible to go back to the original state of the data" (Taylor et al. 2010, 304).[33]

[33] On the same issue, see also Schwerha (2010). For a more detailed discussion of forensic problems in the cloud, See Porcedda and Walden (2011), Sect. 2.

Furthermore, "an organization may not know where the data they are responsible for is located geographically at a particular time, although this may be more of a logical structure, than a geographic one."[34] As a result, evidence collection is much more complicated in the cloud environment than in a traditional one and, in fact, there is not an agreed method to extract data from the cloud. This depends in the first place on whether the data are stored in a private or a public cloud.

In fact, data retrieval in a private cloud seems to be easier for three reasons: (1) data will reside either within the organisation or the cloud provider; (2) potential sources of evidence will be identifiable; (3) key personnel may be interviewed. A public cloud environment, on the other hand, is more dynamic and customisable, "thanks to a seamless interaction of a variety of applications being delivered to the user as if they were accessing just a single site or a logical location".[35] Consequently, identifying where the data are stored exactly is complex and its retrieval in general will be challenging due to three reasons:

1. It will be difficult to assess whether data exist or not, since they are not stored on the physical PC anymore. While traditional documents/files had meta-data, this may not be necessarily true in case of cloud computing documents. That said, it is possible to track the access to the cloud environment, but then it is not easy to investigate modifications done within it, unless modification confirmations were sent to the users' PC, which is subject to the presence or not of audit trails. If the investigation concerns/requires e-mails, then logs of sent/received e-mails from the PC could be used as evidence. Tracking malware, on the other hand, may be very complex.
2. Evidence may only exist temporarily. For example, the duration of registry entries and temporary internet files may be limited to the logged-in session time. In general, the evidence may last less than the time LEAs need in order to act, unless data are retained by cloud providers for some time after deletion by users; in this case, data may be useful for forensic investigations.
3. It may be difficult to conduct an investigation without impacting negatively on cloud users who are not the target of the investigation (Porcedda and Walden 2011; Taylor et al. 2010).

Another important issue is to certify that the PCs were working correctly at the time of the criminal activity. In a cloud computing system, "the cloud manager provides the entry point into the cloud for users and administrators" (Taylor et al. 2010) and should not be liable for the remedies of any wrongdoing done on their service, provided that the cloud manager(s) was not aware of it. In other words, the cloud provider would not be liable for damages/pecuniary remedy/criminal sanctions due to posting data/applications, in cases where the provider is not aware of criminal activity (Van Gyseghem 2008; Directive 2000/31/EC).

---

[34] Taylor et al. (2010), 304.

[35] Ibid.

10.4.2 Data Retention Directive

Among the solutions proposed to facilitate investigation in the cloud is for governments to make arrangements to preserve suspect's data following a request from LEAs to ensure that data does not disappear when needed. In the EU, Directive 2006/24/EC (hereafter the Data Retention Directive 2006) has been adopted precisely to harmonise the Member States' provisions on data retention to make them available for the purposes of the investigation, detection and prosecution of serious crime. Data Retention falls in this section because, in the action for annulment lodged by Ireland (European Court of Justice 2009), the ECJ has confirmed the first pillar nature of the Directive (Van Gyseghem 2008), under the jurisprudence of the essential/ancillary objective (De Busser 2009).[36]

In fact, the obligation to retain certain data falls on the providers of publicly available electronic communications services or of public communications networks, whose definition is laid down by Directive 2002/21/EC. In Article 1 of the Directive 98/34/EC, as amended by Directive 98/48/EC[37] Information Society Services (hereafter ISS) are explicitly excluded from the concept of a publicly available electronic communications service. As a result, whenever a service is principally an ISS, the Data Retention Directive does not apply. Since many services of cloud computing are ISSs, the Directive does not apply to them.

In few cases when it does apply (to cloud computing), the contribution of data retention to detect, investigate and prosecute crime must be assessed in the light of its many shortcomings. While the literature on the subject is vast, given its limited application to the subject at hand it should only be recalled that the Directive is considered poorly conceived under a technical point of view,[38] it would not respect the necessity and proportionality principles that would keep it in line with privacy laws (The Article 29 Working Party 2010b) and the aforementioned action for annulment is simply one of the several lodged. In any case, data retention is a pre-Lisbon Treaty piece of legislation and depending on the developments of the data protection framework revision (European Commission 2010a) and on the Commission's evaluation, it may soon be repealed. This would offer a good chance to address the issues raised by cloud computing services with regards to LEAs, i.e. volatility of evidence and the fact that, by their nature, they may most of the time fall outside the scope of retention.

Another solution proposed is to keep audit trails or other means to record users' activities (Taylor et al. 2010), which relates to, and requires addressing, security issues in the cloud.

---

[36] In detail, the essential objective of the directive is that of regulating the providers' retention of data, whereas data access by LEAs is only the ancillary object, because it is not addressed by the Directive itself (Hijmans and Scirocco 2009).

[37] "Service: any Information Society service, that is to say, any service normally provided for remuneration, at a distance, by electronic means and at the individual request of a recipient of services", amended Article 1(a) 2.

[38] See the work of the Platform on Electronic Data Retention for the Investigation, Detection and Prosecution of Serious Crime, at the page <http://ec.europa.eu/home-affairs/policies/police/police_data_experts_en.htm>.

10.4.3 Data Security Issues

Observing the appropriate level of security in respect to the possible risks would considerably curtail the risk of threats to computer infrastructure, one of the three reasons why LEAs may want to access the data, while helping to preserve the evidence for the other two cases and/or avoiding further incidents deriving from negligence of data control (i.e. hacking into the police systems).

Unprotected data could be easily hacked, lost or damaged in such a way as to ultimately affect the very reason for their storage. It is easily conceivable that a malicious person could be trying to eliminate the evidence, if he/she has the necessary means. Finally, depending on the kind of investigation, it could be relevant to determine responsibility for (the lack of) security.

Data security[39] is one of the basic principles of data protection and should be applied every time personal data are processed (which includes storage). According to its definition (article 17 of Directive 95/46/EC), it implies two factors, namely organisational and technical measures, appropriate to the risks posed by the processing activity, provided these are technically and economically feasible for the controller or the processor—if different—which must in turn be chosen in an accurate manner. These measures include procedures and proper technical means.

Moreover, the article refers to any controller and processor; indeed, data security relates to every data processing, regardless of the fact that it may be a new processing on the same data carried out by a new controller. In fact, recital 30, Articles 10 and 22 of Council Framework Decision 2008/977/JHA (hereafter the Data Protection Framework Decision 2008) provide that LEAs must observe the appropriate security measures when handling data. The same applies to Article 7 of the Data Retention Directive.

As such, data security is a transversal principle, which favours all actors involved. Indeed, a call for increased security has been made not only by the Article 29 Working Party (2009) and the European Data Protection Supervisor (hereafter EDPS 2010), but also by the European Commission (European Commission 2010c) and the ENISA (2009; V. A. 2008). These last two encourage an increased respect for privacy and data protection, for instance by using the so called "privacy by design" principle. Service providers should guarantee appropriate security/confidentiality, even if the client is a data controller (The Article 29 Working Party 2009).

10.4.3.1 Technical and Procedural Measures

Privacy by design (PbD) means to integrate data protection and privacy at an early stage of design and creation of technology, especially technology in risky areas covered also by cloud computing.

---

[39] The Experts Group on Data Retention has published a working paper (not available as of December 2010), on this issue: Series A—Position paper 7—Closer understanding of the term "Data Security" in relation to its application in Directive 2006/24/EC. The paper was adopted on 14 July 2010.

Usually information and communication technology (ICT) manufacturers and providers do not implement PbD, because of a lack of economic incentives, demand or institutional support. Users, too, do not usually question the providers' policies as they assume that their data and privacy are de facto protected (EDPS 2010). Indeed, "since it is difficult to understand what is in a cloud from the outside, users should refrain to think about clouds as big and secured services"; a cautious approach would be needed instead. A good example is provided by the crash of the Magnolia Social Network (Bradshaw 2010), whose data were allegedly lost forever by their owners (but may still be accessible and usable by others).

In any case, the increasing number of breaches shows that the risk is real and that action cannot be delayed. PbD is also very relevant in the area of Justice, Freedom and Security, especially for what concerns the information management system. The current legal framework offers two provisions supporting PbD: one is the already cited Article 17 of the Data Protection Directive, to be read in conjunction with Recital 46 of the same Directive; nonetheless, it is too general and vague. The second one is Article 14.3 of the e-Privacy Directive, which requires the adoption of appropriate protective measures from an early stage in the construction of technologies; however, this has never been applied (EDPS 2010). In this respect, the provisions on security vis-à-vis cloud computing are not so much inadequate, as under-enforced: this still calls for action, albeit of a different kind.

The EDPS proposes to incorporate PbD in the legislation as a general principle and especially in certain areas of ICT. Government implementation of PbD could also stimulate its adoption by the private sector (EDPS 2010).

Examples of appropriate technical security measures, between the cloud provider and user and by the cloud provider itself could be:

1. An adequate information management system to control access to data; this includes the use of audit trails, which allow logs to be kept (and would help at a later stage in investigations);
2. Use of privacy enhancing technology and protection against breaches, for example through the use of patches, encryption etc.;
3. Obligation to segregate data stored;
4. Maintaining a person responsible for security (Gayrel et al. 2010).

Proposed procedural measures include the following:

1. Obligations to audit the system (and keep audit-trails);
2. Cooperation between service providers and Data Protection Authorities (allowing audit of security measures/issuance of recommendations);
3. A security policy expressed in clear language. The terms of service proposed by the cloud computing providers tend to be problematic, with the exception of Intellectual Property rights that are usually well respected. In fact, on most occasions the user does not have any negotiation power and must accept the policies as they are. These often include: limited (if any) liability for the integrity of the data; disrespect of the confidentiality of content; disclaimers against guaranteed provision/continuity of the service; imposed applicable law; and difficult data recovery after termination of services. In addition, providers engage in different levels of

obligation to notify users of data disclosure, typically to LEAs (Bradshaw 2010; Bradshaw et al. 2010).

4. Notification of data disclosure and security breaches. The reviewed e-Privacy Directive 2009/136/EC calls for mandatory notification of security breaches, provided they are likely "to adversely affect their personal data privacy" (i.e. ID theft, reputational loss) and unless encryption measures were enabled. Security breaches are defined as "any breach leading to the destruction, loss, and disclosure of personal data transmitted, stored or processed in connection with the service" (Barcelo 2009, 156).

However, only providers of public electronic communications services are obliged to notify breaches,[40] even if member States can decide to extend the obligation at the national level. In addition, the same providers have to establish procedures to respond to data access requests by LEAs in case of an investigation, or to Data Protection Authorities' information requests on this point.

These measures would also have the positive effect of encouraging the correct applications of the data quality principles, as defined in Article 6 of the Directive 95/46/EC.

10.5 The (Former) Third Pillar. Layer 4 or Data Protection in the Area of Justice, Freedom and Security (JFS)

Attention can be finally turned to processing in the field of police and judicial cooperation. As seen in Sect. 10.2, this falls under the exceptions of data protection rules,[41] therefore the related provisions are *leges speciales*. To date, there are three major families of instruments in the Union concerning the use of data by LEAs.

1. Convention 108, its Additional Protocol (Council of Europe 2001) and the Recommendation 87(15) (Council of Europe 1987): as the first binding international instrument adopted (with the exception of the Recommendation), it established a benchmark for data protection in the former third pillar and still applies to the instruments entered into force prior to the adoption of Council Framework Decision 08/977/JHA;
2. The Data Protection Framework Decision, whose scope is limited, in that it regulates the exchange of data between Member States for all data exchanges which do not fall under a particular, or special, regime;[42]

[40] There has been a fierce political fight on this point, Ibid.

[41] Few Member states have extended the Data Protection Directive to the activities of police and judicial cooperation.

[42] In fact, although pursuant to Article 1 the decision should also apply to "data exchanged between Member States and authorities or information systems established under the former title VI of the Treaty on European Union (TEU)" such as Europol/Eurojust, Article 28 limits substantially this provision.

3. Special regimes regulated in *leges speciales* such as those of Europol (Council Decision 2009b), Eurojust (Council Decision 2009a), Schengen (Regulation 2010), etc. (European Commission 2010b), whose benchmark is the Convention 108, its Additional Protocol and the Recommendation.[43]

As seen in Sect. 10.2, it will be sometime before it is seen how the new rules pursuant to Article 16 will be implemented (Dumortier et al. 2010; also, JFS is an area of shared competence as will be the rules on data protection in the field of CFSP). As a result, the relevant instruments for our discussion are Convention 108, Recommendation (87)15 (because access by LEAs is not specifically addressed by the Convention) and the Data Protection Framework Decision.

There are four main data protection concerns related to data handling by LEAs:

1. The degree to which data protection and privacy rules are respected in the course of collecting, handling and further processing data that are evidence for an investigation.
2. Data transfers to LEAs abroad. This raises concern because states may send data to third countries offering a lower level of data protection.
3. Indiscriminate access of data physically processed in that country (or, more rarely, only retained) by third countries' LEAs which may not offer adequate protection. Concerns are raised when the foreign regime is an autocratic one (for instance, the access of data stored by Google in China).
4. "Purpose creep", i.e. the use of data in an investigation collected for a different purpose, most typically the use of data collected for commercial purposes; this becomes even more worrisome when it entails international transfers (see the case of SWIFT).

10.5.1 Convention 108, the Data Protection Framework Decision and Data Protection Principles

To address the first question, it has to be seen to what extent the Data Protection Framework Decision complies with the essential principles of data protection seen in Sect. 10.2, as applicable in the area of JFS.

a. Purpose Limitation and Legality: Convention 108, Article 5b: *Personal data undergoing automatic processing shall be stored for specified and legitimate purposes and not used in a way incompatible with those purposes*;

Recommendation (87) 15:

[43] Since it does not contain specific rules on data protection, the Council of Europe Convention on Cybercrime (Council of Europe, CETS No.185), which is the only internationally binding instrument existing, does not belong in this category. For this reason, the Article 29 Working Party has criticised it in a working document (The Article 29 Working Party 2001). The same applies to Council Framework Decision 2005/222/JHA on attacks against information systems. For a pertinent discussion on the former instrument, see Porcedda and Walden (2011), Sect. 2.

- Article 2.1: *The collection of personal data for police purposes should be limited to such as is necessary for the prevention of a real danger or the suppression of a specific criminal offence. Any exception to this provision should be the subject of specific national legislation.*
- Article 4: [...] *personal data collected and stored by the police for police purposes should be used exclusively for those purposes.*
- Article 5.5iii (in case of onward transfers): *The data communicated to other public bodies, private parties and foreign authorities should not be used for purposes other than those specified in the request for communication.*

Vs.

The Data Protection Framework Decision:

- Article 3.1: *Personal data may be collected by the competent authorities only for specified, explicit and legitimate purposes in the framework of their tasks and may be processed only for the same purpose for which data were collected. Processing of the data shall be lawful and adequate, relevant and not excessive in relation to the purposes for which they are collected. 2. Further processing for another purpose shall be permitted in so far as: (a) it is not incompatible with the purposes for which the data were collected; (b) the competent authorities are authorised to process such data for such other purpose in accordance with the applicable legal provisions; and (c) processing is necessary and proportionate to that other purpose.*
- Article 11: *Personal data received from or made available by the competent authority of another Member State may, in accordance with the requirements of Article 3(2), be further processed only for the following purposes other than those for which they were transmitted or made available: [...] (d) any other purpose only with the prior consent of the transmitting Member State or with the consent of the data subject, given in accordance with national law.*

Although Article 3 seems to be very restrictive, Article 11.d authorises the possibility of further processing, which seems to be disproportionate vis-à-vis the strict limits envisaged by the Recommendation.

b. Fairness, Transparency and Consent: Convention 108, Article 8: *Any person shall be enabled to establish the existence of an automated personal data file, its main purposes, as well as the identity and habitual residence or principal place of business of the controller of the file.*

Recommendation (87) 15, Article 2.2: *Where data concerning an individual have been collected and stored without his knowledge, and unless the data are deleted, he should be informed, where practicable, that information is held about him as soon as the object of the police activities is no longer likely to be prejudiced.*

Vs.

The Data Protection Framework Decision, Article 16: *1. Member States shall ensure that the data subject is informed regarding the collection or processing of personal data by their competent authorities, in accordance with national law. 2. When personal data have been transmitted or made available between Member States, each*

Member State may [. . .] ask that the other Member State does not inform the data subject. In such case the latter Member State shall not inform the data subject without the prior consent of the other Member State.

Article 16.2 authorises a permanent derogation of the principle of transparency. The recommendation is far more protective in that it states that data subjects should be informed as soon as the outcome of the investigation is not likely to be adversely affected any longer by such notification.

c. Independent Supervisory Authorities: Additional Protocol to Convention 108, Article 1: *Each Party shall provide for one or more authorities to be responsible for ensuring compliance with the measures in its domestic law [. . .].*

Recommendation (87) 15:

- Article 1.1: *Each member state should have an independent supervisory authority outside the police sector which should be responsible for ensuring respect for the principles contained in this Recommendation.*
- Article 6.1: *The supervisory authority should take measures so as to satisfy itself that the public is informed of the existence of files which are the subject of notification as well as of its rights in regard to these files. Implementation of this principle should take account of the specific nature of ad hoc files, in particular the need to avoid serious prejudice to the performance of a legal task of the police bodies.*

Vs.

The Data Protection Framework Decision, Article 25.1:

- *Each Member State shall provide that one or more public authorities are responsible for advising and monitoring the application within its territory of the provisions adopted by the Member States pursuant to this Framework Decision. These authorities shall act with complete independence in exercising the functions entrusted to them.*
- *Each authority shall in particular be endowed with: (a) investigative powers [. . .]; (b) effective powers of intervention [. . .]; (c) the power to engage in legal proceedings where the national provisions adopted pursuant to this Framework Decision have been infringed or to bring this infringement to the attention of the judicial authorities. Decisions by the supervisory authority which give rise to complaints may be appealed against through the courts.*
- *Each supervisory authority shall hear claims lodged by any person concerning the protection of his rights and freedoms in regard to the processing of personal data. The person concerned shall be informed of the outcome of the claim.*
- *Member States shall provide that the members and staff of the supervisory authority are bound by the data protection provisions applicable to the competent authority in question and, even after their employment has ended, are to be subject to a duty of professional secrecy with regard to confidential information to which they have access.*

Article 25 is far more detailed than Article 6 of the Recommendation, and yet it does not require such a positive level of publicity as supervisory authorities should ensure pursuant to the latter.

The combination of Convention 108 and of Recommendation (87)15 offers a stronger protection than the Framework Decision does. Indeed, the latter has already been criticised because of its inconsistencies.[44] Nevertheless, even if the Convention was designed to be technology neutral, the latest developments in computing are much too advanced and, as a result, the text is incapable of addressing the relevant issues in the case of cloud computing (Gayrel et al. 2010). Recital 41 of the Data Protection Framework Decision provides that it does not affect Convention 108, its additional protocol and other Council of Europe instruments in the field of police and judicial cooperation. However, despite the fact that Recommendations (87)15 is mentioned in all *leges speciales* adopted in the field of JFS as a standard setting instrument for data handling by LEAs (i.e. Europol Council Decisions), it is not binding (it contains a set of principles).

As a result, and to address the first concern, the current general data protection legal framework in LEAs is not adequate. While Europol/Eurojust, as *leges speciales*, have a very comprehensive data protection system, questions of a legal nature arise when data are handled by Member States, the 24/7 contact points activated by the G8 or the Cybercrime Convention.

Since the data protection legal framework is under revision, the Data Protection Framework Decision may be repealed before it is able to produce any effect and replaced by a more protective document (possibly better addressing cross-border data exchange issues; The Article 29 Working Party 2009); this is timely to include provisions addressing the problems raised by cloud computing as well.

10.5.2 Transfer/Access of Data by Third Countries' LEAs

When it comes to LEAs, the general rule is to transfer data only to countries ensuring an adequate level of protection (Articles 5.4–5.5iii of the Recommendation(s) (87)15, Article 25 the Data Protection Directive and Article 2 of the additional protocol to Convention 108), subject to very restrictively interpreted exceptions.

However, and to address the second reason of concern, while "in principle" the Data Protection Framework Decision respects the idea (recital 23), in practice both Article 13 on transfers to third state authorities/international bodies and Article 26 (without prejudice to existing instruments) are very permissive.

Yet, these rules do not take into account the fact that third states' authorities may want to access the data without the explicit consent of the state where the data originated. In some US' jurisdictions, for example, providers are obliged to report evidence of child pornography in certain instances. In addition, the ultimate owner of a cloud may be a government, which may therefore have access to all information stored in the cloud (Gellman 2009).

The user is usually unaware of these possibilities, regardless of the terms of service or privacy policies of the cloud provider and the cloud service providers do not often

[44] See, inter alia, Dumortier et al. (2010).

notify users of subpoenas when it is lawful to do so, even if they declare they will do so in their privacy policies.[45]

When the EU accesses those data, it will be bound to respect the EU legislation. However, two issues arise: (1) When (the same) data are processed in several locations, how will jurisdiction be assigned? (2) Third states will not necessarily meet EU data protection standards when accessing cloud data, even if these are EU citizens' data.

On top of this, whereas the substantial respect of the essence of a human right is expected in a democratic society, the same is not true when processing happens on the soil of an autocratic country. There, the authorities may compel data disclosure or even electronic surveillance (Clarke and Stavensson 2010; Gayrel et al. 2010), for instance for enforcement purposes, or on grounds of economic espionage.

This raises special concerns given that public administrations are considering whether to put their computing services in the clouds. To avoid state espionage, states will have to carefully select the providers and the kind of services. As a result, and mentioned in Sect. 10.3, many are voicing the idea to build domestic/mono-jurisdictional clouds only.

This underlines, once more, the need for a concerted international solution.

10.5.3 Purpose Creep: Data Collected by the Private Sector (Commercial Purposes) and International Agreements

The third source of concern is the fact that LEAs have started demanding permanent access to data which have been collected by the private sector (for commercial purposes). Such practice has been growing in the past few years on the basis of the "principle of cooperation" (Dumortier et al. 2010) between law enforcement agents and private companies for investigation purposes.[46]

This practice represents a dangerous attempt against the principle of purpose limitation, in that it intends to be permanent, whereas derogation of data protection rules should be limited in time and scope. It also raises serious concerns in terms of the principle of data quality criteria pursuant to Article 6 of the Data Protection Directive. However, the standard of adequacy required for commercial purposes may well differ from that required for an investigation; alternatively, the data may be inaccurate or old. Even though the correctness of the data may not be essential for, say, behavioural advertising, it becomes crucial when the same data are used as evidence.

Data may also have been collected without the unambiguous and informed consent of the data subject—where required—and as such may be unlawful. The issue was developed in the previous section, together with the Data Retention Directive,[47] which is the most important EU domestic example of data purpose creep.

---

[45] Id., see also Bradshaw et al. (2010).

[46] For a detailed analysis of the public-private partnerships, see Porcedda and Walden (2011), Sect. 2.

[47] For an analysis of the consequences of LEAs' use of evidence obtained unlawfully, see Porcedda and Walden (2011), Sect. 2.

When data retention has an international dimension, (i.e. the collection of data for commercial purposes which are made available to third countries' authorities), the controversial cases such as PNR and the "Society for Worldwide Interbank Financial Telecommunication" (SWIFT 2007) comes into play (EPHR 2010; Lichtblau and Risen 2006).[48] The TFTP Agreement (Agreement between the EU and the US on the processing and transfer of Financial Messaging Data from the EU to the US for the purposes of the Terrorist Finance Tracking Program 2010) is particularly relevant to this discussion for two reasons. On one hand, it represents a good example of the issues of extraterritoriality involving data access with transparency and supervision, which triggered the criticism of the privacy community (EDPS 2007; The Article 29 Working Party 2006). On the other hand, and most importantly, as SWIFT CEO Mike Fish reported, "SWIFT is considered by industry experts to be the pre-eminent example of a secure private cloud for financial services"; for them cloud computing "is about adding additional capabilities that allow more interoperability" (Sibos Issues Thursday 2009).[49]

Even if crucial from a data protection perspective, an analysis of the innovations of the Agreement is not relevant here. What is relevant for this discussion is the fact that "intra-European messages remain in Europe and are no longer mirrored in the United States"(SWIFT 2007), which addresses both questions of extraterritoriality and data transfers and represents a positive step in the matter.

According to Article 4 of the Agreement, "The US Treasury Department shall serve production orders ('Requests'), under authority of U.S. Law, upon a designated provider present in the territory of the United States in order to obtain data necessary for the purpose of the prevention, investigation, detection or prosecution of terrorism or terrorist financing that are stored in the territory of the European Union. [. . .] 4.6 The designated provider shall thereupon provide the data (i.e. on a push basis) directly to the US treasury department. The designated provider shall keep a detailed log of all data transmitted to the US Treasury Department for the purposes of this agreement".

The innovation of the Agreement is therefore the obligation of keeping data relating to EU citizens in the EU and of sending them on a selected basis when strictly needed for anti-terrorism purposes (recital 3). The processing, however, still takes place in the US, where data are retained until needed. This will be the case until the EU creates a TFTP of its own, as envisaged by the Conclusions (Article 2) and Agreement itself (Article 11).

Furthermore, the Agreement provides for independent supervision to ensure its correct implementation, including the articles on purpose limitation (Article 3 of Conclusions and Article 12 of the Agreement). The Agreement represents a substantial step in the direction of the creation of a domestic cloud, although it does not

48 For a synthesis of the facts since the New York Times unveiled the access by US Treasury Department authorities of financial records held by SWIFT 4 years ago see EPHR 2010.

49 See at <http://www.swift.com/about_swift/press_room/swift_news_archive/2010/business_forum/Canadian_Business_Forum_2010.page>.

realise it completely. In fact, from a privacy perspective in cloud computing, the location of the control (and effective processing) is more important than that of simple storage.

10.6 Conclusion

This trajectory has hopefully shown that, from the data (the very essence of this discussion), to their access by LEAs, cloud computing raises questions and highlights shortcomings of the data protection legal framework that can no longer be ignored:

First and second layer: Should the definition of personal data currently be changed to include data which are not personal, and its scope to include data treated for household purposes? Should the cloud provider be treated as a co-controller or external controller? Should obligations on confidentiality be imposed on cloud computing services, regardless of the revision of the definition of personal data?

Third layer: Should the Data Retention Directive apply to Information Society Services? How should the problem of evidence in cloud computing be addressed? Can security laws provide answers to the problem of evidence, without affecting performance? Should PbD and standard procedures be imposed by means of legislation to cloud computing services? For instance, should cloud computing services be obliged to notify breaches (without jeopardising LEAs' activities)?

Fourth layer: How can national LEAs' respect of data protection rules be ensured? Should this be addressed by the new legal framework or would it be sufficient to render Recommendation (87)15 as binding?

Applicable law: Applicability of EU law does not equate with easy enforceability. Is the concept of adequacy still relevant as it is formulated, especially in the light of the issues raised by cloud computing? How should companies' compliance with the law be ensured? Are binding corporate laws the answer? Is an international agreement to address cloud computing issues needed?

Is the obligation to build domestic clouds the only way to avoid the possible drawbacks? Is this feasible without damaging the possibilities offered by cloud computing? Will this obligation prevent, on the other hand, companies from building off-shore data centres (Clarke and Stavensson 2010)? Does the lack of harmonisation between countries risk undermining the advantages of a single jurisdiction—domestic cloud?

Should the EU realize a TFTP agreement of its own with a view to building an entirely domestic cloud? Are companies also willing to implement such systems in practice?

A final point which raises additional questions and has not been addressed so far is the future development of the cloud computing market (Nelson 2009). Will it be dominated by a few powerful super-clouds whose power to build off-shore data centres is uncontrollable or will it be characterised by dispersed small clouds which comply more easily with the law? Will nation States protect their cloud, lock in a cloud or rather cooperate internationally to constrain providers?

Actually, most of the issues raised are simply a radicalisation of existing problems concerning data protection and LEAs and therefore highlight the need for a swift reform of the legal framework, which is currently not up to the task.

Acknowledgments This paper is the result of research carried out at both the CRID (Belgium) and the EUI (Italy). I would therefore like to thank Yves Poullet, Jean-Marc Van Gyseghem, Jean-Philippe Moiny and Giovanni Sartor for the extensive comments and helpful discussions which resulted in substantial improvements of this paper. I am also very thankful to Martyn Egan (EUI) for his thorough and patient linguistic advice.

References

Books and Articles

Armbrust, Michael, Fox, Armando, Griffith, Rean, Joseph, Anthony D., H. Katz, Randy, Andrew, Konwinski, Lee, Gunho, Patterson, David A., Rabkin, Ariel, Stoica, Ion, and Zaharia, Matei. 2009. Above the clouds: A Berkeley view of cloud computing. Technical Report No. UCB/EECS-2009-28. Accessed 10 Feb 2009.

Barcelo, Rosa. 2009. EU: Revision of the ePrivacy directive. *Computer Law Review International* 5:129–160.

Bellanova, Rocco. 2010. The case of the 2008 German-US agreement on data exchange: An opportunity to reshape power relations? In *Data protection in a profiled world*, eds. Paul De Hert, Serge Gutwirth, and Yves Poullet. Dordrecht: Springer.

Bellanova, Rocco, and De Hert, Paul. 2008. Data protection from a transatlantic perspective: The EU and US move towards an international data protection agreement? Study for the European Parliament's Committee on Civil Liberties, Justice and Home Affairs. Brussels.

Bradshaw, Simon. 2010. Cloud computing: Security and privacy aspects and cloud contract. Conference presentation, Ankara.

Bradshaw, Simon, Millard, Christopher, and Walden, Ian. 2010. Contracts for clouds: A comparative analysis of terms and conditions for cloud computing services. Queen Mary School of Law Legal Studies Research (Paper No. 63/201). London.

Clarke, Roger, and Stavensson, Dan. 2010. Privacy and consumers risks in cloud computing. *Computer Law and Security Review* 26 (4): 391–397.

De Busser, Els. 2009. Data protection in EU and US criminal cooperation: A substantive law approach to the EU internal and transatlantic cooperation in criminal matters between judicial and law enforcement authorities. Maklu Uitgevers N.V.

De Hert, Paul, and Gutwirth, Serge. 2009. Data protection in the case law of Strasbourg and Luxembourg: Constitutionalism in action. In *Reinventing data protection?* eds. Serge Gutwirth, Yves Poullet, Paul De Hert, Sjaak Nouwt and Cécile de de Terwangne ,3–44. Springer.

De Hert, Paul, Papakonstantinou, Vagelis, and Riehle, Cornelia. 2008. Data protection in the third pillar: Cautious pessimism. In *Crime, rights and the EU, the future of police and judicial cooperation*, ed. Martin Maik. London: Justice.

Dumortier, Frank, Gayrel, Claire, Poullet, Yves, Jouret, J., and Moreau, D. 2010. La protection des Données dans l'Espace Européen de Liberté, de Sécurité et de Justice. *Journal de Droit Européen* 166:33–46.

Gayrel, Claire, Gérard, Jacques, Moniy, Jean-Philippe, Poullet, Yves, Van Gyseghem, and Jean-Marc. 2010. Cloud computing and its implications on data protection. Paper for the council of Europe's Project on Cloud Computing, Centre de Recherche Informatique et Droit (Namur, March

2010). http://www.coe.int/t/dghl/cooperation/economiccrime/cybercrime/Documents/Reports-Presentations/2079_reps_IF10_yvespoullet1b.pdf.
Gellman, Robert. 2009. Privacy in the clouds: Risks to privacy and confidentiality from cloud computing. Paper prepared for the World Privacy Forum.
Grance, Tim, and Mell, Peter. 2009. The NIST definition of cloud computing (Version 15). http://csrc.nist.gov/groups/SNS/cloud-computing/. Accessed 10 July 2009.
Gutwirth, S., Poullet, Y., Hert, P. de, Terwangne, C. de, Nouwt, S. (Eds.). (2009). *Reinventing data protection?* The Netherlands: Springer.
Hijmans, Hielke. 2010. Data protection and international agreements in the area of law enforcement. Speech delivered at the conference on the area of freedom, security and justice in a wider world. The Hague.
Hijmans, Hielke, and Scirocco, Alfonso. 2009. Shortcomings in EU data protection in the third and the second pillars. Can the Lisbon treaty be expected to help? *Common Market Law Review* 46: 1485–1525.
Hustinx, Peter. 2009. Data protection in the light of the Lisbon treaty and the consequences for present regulations. Speech delivered at the 11th conference on data protection and data security. Berlin.
Hustinx, Peter. 2010. Data protection and cloud computing under EU law. Speech delivered at the third European Cyber Security Awareness Day. Brussels.
Leenes, Ronald. 2010. Who controls the cloud? Revista de Internet. Derecho y Politica 11.
Lichtblau, Eric, and Risen, James. 2006. Bank data is sifted by U.S. in secret to block terror. The New York Times.
Nelson, Michael R. 2009. Cloud computing and public policy. Briefing paper for the ICCP Technology Foresight Forum, Organization for Economic Cooperation Development.
Porcedda, Maria Grazia, and Walden, Ian. 2011. Regulatory challenges in a changing computing environment. Working paper for the conference "Law enforcement in the clouds: regulatory challenges". Brussels, Belgium. http://www.crid.be/cloudcomputing/default.htm. Accessed 24 Feb 2011.
Rodotà, Stefano. 1973. *Elaboratori elettronici e controllo sociale*. Bologna: II Mulino.
Rodotà, Stefano. 2005. Intervista su Privacy e Libertà. A cura di Paolo Conti. Laterza.
Rodotà, Stefano. 2009. Data protection as a fundamental right. In *Reinventing data protection?* eds. Serge Gutwirth, Yves Poullet, Paul De Hert, Sjaak Nouwt and Cécile de Terwangne ,79-80. Springer.
Scheinin, Martin. 2009. Terrorism and the pull of 'Balancing' in the name of security. In Law and security—facing the dilemmas, ed. Martin Scheinin, Florence: European University Institute Working Paper No. 11, 2009.
Schwerha, Jospeh J. IV. 2010. Law enforcement challenges in trans-border acquisition of electronic evidence from cloud computing providers. Discussion paper for the Council of Europe, Strasbourg, France.
Sibos Issues Thursday. 2009. The official daily newspaper of Sibos. Hong Kong. (14–18 Sept 2009)
Solove, Daniel J. 2007. 'I've got nothing to hide' and other misunderstandings of privacy. San Diego Law Review 44, GWU Law School Public Law Research Paper No. 289.
SWIFT. 2007. EDPS glossary. http://www.edps.europa.eu/EDPSWEB/edps/site/mySite/pid/87.
Taylor, Mark, Haggerty, John, Gresty, David, and Hegarty, Robert. 2010. Digital evidence in cloud computing systems. *Computer Law and Security Review* 26 (3): 304–308.
V. A. 2008. Cybercrime and cybersecurity in Europe. The European files.
Van Gyseghem, and Jean-Marc. 2008. 2008eHealth services and directive on electronic commerce 2000/31/EC. In *Proceedings of the HIT@HeathCare 2008 joint event: collection of studies in health technology and informatics* 141: 57–66.

Legal Instruments and Policy Documents

Agreement between the European Union and the United States of America on the processing and transfer of Financial Messaging Data from the European Union to the United States for the purposes of the Terrorist Finance Tracking Program. 2010. Official Journal L 195: 5–14.

Charter of Fundamental Rights of the European Union. 2000. Official Journal C 364: 1–22.

Consolidated versions of the Treaty on European Union (TEU) and the Treaty on the Functioning of the European Union (TFEU). 2010. Official Journal C 83 of 30 March 2010.

Council of Europe. 1950. ETS no 005, Convention for the protection of Human Rights and Fundamental Freedoms, as amended by Protocols No 11 and 14, Rome.

Council of Europe. 1981. CETS No. 108, Convention for the protection of individuals with regard to automatic processing of personal data. http://conventions.coe.int/Treaty/Commun/ChercheSig.asp?NT=108&CM=1&DF=&CL=ENG.

Council of Europe. 1987. Recommendation no R (87) 15 of the Committee of Ministers regulating the use of personal data in the police sector, council of Europe (Police Recommendation).

Council of Europe. 2001. Additional protocol to the convention for the protection of individuals with regard to automatic processing of personal data, regarding supervisory authorities and trans-border data flows, CETS No. 181

Council of Europe. 2001. Convention on Cybercrime, Budapest, CETS No. 105, 23 November 2001.

Council Decision. 2009a. 2009/426/JHA of 16 December 2008, Official Journal L 138:14–32.

Council Decision. 2009b. 2009/371/JHA of 6 April 2009, Official Journal L 121:37–66.

Council Framework Decision. 2008. 2008/977/JHA of 27 November 2008, Official Journal L 350:60–71.

Council Framework Decision. 2005. 2005/222/JHA of 24 February 2005, Official Journal L 69, 16/03/2005, 67.

Directive. 1995. 95/46/EC (Data Protection Directive) Official Journal L 281, 23.11.1995, 31.

Directive. 1998. 98/48/EC of 20 July 1998, Official Journal L 217:18–26.

Directive. 2000. 2000/31/EC of 8 June 2000, Official Journal L 178:1–16 (Directive on Electronic Commerce).

Directive. 2002. 2002/58/EC, Official Journal L 201, 31.07.2002, 37 (Directive on Privacy and Electronic Communications)

Directive. 2006. 2006/24/EC of the European Parliament and of the Council of 15 March 2006 on the retention of data generated or processed in connection with the provision of publicly available electronic communications services or of public communications networks and amending Directive 2002/58/EC Official Journal L 13.04.2006, 105:54–63.

Directive. 2009. 2009/136/EC of 25 November 2009, OJ L 337, 18.12.2009, p. 11–36, without the text 'to be transposed', which is now obsolete.

EDPS. 2007. Opinion on the role of the European Central Bank in the SWIFT case.

EDPS. 2010. Opinion on promoting trust in the information society by fostering data protection and privacy (Opinion on Privacy By Design). 14.

European Commission. 2002. (COM) 2002 0173 final, "Proposal for a Council Framework Decision on Attacks against Information Systems".

European Commission. 2010a. (COM) 2010 0609 final, "A comprehensive approach on personal data protection in the European Union".

European Commission. 2010b. COM (2010)385 final, "Overview of information management in the area of freedom, security and justice". Brussels.

European Commission. 2010c. COM (2010) 0245 final/2 "A Digital Agenda for Europe". Brussels.

European Council—an open and secure Europe serving and protecting citizens. 2010. Official Journal C 115, 4.5.2010, 47, 3.

European Court of Justice. 2009. C-301/06, Ireland vs. Council and Parliament.

European Court of Human Rights. 2002. Goodwin vs. UK (28957/95), judg.

European Network and Information Security Agency (ENISA). 2009. Cloud computing, benefits, risks and recommendations for information security.
European Privacy and Human Rights (EPHR). (2010). Privacy International, the Electronic Privacy Information Center (EPIC) and the Center for Media and Communications Studies (CMCS). (eds.) https://www.privacyinternational.org/article/european-union-privacy-profile.
EU-US Agreement on Extradition and Multilateral Legal Assistance. 2003. Official Journal L 181/34, 19 July 2003.
House of Lords, European Union Committee. 2008. The treaty of Lisbon: An impact assessment. 10th Report Session 2007–2008, 13 March 2008. http://www.publications.parliament.uk/pa/ld200708/ldselect/ldeucom/62/62.pdf.
Regulation. 2001. (EC) No 45/2001, Official Journal L 8:1–21.
Regulation. 2010. (EU) No 542/2010 of 3 June 2010, Official Journal L 155:23–26.
Tanaka Hiroyuki et. al. 2010. Transatlantic information sharing: at a crossroads. Washington: Migration Policy Institute (43 note 140).
The Article 29 Data Protection Working Party. 1998. Working document: Transfers of personal data to third countries: Applying Articles 25 and 26 of the EU data protection directive (WP 12).
The Article 29 Data Protection Working Party. 2001. Opinion 4/2001 on the Council of Europe's draft convention on cybercrime (WP 41).
The Article 29 Data Protection Working Party. 2006. Opinion 10/2006 on the processing of personal data by the society for worldwide interbank financial telecommunication (SWIFT; WP 128).
The Article 29 Data Protection Working Party. 2007. Opinion N. 4/2007 on the concept of personal data (WP 136).
The Article 29 Data Protection Working Party. 2010a. Opinion 3/2010 on the principle of accountability (WP 173).
The Article 29 Data Protection Working Party. 2010b. Report 01/2010 on the second joint enforcement action: Compliance at national level of telecom providers and ISPs with the obligations required from national traffic data retention legislation on the legal basis of Articles 6 and 9 of the e-privacy directive 2002/58/EC and the data retention directive 2006/24/EC amending the e-Privacy Directive (WP 172).
The Article 29 Data Protection Working Party. 2010c. Opinion 1/2010 on the concepts of 'controller' and 'processor' (WP 169).
The Article 29 Data Protection Working Party. 2010d. (WP 170) 2010–2011 work program can be consulted at: http://ec.europa.eu/justice/policies/privacy/docs/wpdocs/2010/wp170_en.pdf.
The Article 29 Data Protection Working Party and The Working Party on Police and Justice. 2009. 'The Future of Privacy': Joint contribution to the consultation of the European Commission on the legal framework for the fundamental right to protection of personal data (WP 168).

Chapter 11
Privacy Self-regulation Through Awareness?

A Critical Investigation into the Market Structure of the Security Field

Carla Ilten, Daniel Guagnin and Leon Hempel

11.1 Introduction

This chapter aims to provide a critical contribution to the ongoing discourse on self-regulation with regard to privacy and data protection (cf. e.g. European Commission 2010). This discourse encompasses the amendment of the EU Data Protection Directive and the related discussion about a principle of accountability (cf. Article 29 Working Party 2010). Underlying these conceptualisations is the assumption that data protection law is generally observed, but could be simplified and even reduced in favour of more self-regulatory approaches which are deemed more efficient.

We would like to raise critical questions about the institutional conditions and frameworks that greatly influence data controllers' potential and motivation for enacting privacy awareness and self-regulation; in other words, the market structures that these organisations operate within. An investigation into organisations' practices is indispensable in order to evaluate these current claims for self-regulation and to lay out the conditions that need to be met if market forces are to be harnessed for privacy and data protection.

The results and conclusions presented were gained in the course of the EU FP7 project "Privacy Awareness through Security Organisation Branding" (PATS). The project inquires into the possibilities of organisational self-regulation in the field of security technology and services by means of branding—understood as a complex, two-sided communication process between companies and stakeholders.[1] Specifically, research from the first three work packages is used. We started out with an analysis of current security regimes and actors, then interviewed representatives of

[1] The PATS project is funded from 2009 to 2012 and involves partners from Germany, the UK, the USA, Poland, Israel and Finland. The findings presented here are mainly based on the outcomes of the German team. The project website can be found at www.pats-project.eu.

C. Ilten (✉)
Centre for Technology and Society (ZTG), Technische Universität Berlin, Berlin, Germany
e-mail: ilten@ztg.tu-berlin.de

S. Gutwirth et al. (eds.), *European Data Protection: In Good Health?*,
DOI 10.1007/978-94-007-2903-2_11, © Springer Science+Business Media B.V. 2012

security organisations in detail about their privacy awareness and practice, as well as conducting a qualitative analysis of security organisations' communications and self-presentations.

The security field can be used as a burning lens to focus particular problems when it comes to the self-regulation of privacy and data protection: while the industry certainly represents a particular case when it comes to actor relationships, our analysis shows which questions need to be asked in order to understand existing structures of, and obstacles to, privacy protection. We argue that powerful obstacles lie in market structures that are obscure rather than a provider of incentives for self-regulation. These findings facilitate further thought about a principle of accountability with regard to the governance of privacy in different industries dealing with (personal) data. It is not enough to look at legal provisions and privacy statements when we want to assess the state of "health" of privacy and data protection in the EU—we need a thorough examination of the patient.

11.2 Security Regimes

The first work package was a research journey of all involved project partners into their respective national empirical fields: mapping the security regimes along the concepts of actors, technology and discourses. For this, we gathered on the one hand quantitative data about the security industry market and developed different qualitative types of security organisations; on the other hand we made a literature review of documents and articles about the development of the security field between 1989 and 2009. This section gives an account of the more general trends we have observed and which focus on the current debate surrounding the regulation of privacy in this sector.

11.2.1 Securitisation

Several discourses on security were identified during our research of current security regimes. A powerful, but creeping discourse concerns the broadening of security both as a term and as a political task. This development has been labelled "securitisation" in the academic discourse and has at the political realm enabled shifts in competences and power (cf. Buzan et al. 1998; for the German security regime, see also Lange et al. 2009; Singelnstein and Stolle 2006). Security is seen as a cross-cutting political issue that needs to be ensured in virtually every social sphere. The notion of a "right to security" propels the pursuit of security to a number one responsibility for the state (Isensee 1983). In Germany, this discourse was first associated with criminal theory but has been utilised by political interests of power extension and centralisation (Busch 2008; Lange and Frevel 2009). Under the title of "security vs. freedom", the shift of the political norm towards measures of securitisation has been discussed and the considerably weakened position of privacy values and other liberties observed (Heinrich and Lange 2009).

The most unquestioned discourse about "new threats" originated in the political realm and is tightly coupled to processes of globalisation and allegedly new forms of war after the end of the Cold War. This discourse has global scope and is taken up by both political and economic actors, especially after 9/11. It is a powerful narrative and justification for securitisation processes in the USA, but in most other countries analysed as well (Lyon 2006; Bukow 2005a).

Another manifestation of the extension of the security notion can be identified in what we called the "network paradigm". Originally coined and used by social scientists in response to socio-technical developments, the "network" term has seen a career beyond compare. It has been appropriated by many scientific communities dealing with organisational structures, politics and economic developments. Management literature has happily taken up the term, and it has become most common in describing social relations. Rooted in the fascination about the Internet and networking technologies in general, the term "network" could be translated with "up to date" or even "futurist". The discourse is used by many, if not all of the actors dealt with here. Yet, it proves most useful to those already most competent when it comes to networking: the companies we have identified as *Systems Integrators* in a security actors typology.

The network paradigm and the rhetoric of "new threats" are tightly coupled: The dissolution of borders, globalisation, new types of conflict or war have been bundled into one image by the 9/11 terrorist attacks in the USA. This focus event, singularly witnessed by millions through extensive media coverage, is probably present before everyone's eyes when "new threats" are mentioned, also in Germany. The invention of the term "Homeland Security" by the US government in the aftermath of the attacks and the instalment of a powerful institution of the same name is the consequence of the "new threat" discourse as well as a medium for safety and security convergence. The Homeland Security department is not only responsible for "Counterterrorism", "Preparedness, Response, Recovery", but also for "Border Security" and "Immigration". It thus includes safety from natural disasters in its security mission and subsumes immigration under the security aspect.

11.2.2 Privatisation

While most telecommunications and internet service providers have unintentionally become part of the security regime, many private actors—companies—benefit from the extension of security in general. A first major trend concerns the rising use of risk management and security measures on the part of companies and industries. Traditional security service companies offered services of locking, guarding and patrolling. With the continued increase in space occupied by industries, more protection has been engaged. Security services have also often been linked with building-related services such as cleaning and other forms of maintenance.

Concerning the notion of security, a qualitative shift has occurred with the introduction of IT in most industrial and service organisations: it has become a security issue and a sector of its own, extending the "security market" vastly. With growing networks and more complex supply chains through outsourcing and lean production, security of business, data, finance, etc. has come to be seen as one issue termed "business continuity". The rescue comes as a comprehensive systems solution from one hand, e.g. the large security service company or the systems integrating company, including risk management, services, and technologies. This development finds its expression in the emergence of a market for security consulting as a stand-alone product. Consultancies take on an intermediary role in the unregulated, diverse and thus confusing security market.

A second development concerns the shift in public and private spaces. Many places have—often unnoticed by the public—become private spaces. Whole infrastructures such as public transport are private, shopping precincts, banks and even streets are in the responsibility of their owners, yet used as, and perceived as, public spaces. The employment of private security services can thus be seen as the "natural" responsibility that comes with property (of space), a kind of "self-help" on the part of those who create these spaces (Feltes 2009; Newburn 2001). To the people who frequent these spaces, and often to the security actors themselves, it is far from clear where the responsibilities lie. At the same time, since security is not the prime function of the organisations using private spaces, it is always in competition with commercial interests. Highly symbolic and visible security measures such as video surveillance thus meet with more approval from the companies than the more expensive security staff. This problem of accountability and legitimacy becomes crucial when privacy and data protection come into view—if security is of secondary importance, privacy is considered to be even less relevant.

The type of outsourcing of security functions commonly perceived as privatisation is the fulfilment of core security functions through private companies in Public Private Partnerships (Morlok 2009). Here, it is not private but public space that is handed over to be secured through private actors. The requirements set by the public agencies are not much higher than otherwise—a point criticised by some actors within the market, because professionalisation processes stay slow. Still, the security service market leaders are prepared for Public Private Partnerships as they themselves are setting higher standards and approaching police quality in terms of education and appearance (von Arnim 1999).

With the blurring of safety and security concepts and functions, actors formerly concentrating on defence (and aviation) step into the civil security market more powerfully. Making intense use of the network paradigm and their experience in real-life missions, these companies now offer comprehensive solutions for the protection of critical infrastructures and crisis management and present themselves as the prime partner for the state when it comes to cooperation with private actors. In this regard, a capacity imbalance of public and private security providers is articulated. While public agencies now use private information infrastructures, they cannot keep pace with the original technological novelties. Large-scale sensitive projects such as the digital telecommunications network for security organisations are implemented by private companies.

To sum up, what is commonly termed "privatisation" is not a mere outsourcing of public functions, but a complex and multi-faceted development. An increase in private space (space privatisation)—industry and business representing an important share—also accounts for the involvement of private actors in security. At the same time, the state encroaches on private assets when security agencies make use of companies' infrastructures. Thirdly, an entirely new sector within security has emerged, adding to the capacity of private actors as compared to state capacities—the field of IT security, a major cross-cutting security issue. Considering these developments, it makes sense to speak first of an extension of the security regime in general—including both public and private actors—, and second of the qualitative extension and quantitative growth of a security market undergoing structural changes. Indeed, the "security market", as heterogeneous as it is, has attracted much attention from economically interested actors, especially in the field of technology.

11.2.3 Networked Security

The institutional vision of "networked security" which connects agencies and includes safety and security is complemented by the security technology oriented use of the term. Perceived changing threats are faced with converging solutions: "Many measures which were initially aimed against organised crime are by now used against international terrorism" (Bukow 2005b). What is more, measures are now aimed at terrorists, burglars and fire at the same time. Security technologies have undergone a process of convergence through digitisation, making new functionalities possible in interconnected systems (cf. Edwards 1996; Zanini and Edwards 2001).

Great hopes are set in the security technology market—mostly from an economic perspective, but from a rhetoric viewpoint and closely coupled to the new understanding of security. The security technology market is booming—at least according to the market overviews available and the self-description of the participants. Still, the market remains completely obscure and mostly arbitrarily defined. All kinds of technologies can be subsumed under "security" if the application indicates it, which is best shown with classic dual-use technologies. Biometric sensors, for example, are quite common in industrial quality management, but have been re-appropriated as a security technology. Security technology development is also generally supported well in terms of funding.

In such a dynamic market, as could be expected, actors try to get their share of the cake. Large economic players play the game—they make the most of existing discourses such as the network paradigm or extended security programmes. Our analysis has shown that many corporate players utilise security extension rhetoric in order to expand their business.[2] Market potential studies and an uncritical use of "new threat" rhetoric become self-feeding mechanisms. Since all technology can be appropriated for security uses, there is a wide field especially through convergence

---

[2] This is what we also found in the analysis of security communication, see Sect. 11.4.

of digital technologies such as IP video and biometrics. Systems integrators benefit from this development.

11.2.4 An Expanding Security Market

Against the backdrop of this general process of securitisation of political, legal and economic regimes and an expanding security market, notions of regulation shift when it comes to the problematic effects of security services and technologies on the people and the public under surveillance. Responsibility for the protection of privacy and data is being transferred to companies with clear for-profit goals and little intrinsic motivation to question the supremacy of security over privacy protection. The underlying assumption of most actors is that legal provisions are clear and sufficient to safeguard the data subjects' privacy and liberties.

There is clearly a contradiction between the goal of "networked" and "total" intelligence pursued and advertised by security companies—the general idea of feasibility and omnipotence—and the public and individual interest to preserve privacy and personal data protection, as well as just having "unobserved" spaces. Yet, when it comes to surveillance, attention focuses mostly on the state as the central actor and potential invader. Decentralised surveillance, delivered by private actors in private spaces such as public transport systems, is harder to discern and grasp in its entirety, or assess with regard to its effects. This is true both for the data subjects and regulating bodies, and the organisations themselves.

The transformation of the security field towards increasingly market-based relations leads to new questions about the governance of privacy and the efficacy of legal provisions (Bennett and Raab 2006). A closer look at the actual, day-to-day practices of security actors is, to this end, necessary. Discussions about new forms of more market-based regulation—"self-regulation"—cannot be led without a clear picture of the context and mechanisms—the market—that these organisations operate within.

While privacy is largely perceived as a "problem", and not an opportunity within the security industry, some developments suggest that there is room for privacy awareness raising within organisations: the targeted professionalisation of the security service market, a trend towards systems solutions including consulting and auditing (risk management), and the branding efforts of globally operating companies. Based on these potential opportunities attached to the hugely enhanced role of the private sector, the PATS project inquired into current levels of privacy awareness among security actors as part of the next research step.

11.3 Security Actors

In this section, we will take a closer look at the actors' practices, attitudes and awareness of privacy. The results presented here are based on 12 in-depth qualitative interviews with stakeholders from security organisations of the different types we discerned in the previous work package: technology producers, service providers, consultancies, research institutions and associations.[3] The main question during this research phase was how privacy is perceived by security actors, and how, in contrast with abstract legal norms, privacy and data protection are actually *practised* in organisational routines and operations. In other words: how does privacy figure in security actors' daily business lives and decisions?

In this section we argue that in practice there is a limited understanding of privacy and often very low awareness. This state of affairs is strongly related to actor constellations and their relationships within markets. These findings lead us to articulate criticism of the current market relationships which represent a less than "perfect" market—in particular, we face substantial problems with regard to the information about security needs and technologies as pointed out in the preceding section.

11.3.1 Organisational Practices

In general, we found a very limited understanding of privacy in security organisations. Privacy is mainly understood as data security—a rather technical understanding of privacy that neglects the democratic value of privacy and the basic principles of data parsimony and sensitivity. Privacy is thus reduced to organisational-technical issues of data processing and storage and is not dealt with on the level of business processes or decisions in general.

Another important practice is the reference to the existence of ISO standards and legal frameworks with the objective of shifting responsibility to those entities. These standards and legal frameworks are used as black boxes when used as an argument for not giving more thought to the related issues: "Why, but there is a data protection law!" The practices and routines regarding privacy and data protection are opaque even to the members of the organisations we interviewed. This becomes problematic when the unquestioning trust in the almost magical workings of legal provisions is accompanied by a reluctance to even discuss the topic—as privacy, so our interview partners argued, had surely been taken care of in some shape or form.

Another dimension of opacity lies in the fact that the organisational structures—which should enhance privacy compliance—depend on the actual practices of each company. For example, it makes a big difference as to whether data protection officers are employed full time or not, how well trained they are in data protection issues and how independently and proactively they can act within their company. As stated in interviews, the qualification of employees is indeed an issue; some actors are still trying to achieve basic legal compliance, which renders active engagement for data protection impossible and sheds a very critical light on ideas of self-regulation.

[3] The interviews were semi-structured, qualitative interviews which lasted from 1 h up to 3 h. All but one interview were conducted face-to-face and recorded. They were then transcribed or paraphrased closely. The analysis was done using the qualitative analysis tool Atlas.ti with a Grounded Theory approach.

In conversations, most of the representatives express their willingness to enhance privacy protection, but they feel that they face the described organisational problems and are limited in their sphere of action, because they have to act according to the needs, more specifically: the demand of the markets. This will be elucidated further in the following.

11.3.2 Privacy Awareness

While there are indeed individuals who wish to enhance the privacy practices within their organisations and who are aware of privacy problems and problematic structures, there is nevertheless a general lack of communication with the public about privacy issues—even when there is a real interest in providing and enhancing privacy within the business model. We found examples of security actors with a strong willingness to improve the privacy situation in relation to services or technologies offered. These interviewees stressed that trust is more important in the long run than instant economic profit, and that they offer data protection education in addition to their security products and services. Yet, according to a technology producer who offered specific Privacy Enhancing Technology (PET) options in combination with an IP camera product, there is little or no demand for these technologies and clients will not buy them as long as it is perceived as a costly "add on". This lack of client interest, along with what one interviewee called a "cat-and-mouse-atmosphere" when talking about data protection issues, seems to lead to a situation where companies do not feel like communicating about privacy in the public domain. It seems like putting oneself in danger for no reason.

This difficult relationship between privacy practice and privacy communication becomes evident when we look at companies that went through privacy scandals. From our interviews, it emerged that data leakage or misuse scandals hit the clients of security (technology) providers, not necessarily the security companies themselves. When misuse becomes publicly known, these organisations mostly show two reactions: either they begin to talk publicly about their privacy efforts or they avoid any (further) publicity about data protection. For the former however—intense communication on privacy efforts—it was reported that organisations try to achieve formal law abidance to "safeguard the management board from claims".[4] This is illustrated by companies that set up entire compliance departments to purify their reputation, notwithstanding the efficacy of these measures. Reputation is an important asset especially in regard to investors' trust, but engagement spurred by this motivation does not surpass a pragmatic attitude towards data protection and privacy. The communication aims to present a good image regardless of the real effectiveness of data protection measures and related practices.

The second common reaction to scandals is the avoidance of further image damage through the avoidance of any communication about privacy related issues, which

[4] See interview 2, line 46.

against the backdrop of the "accountability" discourse seems to be a questionable strategy. Companies that stay silent about their surveillance projects clearly impact their security technology providers' behaviour. Not only are suppliers less than encouraged to enhance their privacy performance, but they are also asked to keep a low profile. This is in stark contrast to ideas of self-regulation or even building a positive image by stressing one's outstanding privacy performance.

11.3.3 The Actors and the Market

To revisit the findings so far: There are intransparent structures which lead to a certain degree of opacity. Responsibility is shifted to institutions such as data protection law or data protection officers, quality standards or—as we will point out in the next section—even technology (e.g. PET). We want to argue here that the market, which is invocated as a source for regulation by the "invisible hand", reflects this opacity and is far from constituting a regulative framework. The current market structures do not relay market pressure or incentives towards more privacy protection to the companies in charge. On the contrary, it seems that the regulating power of the security market weakens privacy as a consequence of the actual relationships.

According to our outcomes we face (1) conflicting interests of different actors, (2) a tendency to hold citizens accountable notwithstanding their constrained possibilities to influence or participate security organisations and their clients' business behaviour, and, maybe most problematic, (3) a total lack of representation of citizens/data subjects and of any information directed towards this group.

The low demand for privacy tools is rooted in the market setup: the clients are interested in (cheap) surveillance technologies, not in citizen rights. It is important to understand the supplier-client relationship here: if we think of clients as those paying for security products and deploying them in their facilities, they provide the demand for security technologies—and are legally held responsible as "data controllers". The suppliers are security technology producers or security service providers offering their products to this market of clients, e.g. public transport companies, airports, other companies or institutions.

Which role does the citizen, public transport passenger, or employee take on in this constellation? The data subject is a client of the security organisations' clients—or even a dependant, e.g. in an employment relationship. The relationship is thus not always a voluntary one based on market forces. Even if we concede consumers some market power in respect of their choice of e.g. surveilled or non-surveilled supermarkets, their power is very low. Sheer selection forces do not go far here; for example, in order to avoid public transport due to the use of CCTV, one has to opt out of the system and use alternative transportation means. It becomes difficult to walk the streets without being captured by any camera, or even realise in whose space—public, private?—one is moving about and whose camera is watching—so in this case, how can consumers possibly exert market influence by pure selection? Accordingly, the actor we expect to demand privacy—the data subject—is utterly uninformed and cannot easily exert influence within the market of security technologies and services. In a sort of pre-emptive move, many interviewees from the security field hold citizens accountable for infringements of their privacy with reference to the fact that they use Google and Facebook—the great icons of voluntary data deluge—and take part in rebate marketing. This attitude suggests that "the horse has already bolted" and is combined with an affirmation of consumers' choice. The assumption that ICT users themselves generally lack privacy awareness is both implicitly and explicitly mentioned, often alleging a generational difference and genuinely new culture of "digital natives" that knows no privacy concept. At the same time the public's and citizens' demand for security is taken for granted and articulated over and over e.g. when it comes to security on public transport where violent events receive a lot of media attention.

In the current communication of the European Commission, the problem of the citizen's burden of being held accountable is addressed with the claim of enhancing the transparency of e.g. privacy notices, replacing opt-outs with opt-ins, and strengthening the power of the users (European Commission 2010).

However, it is questionable as to how internal market regulations can be enhanced to strengthen privacy efficacy when we are facing an utter non-representation of the citizen within the markets. Our findings pertain to the specific case of the security market, but we hold it to be indicative of the general lack of information and transparency when it comes to the much heralded market-based regulation of privacy in other industries (Social Network Sites).

11.4 Security Communication

To round off the perspective we will now give an insight into the security communication of security organisations, based on the analysis of material from security fair, brochures, websites and several issues of a security journal.[5] Notably we find a special mode of communication: the self-representations are strictly oriented to the clients of the specific market. Accordingly the analysis shows which values are communicated and how security is constituted in the security branch (see Fig. 11.1).

11.4.1 Economic Value and Invisibility

The most obvious kind of narratives we find is the presentation of economic values and a general feeling of happiness. The latter is mainly communicated with images of happy people, which are obviously happy because they are secured and protected by technologies and services. Organisations try to communicate that economic value

[5] We collected pictures from stands and brochures at the fair "Security Essen 2010", material from website presentations and ten issues of a security related stakeholder journal between 2009 and 2010. For the analysis, we used Atlas.ti to find common narratives in the self-representations of the organisations, and coded the material using a Grounded Theory approach.

Fig. 11.1 Samsung: Total security solutions—happiness and prosperity

is actually secured through security services and technologies. Economic value is shown both as private home property and in a business context. Remarkably, economic value is sometimes encased with the notion of ethical values such as in the slogan "protecting values".[6] Obviously in the material the threats are hardly shown; yet the economic value and people take centre stage.

For example on the poster from Samsung, smiling well dressed people walk through a stylised financial district. They are happy and busy; they use their cell phones. There is no visible threat; security technology does not even feature in the picture. The threat is completely absent while the slogan is "Total Security Solutions. Beyond your imagination." Only in this slogan is the issue of security made explicit. Yet, no-one appears to take notice of threats or the security technologies. The picture also implies that security is, rather ironically, a precondition for the freedom to move. The message is "freedom through security", meaning that those who are allowed to move have to be "secured" whereas the fact that most people in the world are not allowed to move as they want, and security technologies enhance their exclusion, is not worth mentioning in this poster. Being secured means in this context being scanned and categorised as either a trusted or mistrusted person. Beyond that, this co-constructs the idea that people who are not allowed to move freely are dangerous and have to be excluded.

[6] Quote from a poster of a company named Orfix at "Security Essen 2010".

Fig. 11.2 Bosch: Forensic search in a game-like setting

11.4.2 Total Security and Convergence

At the same time, the "Total Security Solutions" term symbolises another evident narrative; Samsung is offering integrated system solutions, and this sort of product is focusing on the technological promises we find connected with "*networks*" and interoperating systems. We find these lines of argumentation linked to a modern belief in technological possibilities which is not at all aware of critical reflections and the limitations of technologies. It is rather the co-constructing of black boxes that leads to social causes and implications being neglected (see Figs. 11.2 and 11.3).

Continuously we found the theme of the "blessing" technologies, mainly computer analysis tools, which were presented and touted in various ways. We called this theme also *cybernetic*, because it refers to the discourse in the second half of the 20th century, and it seems like a very uncritical dream of almighty computer systems giving men the power and the control over the world. Total security is often presented in a game-like manner at the security fair, when companies want to illustrate the great possibilities provided by technologies in their stands.

The common argumentation line of *cybernetics* is that artificial intelligence promises to regain control over the flood of data. Beyond that we also find a reference to science which strengthens the connection to the modern age and cybernetics: "imageology—the science of surveillance".[7] It is here that technology is constructed both as a data emerging tool and at the same time it gives humans the power to keep an overview through its own intelligence. Consequently it is a double solution

[7] Quote from a Bosch poster at "Security Essen 2010".

Fig. 11.3 Security buyer: Sheltered baby

(seek and control), while humans are neither able to gain all the information nor to keep track over it without technology. In other words, data control supports the vision of crowd control.

Summarising this narrative suggests empowering humans to deploy a ubiquitous surveillance setting which focuses on prevention instead of reaction.

11.4.3 Naturalisation

Imagery of natural settings and natural metaphors is frequently used by the security organisations. In some cases we find a direct comparison with nature such as the "organic" functioning of technologies, systems and organisations. Often, nature is taken as a model for technologies. Many pictures show nature to describe security situations and to construct a certain feeling. For example, on a Honeywell poster there is a picture of a nearly closed shell combined with the slogan "closing the security gaps." Again there is no explicit reference to threats, but a focus on the solution. At the same time it is clearly connoted that threats are a natural problem. The social character of security as a societal concept is completely neglected. Communication of this quality supports an irrationalisation of the discourses. Security is presented as a natural need—and natural facts cannot be discussed.

More subtle than the described nature imagery is the naturalisation of social hierarchies. We find images of mothers protecting their child and of families in their safe home; a clear reference to the "natural" hierarchies of protection.

In all, this stands in line with the argumentation that first, threats are a natural phenomenon and second that security is a natural need which has to be taken care of. Naturalisation is here an argument of determinism, which consolidates the actual relations and neglects social reasons and causes that underlie the challenges facing security.

11.5 Conclusion

We have argued that the market structures in the security field are obscure to the extent that no incentives for self-regulation are perceived by the actors involved. Security actors are clearly interested in making a profit and do not have sufficient intrinsic motivation to kick-start self-regulation. Demand for more attention to privacy would have to be forced upon these actors, but no one currently articulates this demand within the market.

Not only are market relationships indirect, but citizens and the public are rarely even represented in the market at all. Privacy cannot translate into a means of monetary regulation in the marketplace in this set-up. What is more, security companies actively support obscuring discourses about threats and security through their communication strategies of naturalisation and invisibility. Security and privacy are rendered "unspeakable" through these opaque imageries, and public discourse about privacy is further hindered.

This investigation into organisations' practices has shown that current claims for self-regulation need to be backed up by research into the conditions that have to be met if market forces are to be harnessed for privacy and data protection. Institutional conditions and frameworks greatly influence data controllers' potential and motivation for enacting privacy awareness and self-regulation. These structures need to be known in detail in order to make statements about self-regulation prospects and goals in specific sectors.

In particular, internal market regulations cannot be enhanced to strengthen privacy efficacy when we are facing a total non-representation of the citizen or the data subjects within the markets. An important issue to raise within the current self-regulation discourse is thus how, hitherto, under-represented actors can be shifted into a more powerful position within "self-regulating" markets, and which mechanisms need to be implemented in order to make market forces "work" towards privacy protection.

References

von Arnim, Andreas. 1999. Private security companies and internal security in Europe. In *Recht und Organisation privater Sicherheitsdienste in Europa*, ed. Reinhard Ottens, Harald Olschok, and Stephan Landrock, 61–112. Stuttgart: R. Boorberg.

Article 29 Working Party. 2010. *Opinion 3/2010 on the principle of accountability*. Brussels: Article 29 Working Party. http://ec.europa.eu/justice/policies/privacy/docs/wpdocs/2010/wp173_en.pdf. Accessed 28 July 2011.

Bennett, Colin J., and Charles D. Raab. 2006. *The governance of privacy: Policy instruments in global perspective*. 2nd and updated ed. Cambridge: MIT Press.

Bukow, Sebastian. 2005a. Deutschland: Mit Sicherheit weniger Freiheit über den Umweg Europa. In *Europäisierung der inneren Sicherheit,* ed. Gert-Joachim Glaeßner and Astrid Lorenz, 43–62. Wiesbaden: VS Verlag.

Bukow, Sebastian. 2005b. *Internal security and the fight against terrorism in Germany*. Philosophische Fakultät III Institut für Sozialwissenschaften, Humboldt Universität Berlin. http://edoc.hu-berlin.de/oa/conferences/reZgVweZSLueU/PDF/27QEOto3iuZCs.pdf. Accessed 30 July 2011.

Busch, Heiner. 2008. Kein Mangel an Sicherheitsgesetzen. *FriedensForum*. http://www.friedenskooperative.de/ff/ff08/6–61.htm. Accessed 30 July 2011.

Buzan, Barry, Ole Waever, and Jaap de Wilde. 1998. *Security: A new framework for analysis*. Boulder: Lynne Rienner.

Edwards, Paul. 1996. *The closed world: Computers and the politics of discourse in Cold War America*. Cambridge: MIT Press.

European Commission. 2010. *Communication from the Commission to the European Parliament, the Council, the Economic and Social committee and the Committee of the Regions—A comprehensive approach on personal data protection in the European Union*. Brussels: European Commission. http://ec.europa.eu/health/data_collection/docs/com_2010_0609_en.pdf. Accessed 20 July 2011.

Feltes, Thomas. 2009. Akteure der Inneren Sicherheit: Vom Öffentlichen zum Privaten. In *Auf der Suche nach neuer Sicherheit: Fakten, Theorien und Folgen*. 2nd ed., ed. Hans-Jürgen Lange, H. Peter Ohly, and Jo Reichertz, 101–109. Wiesbaden: VS Verlag.

Heinrich, Stephan, und Hans-Jürgen Lange. 2009. Erweiterung des Sicherheitsbegriffs. In *Auf der Suche nach neuer Sicherheit: Fakten, Theorien und Folgen,* ed. Hans-Jürgen Lange, H. Peter Ohly, und Jo Reichertz, 253–268. Wiesbaden: VS Verlag.

Isensee, Josef. 1983. *Das Grundrecht auf Sicherheit. Zu den Schutzpflichten des freiheitlichen Verfassungsstaates*. Berlin: Walter de Gruyter.

Lange, Hans-Jürgen, H. Peter Ohly, and Jo Reichertz. 2009. *Auf der Suche nach neuer Sicherheit: Fakten, Theorien und Folgen*. 2nd ed. VS Verlag.

Lange, Hans-Jürgen, und H. Peter Frevel. 2009. Innere Sicherheit im Bund, in den Ländern und inden Kommunen. In *Auf der Suche nach neuer Sicherheit: Fakten, Theorien und Folgen,* ed. Hans-Jürgen Lange, H. Peter Ohly, und Jo Reichertz, 116–148. Wiesbaden: VS Verlag.

Lyon, David. 2006. 9/11, synopticon, and scopophilia: Watching and being watched. In *The new politics of surveillance and visibility*, ed. Haggerty, Kevin. D. and Ricard V. Ericson. Toronto: Univ. of Toronto Press.

Morlok, Martin, and Julian Krüper. 2009. Sicherheitsgewährleistung im kooperativen Verfassungsstaat. In *Auf der Suche nach neuer Sicherheit: Fakten, Theorien und Folgen,* ed. Hans-Jürgen Lange, H. Peter Ohly, und Jo Reichertz, 331–342. Wiesbaden: VS Verlag.

Newburn, Tim. 2001. The commodification of policing: Security networks in the late modern city. *Urban Studies* 38:829–848.

Singelnstein, Tobias, and Peer Stolle. 2006. *Die Sicherheitsgesellschaft. Soziale Kontrolle im 21. Jahrhundert*. VS Verlag.

Zanini, Michele, and Sean J. A. Edwards. 2001. The networking of terror in the information age. In *Networks and netwars: The future of terror, crime, and militancy*, ed. John Arquilla and David Ronfeldt, 29–60. Santa Monica: Rand.

Part III
Concepts and Prospection

Chapter 12
Privacy Penetration Testing: How to Establish Trust in Your Cloud Provider

Christian W. Probst, M. Angela Sasse, Wolter Pieters, Trajce Dimkov, Erik Luysterborg and Michel Arnaud

12.1 Introduction

In the age of cloud computing, IT infrastructure becomes virtualised, and all aspects of the stack of hardware, platform and software take the form of services. Moreover, these services can be offered by different organisations, which may purchase their capacity from again different organisations. The complexity of who owns, possesses, controls and uses information increases dramatically (Floridi and Turilli 2011).

In this sense, cloud computing forms an instance of the broader concept of de-perimeterisation (Jericho Forum 2005; van Cleeff and Wieringa 2009). De-perimeterisation denotes the disappearing of boundaries around the IT infrastructure of organisations. Whereas information security was previously conceived as separating the trusted *inside* from the untrusted *outside*, such a clear delineation is not possible anymore. The question is what can take its place, i.e., how re-perimeterisation would be possible.

Of course, security has never been completely based on a single perimeter. People working for an organisation would leave the perimeter in their private lives, enabling information to cross the boundary between the organisation and its surroundings. This has become more prominent with the use of mobile devices in the workplace, or "*bring-your-own-device*". Also, the inside of the organisation might not have been completely trusted, as there would always be a chance that people inside the organisation would misuse their credentials for their own benefit. This so-called insider threat has become a substantial area of research (Hunker and Probst 2011).

In this sense, it is not surprising that the notion of a security perimeter has broken down. Developments like cloud computing have only made more explicit that such a concept is untenable, and accelerated the emergence of different security architectures. The original idea that the perimeter is as close to the data as possible (data-level security) cannot be the only solution in cloud-computing environments, as full encrypted processing is not feasible (Pieters 2011b). Instead, we are now

---

M. Arnaud (✉)
Université Paris Ouest Nanterre La Défense, Paris, France
e-mail: Michel.arnaud@u-paris10.fr

S. Gutwirth et al. (eds.), *European Data Protection: In Good Health?*,
DOI 10.1007/978-94-007-2903-2_12,

looking at complicated re-perimeterisation processes, where different layers of protection, in different domains (digital, physical and social) need to be aligned for all organisations involved (Abrams and Bailey 1995). For example, to prevent confidential data from leaving an organisation, we would have to prevent outsiders from gaining access to the building, prevent employees from taking the data home, and check outgoing e-mail traffic. We would need to perform the same checks at the cloud provider. Worse, some attacks may combine weaknesses in different domains to circumvent the carefully crafted multi-perimeters, or "virtual perimeters".

Especially, private and sensitive data requires special protection when being stored or processed in a cloud infrastructure. Organisations want to have some confidence that the benefits of moving to a cloud environment outweigh the risks. People accept risk and uncertainty in exchange for an expected benefit, but as the cloud infrastructure is not transparent for the user, this requires trust in the providers and their security policies. The more sensitive the data in question is, the better and stronger guarantees are required when the data is being stored or processed in a cloud infrastructure—or more trust. Because of the difficulties of cross-organisational security assessment, this trust may be hard to justify.

The questions are thus how to empower cloud users to develop trust in cloud infrastructures, how to align security policies to form a reliable perimeter within one's own organisation, and how to trust and/or verify the security measures in place in other organisations in the cloud network? After discussing security challenges in the cloud in the next section, we first look at the question of trust into cloud infrastructures in Sect. 1. This leads to the suggestion of PuPPeT, a public privacy penetration-testing agency, in Sect. 4. In this section we also discuss how to test security policies and how to verify security measures; since the suggested agency will have to act across organisations, we introduce cross-domain methods for security testing and for modelling organisational security. The present contribution brings together these different factors in securing data in the age of the cloud, for which open questions are discussed in Sect. 1, followed by conclusions in the final section.

To simplify discussion, we will in the following use the term "cloud operator" for organisations offering a cloud infrastructure, the term "cloud user" for organisations running cloud applications operating on their customers' data, and "data owner" for organisations and individuals using cloud applications.

12.2 Security Challenges in the Cloud

When considering the security impact of adopting a cloud-computing environment, opinions regarding the exact nature of the "cloud threat" differ quite substantially.

Some state that there is really nothing new under the sun, and that, especially with respect to a private cloud environment, the security-related issues are the same as those existing in a "non-cloud" enterprise today (Robinson et al. 2011). Some state that, because of the nature of the cloud itself (i.e., difference in scale of entry points to potentially be subject to attacks), the security risks are clearly of a different nature or type (Jansen and Grance 2011). Others talk more about a difference in scale, not in type of threat (Mitra and Mallik 2010).

However, when the exact nature of challenges in the cloud needs to be quantified, there is one thing almost everyone agrees upon: cloud computing does pose a number of real challenges in terms of security, privacy, and trust, both for cloud providers and cloud users.

Indeed, because cloud computing grew out of an amalgamation of technologies, e.g., virtualisation, Web 2.0 and service-oriented architecture, the related security, privacy and trust issues should be viewed as already known problems in a new setting. However, it is the importance of their combined effect that should not be underestimated. Therefore, in order to propose an appropriate response to the threats related to cloud computing, it is necessary to first understand and define properly the exact challenges in this regard.

In general, when people talk about ensuring security, they refer to integrity, access and availability of data, as well as confidentiality and non-repudiation. Privacy, on the other hand, embraces much more; it is often seen as primarily being about compliance with applicable data protection laws and regulations relating to, e.g., data transfer or location, purpose of processing and data subject rights of access and control. But privacy is much more than data protection, for example, it is also about observable behaviour and anonymity. One could say that data protection only provides the means of protecting privacy, but they need to be used in the right way.

When addressing privacy in the cloud, two aspects must be distinguished: on the one hand, applications running in the cloud should protect the privacy of the data they process; on the other hand, cloud providers should protect the data that is stored or that is processed on their infrastructure. These requirements are not new; the first one is the same as privacy protection in every other application, and the second one is the same as for regular hosting companies. In cloud computing, the risk just is amplified by the multitude of outsourced components and, for example, the uncertainty about location of data.

Therefore, the above concepts need to be further refined and clarified in order to be fully understandable in the cloud context. We propose to add the following clarifications to the existing concepts. Please note that some of these can apply many times but for sake of clarity, we have listed them only once. They are also valid for both cloud users as well as cloud providers.

12.2.1 Security Challenges and Granularity

Security challenges in relation to the cloud environment can (non-exhaustively) be categorised as lack of control on the provider's resources, increased exposure of internal infrastructure via new technologies/interfaces, insufficient adaptation of application/platform security and development lifecycle, unclear ownership of security tasks and lack of cloud specific security standards, to list some.

The above demonstrates that the main security challenge can be translated into one of *granularity*. In other words, in order to understand the full scope of the cloud security challenges, one needs to identify at which level of granularity one can identify the relevant security threats. This will largely depend on criteria such as, e.g., the type of data concerned, the scale of outsourcing, the number of third parties involved, the architecture/technology used, etc. Another important factor is the extent in which cloud providers offer customised services as opposed to standardised ones. The customised approach will allow to better master the security issues in a more adapted manner, also addressing the issue of attribution of responsibilities between the different parties involved.

12.2.2 *Privacy and Accountability*

Data privacy generally refers to a bundle of legal/contractual rights and obligations related to the collection, processing and sharing (transferring) of personal information. Although several definitions exist, one of the most comprehensive definitions of personal information is included in the so-called 1995 European Data Protection Directive 95/46/EC:

> Personal information is any information relating to an identified or identifiable natural person ('data subject'); an identifiable person is one who can be identified, directly or indirectly, in particular by reference to an identification number or to one or more factors specific to his physical, physiological, mental, economic, cultural or social identity.

One could argue that this definition means for cloud computing that most data stored in the cloud will be personal information, resulting in the above-mentioned directive being applicable. This means that somebody in the conglomerate of maybe several cloud users and cloud providers collectively processing the data is responsible for protecting its privacy. However, this responsibility may be hard to assign in practice.

Typical privacy issues that are mentioned in connection with a cloud environment are data localisation and applicable law, data breach/leakage issues and data transfers. Clearly different concerns exist when outsourcing customer data to the cloud versus, for example, outsourcing an organisation's business records.

Even though the current privacy legislative framework is far from ideal, and even though often very divergent privacy laws and regulations exist, rendering difficult the handling of data in the cloud, in reality all of these hurdles are not insurmountable. They can indeed be summarised in the challenge of "accountability". Given the volume and/or location of the different cloud service providers, it will be crucial to establish clear rules regarding the (contractual) responsibilities for privacy compliance by any of the parties in the (cloud) chain. As such, and using the terminology of the data protection regulations, clearly identifying the data flow as well as the roles of each data controller, processor and sub-processor, and where they/the data are located/restricted to, will go a long way in ensuring compliance with the applicable privacy laws and (contractual) rules.

12.2.3 Trust and Transparency

Finally, one of the most difficult challenges in cloud computing is to enable customers to develop trust in a cloud environment. In a cloud environment, one of the key questions from individuals and companies is: can I trust the cloud with my data? To answer this question, we need to first examine what "trusting" the cloud means. We only need trust in situations with risk and uncertainty—people accept risk and uncertainty in exchange for an expected benefit.

With cloud computing, the expected benefit for the user of a cloud computing service, e.g., a medium-sized enterprise, is reducing cost and increasing reliability. The risks associated with cloud computing include availability and integrity (*Will I always be able to access the data when I need them, and will they be the data I stored?*) and confidentiality (*Might someone working at the cloud provider or another client get access to my customer's personal data?*). Uncertainties surrounding cloud computing include questions such as whether the provider will do what they promise (such as not transferring the data outside the EU without explicit consent), and whether there is any redress and restitution if they fail to deliver.

When deciding whether to trust someone, humans usually consider two qualities: the trustee's *ability* and *motivation* to deliver their side of the transaction. In terms of ability, cloud providers argue that data storage and processing is their core competence, which means they are better equipped to keep data secure than most of their customers—*trust us, we're the professionals*. In terms of motivation, Pearson and Charlesworth (2009) argue that cloud computing providers should be highly motivated to safeguard their customers' data, since their reputation depends on it. In a system where customers and providers can trust each other to deliver what their transaction partner expects, all parties can expect to benefit (Riegelsberger et al. 2005). So—is it time to stop worrying, and learn to trust the cloud?

Taking the above into account, the trust challenge can be summarised in one word: transparency. Indeed, establishing a level of trust or confidence about a cloud environment largely depends on the ability of the cloud provider to demonstrate clearly and upfront (and on a regular basis thereafter) the provision of security and privacy controls required to protects the client's data and applications, as well as any evidence to demonstrate the effectiveness of such controls.

12.3 A Pragmatic Trust-based Approach

To pick up our earlier question—maybe it *is* time to stop worrying and trust the cloud. But given that enterprises turn to cloud computing to save money, it makes sense for cloud providers to feel compelled to compete on price. Such competition could lead to cloud providers trying to save on parts of the services that are regarded as non-essential. Whilst customers would note problems with availability in day-to-day usage, effective security and privacy protection manifest themself as *absence* of security breaches. A cut in expenditure on protecting security and privacy does not necessarily lead to a breach—or at least not immediately. So, there is a likely temptation to save on this protection to offer more competitive prices. Once there has been a breach, a cloud provider's reputation will suffer, but by then, for the cloud user that entrusted its data to that cloud provider, the damage to the enterprise, and its customers, is done. So the question is—how can cloud users tell apart the cloud service providers that take good care of their data and safeguard their customers' privacy, and those that do not?

Unfortunately, there are no reliable trust signals that would allow cloud users to infer whether cloud providers are acting as they promise (Riegelsberger et al. 2005). This means that, rather than trusting cloud providers, they have to put assurance mechanisms in place, such as contracts, inspection of facilities, and testing the security and privacy protection measures in place (Flechais et al. 2005). However, such assurance mechanisms introduce cost for both the cloud user and the cloud provider—meaning neither can reap the full financial benefits of a trust relationship.

So, the answer to our earlier question is that we can learn to trust the cloud, but not without investing in the necessary assurance mechanisms. To be effective, these mechanisms need to address the challenges introduced in the previous section.

Adopting a granular approach means demanding a more customised service adapted to the "sensitivity" level of the data processed or services requested. Companies should not only employ specific security controls to verify the correct functioning of the various subsystems in the cloud environment, they should also ensure strong and adapted security management practices adapted to their changed role.

Current privacy laws and regulations are what they are: they are far from ideal and should be further improved and more harmonised. Meanwhile, both cloud users and cloud providers need to comply with this current legal framework. In order to achieve this, both need to clearly attribute accountability to each of the intermediaries for compliance with all relevant (contractual and legal) rules related to, e.g., location of data, data transfer, data usage and data breach procedures in relation to its role and responsibility in the cloud "chain".

Finally, cloud providers must be able to demonstrate in a clear and transparent way that they implement the above-mentioned assurances and approach. At the same time, the cloud user must accept that its role (especially that of its IT department) has changed and entails certain governance responsibilities as well. As such, a trustworthy relationship can be created, not by assuring that everything can be guaranteed in a bullet-proof fashion, but by ensuring that a flexible framework exists whereby data will be protected in a manner consistent with agreed upon policies, procedures and contractual arrangements and that adequate redress or alert procedures are in place. To support this process, we suggest PuPPeT, a public privacy penetration-testing agency.

12.4 PuPPeT—A Public Privacy Penetration-Testing Agency

It is obvious that the principles or concepts of granularity, accountability and transparency apply to any of the above-mentioned security, privacy or trust challenges and are highly intertwined. We believe that they are key to ensuring a properly configured, well balanced and secure cloud environment, thereby allowing both the cloud user as well as the cloud provider to fully exploit the potential benefits of the cloud. They also illustrate that securing the cloud is not only a matter of mere technology, but also a combination of people, processes and technology.

Institutional safeguards, such as regulation, could offer protection, but regulation always lags behind technology, and has not caught up with cloud computing (Pearson and Charlesworth 2009). Additionally, cloud computing is an international business, which means that it is often beyond the regulator's reach. One approach is to rely on self-regulation of markets (Hirsch 2011).

Pearson and Charlesworth make a compelling argument that the solution for this problem is accountability of the cloud provider to their customer enterprises (cloud users). In the case of privacy, the elements for accountability for privacy are (Pearson and Charlesworth 2009):

1. *Transparency*: informing data owners how their data is handled in the cloud, and who has responsibility for which parts of processing;
2. *Assurance*: through privacy policies;
3. *Responsibility*: must be clearly allocated, and taken actively by the cloud provider (rather than relying on enforcement by regulators or cloud users); and
4. *Policy compliance*: rather than following the letter of policies, cloud providers must strive to achieve a proportionate and responsive process for reacting to context-dependent privacy risks.

Pearson and Charlesworth further suggest that these privacy-protecting controls should be built into different aspects of the business process, and cloud users and cloud providers must work together over time to develop robust routine protection of privacy in the cloud. This approach mixes trust and assurance, but remains very much on the assurance side, meaning that the cost for both sides remains substantial.

To overcome this, we suggest PuPPeT, a public privacy penetration-testing agency. We envision PuPPeT to be a more economic alternative to the process sketched above. The agency would award a trust symbol for cloud computing providers that cloud users and data owners can use to make an informed decision about whether or not to trust a cloud provider (Balboni 2009). To award the trust symbol, the agency would perform unannounced security audits and checks—a kind of "privacy penetration-testing".

The agency would be funded by enterprises using cloud computing, but be cheaper than traditional assurance through contracts. It would provide an incentive to keep cloud providers honest in the face of price competition, and is likely to detect problems before they lead to a privacy breach. If enterprises have to pay more for this service for the more sensitive data they place in the cloud, it would provide an incentive for them to minimise the amount of sensitive data they put out there, and thereby limiting the amount of risk they take on behalf of their clients.

The biggest issue is how the agency can actually test whether a cloud provider complies with privacy laws. The rest of this section will discuss some aspects of testing socio-technical aspects of security, but this is only part of the story. The other part is an evaluation of the infrastructure, processes in place, etc. One important requirement is that the agency must ensure, that these evaluations actually arc conducted, and repeated at random intervals to ensure the results' validity. The results of agency evaluations must be available publicly, to allow cloud users and data owners to access, e.g., comments and development of evaluations.

It is important to note that the agency would only be able to test and evaluate the security and privacy measures in place at a cloud provider. Questions such as local jurisdiction being able to force a provider to give access to data might be noted in the agency's report, but per se cannot be part of the seal-decision process, since they are independent of the quality of privacy measures.

Other privacy-relevant questions that are related to the application run by the cloud user on the cloud provider's infrastructure cannot be part of the evaluation either.

12.4.1 Socio-Technical Security Testing

When an organisation decides to work together with a cloud provider, thereby investing a certain amount of trust as described above, the organisation needs to adapt its security and privacy protection measures to accommodate for the new scenario that non-organisation owned premises become part of the organisation's premises, and that non-organisation staff becomes enabled to access the organisation's data. These scenarios did already exist before cloud computing, e.g., with hosted computing and outsourcing, but the promise of cloud computing is that outsourcing becomes an easy-to-use service, and that data can relocate between different machines, countries, continents and (at some point) also providers, without the data owner noticing.

To protect their resources, organisations usually develop security and privacy measures in a top-down manner. The high-level policies describe the desired behaviour of the employees (social domain), the physical security of the premises where the employees work (physical domain), and the IT security of stored and processed information (cyber domain). After the high-level policies have been designed, the individual departments, often with help of a company-wide security department, refine these policies into implementable, low-level policies. These should be enforced through physical and digital security mechanisms as well as employee training. For example, to make sure that data stored on laptops does not end up outside the organisation, policies may be put in place on encryption, physical access to offices, as well as bringing in guests.

Assessing whether the organisations' policies address all identified threats, and whether they are correctly enforced, consists of two steps: *auditing* and *penetration testing*. During the auditing process, auditors systematically check whether proper security policies are defined at all levels and ensure that the policies in place address the identified threats. After the auditing process, penetration tests are used to check whether the policies are enforced without error and whether the policies follow the design specifications.

Both auditing and penetration testing are mature fields in information security and follow methodologies that aim for reliable, repeatable and reportable results.

To address cloud computing they must be extended, e.g., to implement the privacy-penetration testing suggested above. However, the attention paid to the physical and social domain by these methodologies is limited. Unfortunately, adversaries do not limit their actions only to the digital domain, but they use any weak link they can find regardless of the domain. The lack of methodologies for auditing and testing the alignment of security policies across all three domains makes organisation vulnerable to attacks where the adversary combines physical, digital and social actions to achieve a goal. These cross-domain attacks are even more significant in cloud-computing environments than in standard IT infrastructures, since an organisation's perimeter now includes the cloud provider's premises, its IT infrastructure and staff, all providing new attack vectors into the system.

These problems are further aggravated when organisations have to deal with distributed perimeters or the aforementioned de-perimeterisation caused by cloud-computing infrastructures. In these cases policies need to address much more complex scenarios, since the different *domains* now need to be considered in different *perimeters* as well. The same holds for auditing and penetration testing of policies.

A typical example for an attack that cannot easily be found by evaluating policies only at one level is the so-called "road apple attack":

> An attacker leaves a number of dongles with malicious software in front of the premises of an organization. Employees will take dongles, some of them will plug them into their computer, some of which will not be sufficiently protected, and on some of the thus infected machines the malicious payload will find interesting data to encrypt and send with the employee's credentials.

This attack clearly combines elements from different domains (unawareness of employee, inability to check for dongles, inability to check encrypted traffic) that make it hard to detect, but also hard to audit for. To mitigate this attack we need to apply a combination of policies that are coordinated between different stakeholders. Whilst the likelihood of an attack like this on a cloud provider hopefully is rather small (after all, *they are the experts*), a cloud user itself might imagine its data is safe in the cloud, but might still be attackable since the data needs to be transferred to the local machines to work on them.

Once low-level policies have been defined, they need to be enforced using security mechanisms, and this might result in mistakes. Technicians might put the wrong lock on a door, an employee might ignore or forget some of the policies, or some computers might be misconfigured and, for example, might still accept remote connections. Therefore security departments need to be able to test whether the security policies are properly implemented. These tests should include attempts of gaining physical access to restricted areas, as well as attempts in tricking employees to violate a policy (Dimkov et al. 2010b).

Whilst these tests already are hard to apply in a thorough way for traditional scenarios, issues get worse when we consider cloud computing and its additional challenges. We can expect privacy-penetration testing to work well for testing cloud providers' compliance with privacy regulations to a certain extent as discussed above; however, because of likely differences in tools, languages and ontologies used in different organisations, it will in general be impossible to test the alignment of the

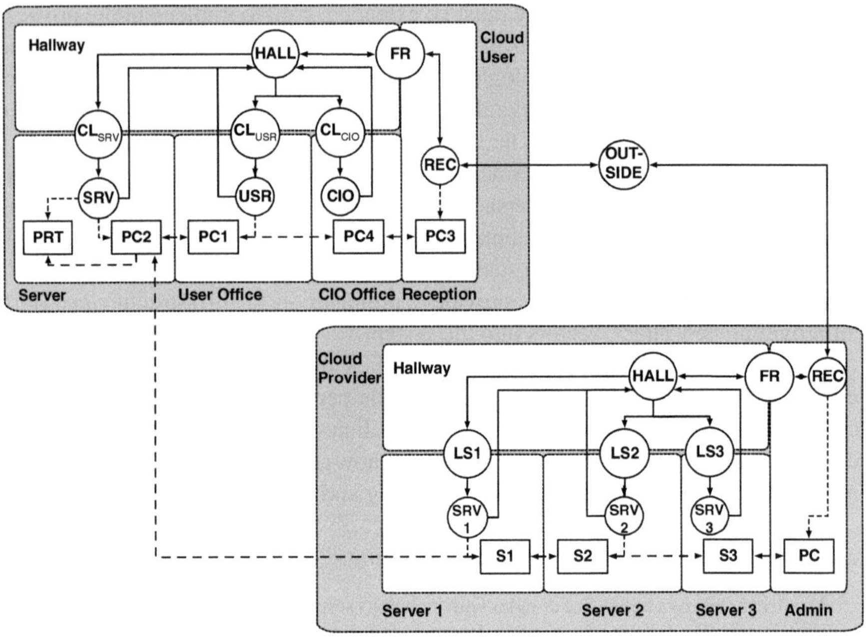

Fig. 12.1 A system model based on ExASyM (Probst and Hansen 2008) illustrating how the physical level (*solid lines*) and the cyber level (*dashed lines*) interact. The model combines a company (*upper left*) with a cloud provider (*lower right*), and represents physical and cyber infrastructure. This model forms the basis of generating attacks based on policies and access control specifications, and can be used for guiding privacy penetration testing

providers' policies with the organisation's policies, and whether the former are in accordance with the latter. On the other hand, providers who are willing to cooperate with organisations to conduct social penetration testing as described above may be able to obtain higher ratings in a quality evaluation.

In the end, the cloud user will need to trust to a certain extent in the cloud provider's will and credibility to enforce certain policies—the goal must be to minimise the gap between the real risk faced by the organisation and the risk it is willing to accept (Probst and Hunker 2010). The suggested public penetration-testing agency is one tool for organisations to evaluate how big a risk they need to take, or how much trust they can have in their cloud provider.

12.4.2 Socio-Technical System Models

To allow for systematic approaches to testing of information infrastructures, including cloud-computing service architectures, we need models for describing the interesting aspects of the system in question. In the penetration test described above, such systems models (Fig. 12.1) can be used to automatically develop attack scenarios to be executed in the tests. The benefit of this approach is that it takes into account the actual system of technologies, physical infrastructures such as buildings, and people in a systematic way (Dimkov et al. 2010a,b; Probst et al. 2006).

System models are specific tools within the framework of organisational security-policy alignment; making sure security policies adequately address the goals they were put in place for. In cloud scenarios, this involves alignment of policies *between* organisations as well. Policy alignment aims at ensuring that policies are aligned horizontally, with policies at the same abstraction level, and vertically, with policies at different abstraction levels.

When defining a set of high-level policies, two problems arise: the policies might conflict with each other, or there might be situations for which no policy is defined, resulting in threats not being addressed. Horizontal alignment of policies aims at assuring that high-level policies are consistent and address as high a percentage of threats as possible. When introducing new policies they need to be checked for consistency with existing policies, and for adequacy in protecting against the attacks they were meant to address.

Ideally, high-level policies and low-level policies should allow and forbid the same behaviour. Vertical alignment of policies aims at refining high-level policies to low-level policies whilst ensuring that the latter faithfully implement the former. It is this vertical alignment of policies that system models aim to address, by testing the infrastructure with its low-level policies against the targets expressed by the high-level policies. It is then for example verified if, within the constraints represented by the low-level policies, it is possible for sensitive data to leave the premises. By describing policies in system models, it can thus be verified whether higher-level policies are satisfied. When low-level policies allow behaviour that violates a high-level policy, an attack scenario is produced. Such an attack scenario can then be used as input for the penetration tests.

Although the low-level policies developed in the departments may be complete when restricted to a single domain, when combined with policies from other security domains the combination may not necessarily be complete as well. Thus, a number of actions allowed in one domain may lead to an attack when combined with allowed actions from other domains. In order to support attack scenario generation, models need to be able to describe not only the technical aspects of the system, such as infrastructure, policies, and access-control specifications, but also sociological aspects, such as potential behaviour of employees (Probst and Hansen 2008; Dimkov et al. 2010a; Pieters 2011a). Using this additional information, attacks on the modelled infrastructure can be generated. These represent misalignments between high-level policies and low-level policies.

Using models that include likelihood of certain events to happen, it becomes possible to include descriptions of the less predictable human behaviour into reasoning. The models can then be used to estimate the *risk* of attacks, namely the probability of success with the losses incurred when the attack succeeds, and attacks can be ranked based on the risk. The losses incurred, often called the *impact* of an attack, can be calculated based on the value of the information assets that are affected.

The important benefit of using models and tools for generating attacks is twofold. First, tools can explore also large system specifications in systematic ways, and guarantee a thorough investigation, resulting in a list of possible attacks. Second, and this is especially important when considering cloud computing, one can combine models from different sources to obtain a holistic view of the overall system. This guarantees that the penetration tests performed by the PuPPeT agency cover all possible weaknesses.

Again this is a special problem when considering policies that are defined within an organisation, policies that are defined at the cloud provider, and policies that are defined between the two. It is because of the increased possibilities for misalignment in a multi-organisational context that inter-organisation penetration testing becomes even more important. This means that cloud providers could provide a model of their system, which could then be used by privacy penetration testers to guide the testing process. Since such a system model should be considered sensitive information, we expect that it either is shared only between the cloud provider and the agency performing the privacy penetration testing, or it is developed by the agency as part of a kind of certification process.

12.5 Open Questions

So far we have discussed security and privacy protection-related issues in the cloud and cloud-based applications as faced by cloud providers and cloud users. One of the biggest problems is that of trust into the protection of sensitive data, and an awareness of privacy issues when storing data in the cloud. The suggested privacy penetration-testing agency PuPPeT could help in addressing these points. Of course, there still remain a lot of open questions with respect of protection of privacy of data stored in the cloud.

There seems to exist an inherent contradiction between the rationale of cloud computing—to compute on data where it can be done more efficiently and therefore cheaper—and the requests to ensure privacy of personal and confidential data. Constant tracking to avoid *any* leak or abusive use is technically unfeasible. The big goal remains to combine the seemingly contradicting goals of reducing costs and ensuring security and privacy. As discussed before, the incentive to be able to save money often will be more tempting than the obligation to protect privacy. To overcome this, adjusting the cost for obligatory privacy penetration testing based on the sensitivity of data may be a promising approach.

Data location is an important issue for legal protection. The European Commission seems inclined to keep personal data being processed in cloud computing on European territories, but the question is, how feasible this requirement is—making the location of data redundant is one of the big promises of the cloud, and in most frameworks it is difficult at best to limit data's location (van Cleeff et al. 2010). This is also difficult from an auditing point of view.

However, even if we were able to solve these two problems, there remain other, equally important questions. Once data is stored in the cloud, how do we secure cloud-computing systems against breaches? Because of the stack of technologies used in cloud-computing infrastructures, they also offer new, increased attack surfaces, and as before we need to develop security procedures that can mitigate the threats resulting from these.

The ultimate goal, however, must be to enable end users to have confidence that their data is protected when being stored in the cloud—either by them or by organisations. To this end, there is a significant need for privacy frameworks for cloud applications that ideally should embrace different cloud providers. In the long run this would help to ensure that storing data in the cloud could be considered safe.

A public cloud that offers on-demand services to a wide population of users must take relevant compliance mandates with utmost responsibility to minimise the risk of breaches of data privacy—or risk loss of business due to bad publicity and lack of trust. To achieve this high level of data protection, identity management technologies such as authentication, authorisation, user management, compliance and others are paramount:

- Users must be strongly authenticated to validate their identity;
- Web-based Identity Federation to ease the authentication process should be available;
- Up to date access rights must be checked against cloud application's access control policies;
- All user interactions must be logged to ensure non-repudiation;
- User accounts must be de-provisioned in a timely manner;
- Dormant accounts must be identified and removed quickly; and
- Access permissions must be certified on a continuous basis.

To date, many of these points require explicit actions, which results in untimely responses and consequently vulnerabilities.

Future research clearly should address these points, and try to proactively develop protection and detection mechanisms. We expect to see an increasing number of vulnerabilities in cloud computing that we need to be able to handle. On the one hand systems will be ported to the cloud, which have not been developed for cloud computing, thereby being exposed to threats that were not relevant in the original development. On the other hand, once we know how to address infrastructure vulnerabilities, we expect to observe new threats on the application level, threats that are enabled by the cloud infrastructure.

12.6 Conclusions

Cloud computing is offering new opportunities for public and private organisations to get access to IT infrastructure. A traditional, cloud-based environment offers quick and cost-effective access to technology using a browser. This brings agility to enterprises and improved satisfaction to end users, whilst lowering overall costs. We have argued above that this promise of cost reduction is very likely to result in increased uncertainty about security as well. This is the case since reduced cost is easy to identify (by reading ones invoice), whilst reduced security and privacy protection first becomes obvious once a breach has happened.

We therefore see the need of strict controls and rules to be applied in cloud computing to meet the requirements for efficient personal data protection. European laws must evolve to regulate this new computing approach, and we propose PuPPeT, a privacy penetration-testing agency, to facilitate this.

An independent agency would be an important step in the right direction. It would signal industry and their clients that public agencies realise the risks and take them serious. It would signal customers, by means of a privacy seal, whether or not they should trust cloud providers. And it would signal end users whether the companies they interact with use trustworthy providers. The combined effect of these signals would be that customers are empowered to decide whether or not they are willing to trust in a cloud provider.

Whilst we strongly believe that the suggested privacy seal issued by a privacy penetration-testing agency is an important step in the right direction, a word of warning seems in order. As Edelman (2011) has noted, among online shops the services accredited with trust certificates are more than twice as likely to be untrustworthy as uncertified sites. This is why we believe that a European agency with comprehensible, publicly documented standards and publicly available testing results is essential in guaranteeing privacy of data stored in the cloud.

Another challenge of cloud computing is the increased amount of third-party infrastructure that organisations need to rely on. By letting go of the infrastructure, managing security risk becomes thus an even more important task than before, requiring a joint effort between the client and cloud provider. Here the modular approach described above might be a viable solution to enable companies to evaluate the risk of including a certain provider's infrastructure.

References

Abrams, Marshall D., and David Bailey. 1995. Abstraction and refinement of layered security policy . In *Information security—An integrated collection of essays*, ed. Abrams, Marshall D., S. Jajodia and H.J. Podell, 126–136. New York: IEEE Computer Society Press.

Balboni, Paolo. 2009. *Trustmarks in e-commerce*. The Hague: Cambridge University Press.

van Cleeff, André, Wolter Pieters, and Roel J. Wieringa. 2010. Benefits of location-based access control: A literature study. *Proceedings of the 3rd IEEE/ACM International Conference on Cyber, Physical and Social Computing (CPSCom 2010)*. Hangzhou: IEEE Computer Society.

van Cleeff, André, and Roel J. Wieringa. 2009. Rethinking de-perimeterisation: Problem analysis and solutions. *IADIS International Conference Information Systems*, 105–112. Barcelon: IADIS.

Dimkov, Trajce, Wolter Pieters, and Pieter H. Hartel. 2010a. Portunes: representing attack scenarios spanning through the physical, digital and social domain. *Proceedings of the Joint Workshop on Automated Reasoning for Security Protocol Analysis and Issues in the Theory of Security*

(ARSPA-WITS'10). Revised Selected Papers, 112–129. Lecture Notes in Computer Science (6186). Springer Verlag.

Dimkov, Trajce, André van Cleeff, Wolter Pieters, and Pieter H. Hartel. 2010b. Two methodologies for physical penetration testing using social engineering. *Proceedings of the Annual Computer Security Applications Conference (ACSAC)*, 06–10 Dec 2010, Austin, 399–408.

Edelman, Benjamin. 2011. Adverse selection in online "trust" certifications and search results. *Journal Electronic Commerce Research and Applications* 10, (1):17–25.

European Data Protection Directive. 1995. Directive 95/46/EC of the European Parliament and of the Council of 24 October 1995 on the protection of individuals with regard to the processing of personal data and on the free movement of such data. *Official Journal* L 281.

Flechais, Ivan, Jens Riegelsberger, and M. Angela Sasse. 2005. Divide and conquer: The role of trust and assurance in the design of secure socio-technical systems. *Proceedings of the 2005 Workshop on New Security Paradigms, NSPW' 05*. New York: ACM.

Floridi, Luciano, and Matteo Turilli. 2011. Cloud computing and its ethical challenges. Paper presented at the Workshop on New Ethical Responsibilities of Internet Service Providers. Hatfield.

Hirsch, Dennis D. 2011. The law and policy of online privacy: Regulation, self-regulation, or co-regulation? *Seattle University Law Review* 34 (2). http://ssrn.com/abstract=1758078. Accessed 1 Sept 2011.

Hunker, Jeffrey, and Christian W. Probst. 2011. Insiders and insider threats, an overview of definitions and mitigation techniques. *Journal of Wireless Mobile Networks, Ubiquitous Computing, and Dependable Applications* 2 (1): 3–25.

Jansen, Wayne, and Timothy Grance. 2011. Guidelines on security and privacy in public cloud computing, Draft NIST Special Publication, National Institute of Standards and Technology.

Jericho Forum. 2005. Jericho whitepaper. http://www.opengroup.org/projects/jericho/uploads/40/6809/vision_wp.pdf. Accessed 1 Sept 2011.

Mitra, Sramana, and Saurabh Mallik. 2010. Thought leaders in cloud computing: Interview with Mark White, CTO of Deloitte (Part8). www.sramanamitra.com. Accessed 1 Sept 2011.

Pearson, Siani, and Andrew Charlesworth. 2009. Accountability as a way forward for privacy protection in the cloud. Proceedings of the 1st International Conference on Cloud Computing, CloudCom' 09. Berlin: Springer.

Pieters, Wolter. 2011a. Representing humans in system security models: An actor-network approach. *Journal of Wireless Mobile Networks, Ubiquitous Computing, and Dependable Applications*, 2 (1): 75–92.

Pieters, Wolter. 2011b. Security and privacy in the clouds: A bird's eye view. In *Computers, privacy and data protection: An element of choice*, ed. Serge Gutwirth, Yves Poullet, Paul De Hert and Ronald Leenes, 445–457. Dordrecht: Springer.

Probst, Christian W., Rene Rydhof Hansen, and Flemming Nielson. 2006. Where can an Insider attack. Proceedings of the 4th international conference on Formal aspects in security and trust, FAST' 06. Springer.

Probst, Christian W., and Rene Rydhof Hansen. 2008. An extensible analysable system model. *Information Security Technical Report*, 13 (4): 235–246.

Probst, Christian W., and Jeffrey Hunker. 2010. The risk of risk analysis and its relation to the economics of insider threats. In *Economics of information security and privacy*, ed. Tyler Moore, David Pym and Christos Ioannidis, 279–299. Springer.

Riegelsberger, Jens, M. Angela Sasse, and John D. McCarthy. 2005. The mechanics of trust: A framework for research and design. *International Journal of Human-Computer Studies* (Elsevier) 62 (3): 381–422.

Robinson, Neil, Lorenzo Valeri, Jonathan Cave, Tony Starkey, Hans Graux, Sadie Creese, and Paul Hopkins. 2011. The cloud: Understanding the privacy and trust challenges, RAND Europe, Technical Report, 2011.

Chapter 13
Review of the Data Protection Directive: Is There Need (and Room) For a New Concept of Personal Data?

Mario Viola de Azevedo Cunha

13.1 Introduction

The entry into force of the Lisbon Treaty brought many changes to the European Union (EU) legal framework and some of them have a direct impact on data protection. From an institutional point of view, the abolition of the pillar structure allows a greater involvement of the European Parliament in the area of the former third pillar. According to Article 87(2) of the Treaty on the Functioning of the EU, the Council and the Parliament may adopt—on the basis of the ordinary legislative procedure—measures related to "*the collection, storage, processing, analysis and exchange of relevant information*" between the police authorities of member states. For instance, the decision of the European Parliament rejecting the SWIFT agreement with the U.S. on the transfer of banking data to U.S. authorities in the fight against terrorism shows that the Parliament will not be afraid to exercise its new veto power (European Parliament 2010).

Furthermore, the entry into force of the Lisbon Treaty also marked a turning point in relation to the protection of the right to privacy and personal data. These two fundamental rights are expressly recognised by Articles 7 and 8 of the Charter of Fundamental Rights. The Lisbon Treaty included an express reference to the Charter in Article 6 of the Treaty on European Union (TEU). According to Article 6(1) TEU, "*The Union recognises the rights, freedoms and principles set out in the Charter of Fundamental Rights of the European Union of 7 December 2000, as adapted at Strasbourg, on 12 December 2007, which shall have the same legal value as the Treaties*". Thus, despite the fact that the text of the Charter has not been included in the Treaty of Lisbon, it has become a binding document. Moreover, Article 51(1) of

---

Mario Viola holds a PhD in Law and a Master of Research in European, International and Comparative Law from the European University Institute and an LLM in Private Law from Rio de Janeiro State University.

M. Viola de Azevedo Cunha (✉)
European University Institute, Via Boccaccio 121, 50133 Firenze, Italy
e-mail: Mario.cunha@eui.eu

S. Gutwirth et al. (eds.), *European Data Protection: In Good Health?*,
DOI 10.1007/978-94-007-2903-2_13,

the Charter establishes that it has to be taken into account by European institutions when carrying out their legislative activities.

Finally, a new legal basis for the regulation of the processing of personal data was introduced by the Lisbon Treaty. Article 39 of the TEU requires the adoption of a more comprehensive instrument for data protection (European Commission 2010). Nevertheless, Declaration 21 on the protection of personal data in the fields of judicial and police cooperation in criminal matters, annexed to the final act of the intergovernmental conference which adopted the Treaty of Lisbon, recognises that

> *specific rules on the protection of personal data and the free movement of such data in the fields of judicial cooperation in criminal matters and police cooperation, based on Article 16 of the Treaty on the Functioning of the European Union may prove necessary because of the specific nature of these fields.*[1]

In the same sense is the abovementioned Article 39 of the TEU, with regard to the areas of Common Foreign and Security Policy, which authorises the Council to

> *adopt a decision laying down the rules relating to the protection of individuals with regard to the processing of personal data by the Member States when carrying out activities which fall within the scope of this Chapter, and the rules relating to the free movement of such data.*[2]

This is a great opportunity for the adoption of single concepts and legal instruments which would provide general rules for all types of processing of personal data within the EU. In fact, it is not by chance that the European Commission recently launched a public consultation on the Commission's comprehensive approach on personal data protection in the EU (European Commission 2010) and announced that it would put forward new (and comprehensive) legislation this year (Reding 2011, 5).

In this scenario, the concept of personal data is of pivotal importance. In effect, it is the foundation of the EU legislation in the field of data protection and has to be in the centre of any discussion about the future of the Directive or of the comprehensiveness and coherence of the EU legislation in this matter (European Commission 2010, 5).[3]

The trend present in the vast majority of laws concerning the protection of personal data, including Directive 95/46 is to consider anonymous or statistical data as the exact opposite of personal data and, therefore, not subject to the law which regulates the protection of personal data (Walden 2002, 235).[4]

In a time where "*Information has become the new raw material of the world economy*" (Kuner 2003, 29) and of the fight against terrorism and organised crime,[5] a harmonised concept of personal data can play an important role in preventing undesirable barriers to free flow of personal information within the EU, which is

---

[1] Available at http://eur-lex.europa.eu/LexUriServ/LexUriServ.do?uri=OJ:C:2010:083:0335:0360:EN:PDF, Accessed 3 Aug 2011.

[2] This article is included in Chap. 2 of the Treaty, entitled 'Specific Provisions on the Common Foreign and Security Policy'.

[3] "The concept of 'personal data' is one of the key concepts for the protection of individuals by the current EU data protection instruments (. . .)."

[4] In this sense are the Spanish and Austrian Data Protection Laws (see Walden 2002, 235)

[5] See, for instance, the debates about the PNR and SWIFT Agreements.

one of the main objectives of Directive 95/46/EC.[6] Differences in the concept of personal data in the member states can create situations where a specific processing of data would be covered by the data protection legislation in one state and not covered in another, leading to the creation of constraints for the development of many activities which rely on personal information, especially in a cross-border scenario, where data controllers would have to adapt themselves to different legal requirements. As highlighted by Nugter, the lack of convergence of data protection laws in a transnational scenario, as the one of an economic bloc, "*leads to additional costs, administrative and organizational problems, or may even lead, though in practice only occasionally, to a total prohibition*" of data flow amongst the countries involved and, of course, "*creates uncertainty for those who are dependent on the free flow of personal data*" (Nugter 1990, 320).

Taking into account this scenario, this chapter initially analyses the concept of personal data as provided for by Directive 95/46 and the views of the Article 29 Working Party and of the European Data Protection Supervisor regarding the concepts of personal data and anonymous data. Then, it concentrates on the experiences of France, Italy and UK, seeking to identify the differences in the concept of personal data in these member states. After carrying out these analyses, the chapter proposes some changes on the concept of personal data which could be incorporated in the review of the General Data Protection Directive.

It is important to highlight that the current EU concept of personal data was included in Directive 95/46/EC more than 15 years ago and since then many technological developments were put in place, including new techniques in the field of re-identification of anonymised data, which by itself would justify a reassessment of the definition of personal data. Moreover, the experiences of the EU member states suggest that an improvement of the concept of personal data provided for by the Directive is needed, not only to meet the challenges posed by new technologies but also by other situations that have surfaced with the practice of the member states in enforcing data protection legislation.

13.2 The Concept of Personal (and anonymous) Data in the EU[7]

Before starting the analysis of the selected EU member states, it is important to discuss the concept of personal data as provided for by Directive 95/46/EC. Article 2(a) of the Directive defines personal data as "*any information relating to an identified or identifiable natural person ('data subject')*" considering an identifiable person

[6] The main purposes of Directive 95/46/EC are: "*(1) to allow for the free flow of data within Europe, in order to prevent member states from blocking inter-EU data flows on data protection grounds; and, (2) to achieve a harmonized minimum level of data protection throughout Europe*". (Kuner 2003, 27)

[7] Some parts of this section were included in a previous work, published in Italian (see Viola de Azevedo Cunha et al. 2010).

"*one who can be identified, directly or indirectly, in particular by reference to an identification number or to one or more factors specific to his physical, physiological, mental, economic, cultural or social identity*" (Kuner 2007, 92).[8]

The Article 29 Working Party (2007, 6) in its opinion on the concept of personal data considered that:

> *(...) the concept of personal data includes any sort of statements about a person. It covers "objective" information, such as the presence of a certain substance in one's blood. It also includes "subjective" information, opinions or assessments. This latter sort of statements make up a considerable share of personal data processing in sectors such as banking, for the assessment of the reliability of borrowers ("Titius is a reliable borrower"), in insurance ("Titius is not expected to die soon") or in employment ("Titius is a good worker and merits promotion").*

Besides, it is also important to define the concept of anonymous data, both for statistical or data protection purposes. The Directive, despite not defining anonymous data, adopted the position that "*the principles of protection shall not apply to data rendered anonymous in such a way that the data subject is no longer identifiable*".[9] The doctrine points for a definition in the same direction, considering as anonymous data that which could not be linked to an identified or identifiable individual (Gediel and Corrêa 2008, 144) and some EU member states have adopted legislation which goes in the same direction.[10] The characterisation of anonymous data, however, is fundamental, in the sense that information which cannot be related to an identified or identifiable person lead to the non-application of data protection rules. Many member states have included in their national legislations a procedure called anonymisation of personal data, which is requirement for the "free" processing of such data in specific circumstances (French Act 1978).

Nevertheless, the distinction between personal data and anonymous data is not absolute and the validity of anonymous data as an exception to the application of data protection rules has been put into question (Information Commissioner's Office 2001, 13),[11] since through some modern techniques of data processing it is possible to recover the link between the anonymous data and the respective data subject (Murray 1997, 63).[12] Back in 2000, a research conducted by the computer

---

[8] "The requirement that the data relate to an 'identifiable' person in the General Directive similarly means that a set of data which, taken together, could be matched to a particular person, or at least make identification of that person considerably easier, is considered 'personal data'."

[9] Recital 26 of Directive 95/46/EC.

[10] Article 4(1)(n) of the Italian Personal Data Protection Code (Legislative Decree no. 196 dated 30 June 2003) considers anonymous data as "*any data that either in origin or on account of its having been processed cannot be associated with any identified or identifiable data subject.*"

[11] "*The Commissioner considers anonymisation of personal data difficult to achieve because the data controller may retain the original data set from which the personal identifiers have been stripped to create the 'anonymised' data.*"

[12] Regarding the anonymisation of genetic data the situation is even more complicated (see Murray 1997, 63). "*If a database contained sufficient information about the sequence, even if the person's name were not attached to the file, it might be possible to identify the individual whose sequence it is, in a manner similar to the method of genetic fingerprinting. So, although the practise of removing*

scientist, Latanya Sweeney proved that through the combination of three "anonymous information" (zip code, sex and birthday) it was possible to identify 87% of North American citizens (Sweeney 2000). In a recent research, Paul Ohm argues that anonymisation techniques are misleading, because there is a wide range of possibilities of re-identification of personal data, that is, through the use of statistical and mathematical techniques and the combination of different databases it is possible to link the anonymous information to a data subject (Ohm 2009). According to Ohm, every information can become personal information if combined with other relevant information (although anonymous).

It is worth noting that many IT systems are built taking into account the fact that the processing of personal data after the use of anonymisation techniques would be free from the incidence of the data protection legislation. Such approach, on the one hand, could be considered as the application of Privacy by Design principles, but, on the other hand, it could lead to many risks, because data considered as anonymous would be processed until being linked to their data subjects.[13]

The importance of this subject led to a discussion about the distinction between anonymous data for statistical purposes and anonymous data for data protection purposes. The Article 29 Working Party in its Opinion 1/2008 on data protection issues related to search engines stated that for data to be considered as anonymous and, therefore, out of the scope of the General Data Protection Directive, the anonymisation has to be complete and irreversible, in a manner that renders the data subject unidentifiable (Article 29 Working Party 2008, 20).[14] Such complete anonymisation, however, even if possible, is a hard task, both from a technical viewpoint and in terms of adequacy to data protection rules (Walden 2002, 226).[15]

The European Data Protection Supervisor in some of his opinions developed a clear distinction between "anonymous" or "anonymised" data, for data protection purposes, and "statistical anonymous data". The former would be "*any information relating to a natural person where the person cannot be identified, whether by the data controller or by any other person, taking account of all the means likely reasonably to be used either by the controller or by any other person to identify that individual. Anonymised data would be anonymous data which previously referred to an identifiable person, no longer identifiable.*" From a statistical point of view, anonymous data "*are data for which no direct identification is possible. This definition implies that the possibility of indirect identification would still qualify the*

---

identifying information is usually thought to confer anonymity by making records impossible to trace to an individual, that may not be the case with records containing significant chunks of DNA sequence data."

[13] An example of the risks is the software produced by Phorm, called WebWise, which was hardly criticized by data protection advocates (see Clayton 2008).

[14] In its opinion 4/2007, the Article 29 Working Party presents a definition of anonymous data that take into account 'the means likely reasonably to be used' for the identification of the data subject (see Article 29 Working Party 2007, 21).

[15] "*Achieving effective anonymisation may be a challenging task, from both a technical and compliance perspective. Sophisticated data analysis and data mining techniques on supposedly anonymous data may eventually yield data that does 'directly or indirectly' relate to a specific individual (. . .).*"

data concerned as anonymous from a statistical point of view, but not necessarily from a data protection point of view" (European Data Protection Supervisor 2008, 4; 2007, 4).[16]

Regarding the issue of a person being "*identified or identifiable*", the Article 29 Working Party, in its opinion on the concept of personal data (Article 29 Working Party 2007, 12), tried to establish a distinction between these two concepts. For the Working Party "*a natural person can be considered as 'identified' when, within a group of persons, he or she is 'distinguished' from all other members of the group. Accordingly, the natural person is 'identifiable' when, although the person has not been identified yet, it is possible to do it (that is the meaning of the suffix '-able').*"

Directive 95/46/EC itself, in its recital 26, presents an indication on how to determine whether an individual is identifiable or not:

> *(...) to determine whether a person is identifiable, account should be taken of all the means likely reasonably to be used either by the controller or by any other person to identify the said person (...)*

This reasonableness referred to by the Directive is the key element in establishing the threshold for considering a data as related to an identified or identifiable person and, as a consequence, personal data.[17] As highlighted by the Article 29 Working Party (2007, 13), "*This means that a mere hypothetical possibility to single out the individual is not enough to consider the person as 'identifiable'*".

Recommendation R (97)5 of the Committee of Ministers of the Council of Europe, on the protection of medical data, in its Article 1(1) states that data cannot be considered identifiable "*if identification requires an unreasonable amount of time and manpower*" (Committee of Ministers of the Council of Europe 1997). These criteria of time and manpower were taken into account by the German Data Protection Law:

> *(...)'depersonalized' if a person can only be identified with 'a disproportionate amount of time, expense and labour'(...).*[18]

As Walden argues, "*The issue of effective anonymisation is ultimately one of fact for the regulator or courts to decide in the particular circumstances, although a statutory definition for anonymous data would provide criteria against which such an assessment could be made*" (Walden 2002, 227).

This is an aspect which is missing in the Directive and that should be included in the discussions regarding its review. A possible solution would be to follow the

[16] "*18. The same analysis occurs with the notion of anonymity. Although, from a data protection view, the notion of anonymity would cover data that are no longer identifiable (see recital 26 of the Directive), from a statistical point of view, anonymous data are data for which no direct identification is possible. This definition implies that indirect identification of data would still qualify these data as anonymous, from a statistical point of view.*"

[17] According to the Article 29 Working Party, the reasonableness is "*Another general limitation for the application of data protection under the Directive.*" (see Article 29 Working Party 2007, 5).

[18] *Apud* Ian Walden, *Op. cit.*: 226. The cost of the identification was recognised by the Article 29 Working Party as one of the factors to be taken into account when analysing whether an individual is identifiable or not (Article 29 Working Party 2007, 15).

same criteria established by both the German Data Protection Law and Recommendation R (97)5 of the Committee of Ministers of the Council of Europe, that is, if the necessary measures to identify the link between a person and a certain datum are disproportionate, this datum will not be considered as personal data but as an anonymous one for the purposes of the application of data protection rules (Sarmento e Castro 2005, 72).

In the next sections, the experience of the selected member states will be analysed.

13.3 The Concept of Personal Data in the Selected EU Member States

Firstly, it is important to justify the choice of Italy, France and UK as case studies. Although these three countries have implemented EU legislation in the field of data protection, the way they have incorporated these rules vary from each other, as will be seen in the following topics. In addition, the UK is a common law country, which explains the fact that it has the richest judicial experience, while Italy and France have a continental system, providing a different analysis perspective.

13.3.1 Italy

Article 4(1)(b) of the Italian Data Protection Code (*Codice per la protezione dei dati personali*) establishes that "*personal data shall mean any information relating to natural or legal persons, bodies or associations that are or can be identified, even indirectly, by reference to any other information including a personal identification number*" and Article (1)(i) stresses that "*data subject shall mean any natural or legal person, body or association that is the subject of personal data*" (Bianca and Busnelli 2007, 74; Garante per la protezione dei dati personali 2004, 61).

This definition goes beyond the one of Directive 95/46/EC; firstly, because it includes both natural and legal persons (Bianca and Busnelli 2007, 25)—the Directive only considers natural persons[19]—and secondly, because it does not specify any factor to which the information should be related (Bianca and Busnelli 2007, 52), as does the Directive when it establishes that an identifiable person "*is one that can be identified, directly or indirectly, in particular by reference to an identification number or to one or more factors specific to his physical, physiological, mental, economic, cultural or social identity*".[20] It is important to notice, however, that some Italian courts have been deciding that only data that is objective can be considered as personal data. This is the case, for instance, of a decision adopted by the Court of Rome in a case where an employee requested access to the evaluation forms filled in

[19] See Article 2(a) of Directive 95/46/EC.

[20] See Article 2(a) of Directive 95/46/EC.

by his employer. The Court refused the request on the grounds that "*The indicated evaluation, as a consequence of its subjectivity*", could not be considered as personal data (Tribunale di Roma 2000). Furthermore, the Italian Code also applies to data related to deceased persons.[21]

In relation to anonymous data, the Italian Code, although presenting a definition of anonymous data in its Article 4(1)(n), does not take into account the idea of reasonableness contained in the Directive (Bianca and Busnelli 2007, 54). According to the Italian Code "*'anonymous data' shall mean any data that either in origin or on account of its having been processed cannot be associated with any identified or identifiable data subject*".[22]

13.3.2 France

The French Act defines personal data as "*any information relating to a natural person who is or can be identified, directly or indirectly, by reference to an identification number or to one or more factors specific to him. In order to determine whether a person is identifiable, all the means that the controller or any other person uses or may have access to should be taken into consideration*",[23] in other words, in France all possible means to identify a person can be considered independent of the amount of time, expense and labour to be applied in this effort (Laffaire 2005, 43).[24] Nevertheless, the French Court of Cassation (*Cour de Cassation*), in a case involving information regarding the place of residence of a famous princess and a meeting with her husband in a restaurant, the Court concluded that the 'Triviality' of some information was likely to exclude the invasion of privacy (Mallet-Poujol 2006, 5).

Moreover, although Article 2 does not include expressly deceased people in the definition of personal data, it can be inferred from Article 40, paragraph 5 that this category of persons is also included in the definition, since it recognises the right of heirs of the deceased person to demand of data controllers that they "*take the death into account and update the data accordingly*",[25] and from the fact that it did not

---

[21] Article 9(3) of the Italian Personal Data Protection Code.

[22] The Italian Code of Conduct and Professional Practice applying to processing of personal data for statistical and scientific purposes lists in its Article 4(1) some means that can be considered as reasonable for identifying a data subject. www.garanteprivacy.it/garante/doc.jsp?ID=1115480. Accessed 23 Dec 2010.

[23] Article 2, 2nd paragraph of the French Act 78–15.

[24] The French Act, as the British one, did not adopt a definition of anonymous data, although in some of its provisions it refers to this kind of data. See, for instance, Article 25, § 1 of the French Act 78–15.

[25] Article 100 of the French Decree n 2005–1309 requires that "*the heir of a deceased person who wishes to update the data concerning the deceased*" proves "*his capacity as heir by producing an attested affidavit or a family record book.*" In overseas regions such proof can be made through any means (Article 111, § 9 of the same Decree).

include the words "living individuals" as did the UK Data Protection Act 1998.[26] Nonetheless, in a famous case involving the book "Le grand secret", which referred to a disease the former French President François Mitterrand had at the end of his life, the Court of Cassation decided that the right to take legal action regarding the violation of private life ends with the death of the data subject, the only holder of this right (Mallet-Poujol 2006, 5). Such decision limits the scope of personal data, since it limits the exercise of rights, by the heirs of the data subject, in terms of violations of privacy (and data protection).

Furthermore, despite the fact that the Act does not include information regarding legal persons in the definition of personal data (Bensoussan 2008, 19),[27] the French Authority in some of its decisions has been dealing with the processing of such data as it was amongst its competences.[28] In addition, some courts in France have already recognised the protection of privacy to legal entities (Lacoste 2008, 188).

13.3.3 The United Kingdom

First of all, it is important to highlight the fact that the UK is the country among the three I am analysing in this study which has the richest judicial experience in terms of data protection, especially with regard to the concept of personal data, making the analysis of this country longer than the previous ones.

In terms of legal concepts, Sect. 1.1 of the Data Protection Act 1998 defines personal data as data relating to "*a living individual who can be identified—a) from those data, or b) from those data and other information which is in the possession of, or is likely to come into the possession of, the data controller.*" The Act also adds that "*any expression of opinion about the individual and any indication of the intentions of the data controller or any other person in respect of the individual*" are comprised by the definition of personal data.

To verify whether there is or not a processing operation of personal data, it is important to understand what the Act means by "*likely to come into possession*". Although there is no clear criterion in the law to verify this "*likely to come into possession*" (Information Commissioner's Office 2001, 12), the Information Commissioner (ICO) takes the view that it is not necessary that the identifying data are or are likely to come under the physical control of the data controller to consider that such data are covered by the concept of personal data (Information Commissioner's Office 2001, 13). To illustrate, the ICO gives the example of the relationship between a data controller and a data processor, where the latter will receive data from third

[26] See Section 1(1) of the UK Data Protection Act 1998.

[27] "*La protection de libertes fondamentales ne concerne que les personnes physiques. Les personnes morales sont exclues du regime de protection.*"

[28] In that sense, see *Autorisation Unique* n AU-003 and *Autorisation Unique* n AU-024 of the French Data Protection Authority. http://www.cnil.fr/en-savoir-plus/deliberations/autorisations-uniques/. Accessed 4 Jan 2011.

parties and will process such data in accordance with the instructions of the data controller. The controller will not have direct access to identifying data, although this data is in the possession of the data processor. Therefore, according to the ICO, in such a situation, the data controller could not argue that the identifying data is not in his possession or likely to come (Information Commissioner's Office 2001, 13).

I agree with the ICO to the extent that the data has not to be in the possession of the data controller, but in the given example, it seems to me that such data is likely to come into the possession, or at least is available, to the controller without much effort. In that case, I would say that the view of the ICO regarding "*in the possession or likely to come into possession*" follows the position adopted by Directive 95/46/EC when it refers to an "*identifiable natural person*".

It is important to notice that the UK Act, similar to the Italian Code, did not adopt the idea of reasonableness contained in Directive 95/46/EC as a limitation for the application of the data protection legislation.[29]

Another aspect of the definition of personal data in the UK is the extension of the concept to include "*expressions of opinion or intention*". According to the ICO, it means that any data controller would have to disclose not only the data it has about a specific data subject, but also its opinions about such person or its intention to offer or decline something to such a person. An example would be the case of the employer that has to disclose his/her opinions about his/her employees and also "*any intention to offer or decline promotion on the basis of those opinions*" (Information Commissioner's Office 2001, 15, 2009, 22). It would force, for example, insurance companies to disclose medical opinions about an individual regarding his/her risks in terms of insurance, and also the impact these opinions would have in denying coverage, leading to a completely different outcome than the one adopted in Italy, where the advice given by medical doctors to their clients (insurance companies) does not have to be disclosed to data subjects (Garante per la protezione dei dati personali 2007).[30]

With regard to the scope of the concept of personal data, it is clear that data subjects will only be living individuals; deceased persons are not covered (Information Commissioner's Office 2001, 11)[31]—and also only natural persons and not legal

[29] It also did not include a definition of anonymous data.

[30] The Italian Data Protection Authority recognises that in such cases there are personal data not only in the identification part of the opinion, but also in the conclusions and evaluations of the medical expert of the insurance company, and, indeed, Article 7 of the Data Protection Code applies to the evaluation and conclusions of the expert. Nevertheless, it does not mean that full access has to be given: information related to counselling given by the expert to the insurance company concerning the decision of paying or not paying an indemnity or the strategy in a future legal claim are not included.

[31] "*The Act is only concerned with living individuals and so if the subject of the information is dead, then the information cannot be personal data.*" In the same sense, see Information Commissioner's Office (2009, 26).

entities (Information Commissioner's Office 2001, 16).[32] As Peter Carey pointed out:

> *There are two points to be made here (...). The first is that the data must relate to a living person—once a person has died her rights under the legislation cease. The second is that the definition applies only to individuals. A database containing names and addresses of limited companies is therefore not caught. However, where such a database includes names of officers or employees within the company (e.g. contact names) it will fall within the definition of personal data.* (Carey 2000, 12)

The only exceptions, in the view of the UK Commissioner, would be the cases of small partnerships and of a sole trader, where the information about the business is indistinguishable from the partner's information (Information Commissioner's Office 2001, 11). Therefore, the processing of data related to an identified "deceased" person or to an identified "legal" person would not be included in this concept and would not be covered by the Data Protection Act 1998 (Information Commissioner's Office 2007, 3). Nonetheless, the Consumer Credit Act 1974, as amended by the Consumer Credit Act 2006, grants some similar rights than the ones regarding data subjects—access, correction, erasure[33]—to partnerships[34] and unincorporated bodies of persons[35] when dealing as consumers with credit reference agencies. In these cases, however, the supervisory authority will not be the ICO but the Office of Fair Trading (OFT).[36]

Furthermore, the England and Wales Court of Appeal, in a ruling of 8 December 2003, in the famous case "Durant", gave a narrow interpretation to the concept of personal data, adding more limits for the application of the Act. According to the findings of the Court:

> *(...) not all information retrieved from a computer search against an individual's name or unique identifier is personal data within the Act. Mere mention of the data subject in a document held by a data controller does not necessarily amount to his personal data. Whether it does so in any particular instance depends on where it falls in a continuum of relevance or proximity to the data subject as distinct, say, from transactions or matters in which he may have been involved to a greater or lesser degree. It seems to me that there are two notions that may be of assistance. The first is whether the information is biographical in a significant sense, that is, going beyond the recording of the putative data subject's involvement in a matter or an event that has no personal connotations, a life event in respect of which his privacy could not be said to be compromised. The second is one of focus. The*

[32] "*A data subject must be a living individual. Organisations, such as companies and other corporate and unincorporated bodies of persons cannot, therefore, be data subjects.*"

[33] See Sects. 158 and 169 of the Consumer Credit Act 1974. In the same sense see The Consumer Credit (Credit Reference Agency) Regulations 2000.

[34] "Partnership" means "*a partnership consisting of two or three persons not all of whom are bodies corporate*" (Sect. 189(1)(a) of the Consumer Credit Act 1974 as amended by the Consumer Credit Act 2006).

[35] "Unincorporated body of persons" means "*an unincorporated body of persons which does not consist entirely of bodies corporate and is not a partnership*" (Sect. 189(1)(b) of the Consumer Credit Act 1974 as amended by the Consumer Credit Act 2006).

[36] See Sects. 38–54 of the Consumer Credit Act 1974 as amended by the Consumer Credit Act 2006.

> *information should have the putative data subject as its focus rather than some other person with whom he may have been involved or some transaction or event in which he may have figured or have had an interest, for example, as in this case, an investigation into some other person's or body's conduct that he may have instigated.* (England and Wales Court of Appeal 2003, § 28)

In this case, there was also a discussion about the definition of "*a relevant filing system*" for the situations where personal data were not processed by automated means, but only by manual ones, and the Court of Appeal created some criteria to identify when the Act should apply to those instances of manual processing of personal data. Following the Court's criteria, a manual filing system would be a "relevant" one if "*1) relate to individuals; 2) be a 'set' or part of a 'set' of information; 3) be structured by reference to individuals or criteria relating to individuals; and 4) be structured in such a way that specific information relating to a particular individual is readily accessible*" (England and Wales Court of Appeal 2003, para 46).

Such scope defined by the Court of Appeal for both the concept of personal data and the "*relevant filing system*" contradicts the one provided for by Directive 95/46/EC (Nouwt 2009, 283). The Directive does not impose any limits for the data to be considered as personal data, being enough that it is related to "*an identified or identifiable natural person*".[37] Therefore, in applying the decision of the Court of Appeal, the UK is not complying with—or better—has not adequately implemented EU legislation, in this case the referred Directive.

Despite the fact that such decision is binding in terms of common law, the ICO, after the adoption by the Article 29 Working Party of its Opinion on the concept of personal data, published a document entitled "What is personal data?—A quick reference guide", which goes in a different direction than the one adopted by the Court of Appeal (Information Commissioner's Office 2008).[38] According to this document, the intention of the data controller in identifying the data subject is not the utmost issue as in the findings of the Court.

---

[37] Article 2(a) of Directive 95/46/EC (see Article 29 Working Party 2007, 13). "*Concerning 'directly' identified or identifiable persons, the name of the person is indeed the most common identifier, and, in practice, the notion of 'identified person' implies most often a reference to the person's name.*" In the same sense are the findings of the European Court of Justice (2003) in the Lindqvist case. C-101 § 24.

[38] The Information Commissioner formulated eight questions to help data controllers in identifying if a certain data is personal data. If the answer for one of those questions is affirmative, it is likely that the processing into question involves personal data in the Commissioner's view. The questions are the following: 1) Can a living individual be identified from the data, or, from the data and other information in your possession, or likely to come into your possession? 2) Does the data 'relate to' the identifiable living individual, whether in personal or family life, business or profession? 3) Is the data 'obviously about' a particular individual? 4) Is the data 'linked to' an individual so that it provides particular information about that individual? 5) Is the data used, or is it to be used, to inform or influence actions or decisions affecting an identifiable individual? 6) Does the data have any biographical significance in relation to the individual? 7) Does the data focus or concentrate on the individual as its central theme rather than on some other person, or some object, transaction or event? 8) Does the data impact or have the potential to impact on an individual, whether in a personal, family, business or professional capacity?

Nonetheless, the Court decision has a binding effect up to the moment when there is a new decision of the respective court or of a higher court which changes its view. As stated by the ICO

> *The Commissioner can only give general guidance; the final decision in case of dispute is a question for the courts. (...) It is not possible for the Commissioner to state categorically whether or not certain types of information or files are caught by the Act although it is recognised that there are certain areas of business where the question of whether manual information falls within the definition will be of particular significance, e.g. personnel files.* (Information Commissioner's Office 2001, 9)

It is clear, hence, that Directive 95/46/EC went further than the UK Data Protection Act 1998 regarding the definition of personal data, "*by not including the words 'likely to come into the possession of', thus rendering an encrypted database personal data where the key exists anywhere in the world, however, unlikely it is that the key would come into the possession of the data controller*" (Carey 2004, 15).

13.4 'New' Concept of Personal Data

As demonstrated in this chapter, the concept of personal data is the main foundation of the EU legislation in the field of data protection.

Furthermore, the description of the concept of personal data carried out in the previous section, shows that there are significant differences in relation to this concept among the selected member states. In Italy and France, it includes deceased persons, which does not happen in the UK where just "living individuals" are covered by the Data Protection Act of 1998. This difference creates a situation where a processing activity as the one regarding the beneficiaries of a life insurance policy, just to give an example, will not be supervised by the Data Protection Authority, because the concept of personal data does not consider information related to deceased people as personal data.

Moreover, as previously shown (Sect. 13.3.3), the UK Act has reduced the scope of the concept of personal data by adding the expression "likely to come into possession", as a requirement for the information to be considered as personal data. Furthermore, the England and Wales Court of Appeal has tightened even more the scope of the concept as discussed above (Sect. 13.3.3). As a consequence, the UK Data Protection Framework does not fully comply with Directive 95/46/EC.

These different approaches concerning the concept of personal data can lead to a barrier for the free movement of services, since it would cause providers of certain activities to deal with different rules in different member states and possibly creating problems for the free movement of services, because in some states a processing carried out could involve information considered as personal data and, in others, the same information would not be considered as such. This is of particular importance in a cross-border scenario, as in the EU, where the trade between member states intensifies the flow of personal data, making it necessary to establish a uniform trade environment in order to guarantee the protection of personal data within the bloc, avoiding different levels of protection between member states (Doneda and Viola de Azevedo Cunha 2010, 366).

Another aspect for the review of the EU Data Protection Legal Framework is the extension of the concept of personal data to include also information related to legal persons. Among the analysed member states, only Italy has adopted such a position, making all processing of data related to individuals or legal persons subject to the Data Protection Legislation.[39]

It could be argued that physical, physiological, and mental data are only related to natural persons; nonetheless, the other kinds of data mentioned in the Directive can obviously belong to legal persons and, for such a reason, it seems that the Italian Legislator adopted a better solution, since problems concerning the processing of these data can affect both legal and natural persons.

Furthermore, in many member states the protection given by the Data Protection Legislation to natural persons is usually extended to legal persons through other pieces of legislations, as for example the UK Consumer Credit Act 1974 (as amended by the Consumer Credit Act 2006), which grants the rights of access, correction and erasure to partnerships and unincorporated bodies of persons, giving supervisory powers to the OFT and not to the ICO. This is another reason for adopting a general concept of personal data which includes legal persons, that is, to avoid different interpretations/positions of different supervisory authorities dealing with similar cases. It can be a case, for example, where the ICO and the OFT will analyse complaints dealing with the same processing activity, one from a data subject and the other from a legal entity, and will have different outcomes, one considering the processing fair and the other unfair. This situation becomes even more complicated in a cross-border scenario as, for example, in the e-justice initiative aiming at interconnecting national insolvency registers (see Buttarelli 2009). These differences in the concept of personal data could create considerable problems for the intended interconnection of databases.

Moreover, as highlighted in the analysis of the French Data Protection Framework (Sect. 13.3.2), in some situations it is difficult for the supervisory authority not to deal with processing activities involving data related to legal entities, as did the CNIL in some of its general authorisations.

It can be seen, hence, that there are slight differences in the concept of personal data adopted by the selected member states, which can have important consequences, especially in a cross-border perspective. Just to illustrate, a database containing information related to legal persons could be processed in a member state without the incidence of data protection rules but in others such rules would apply.

A possible solution seems to be the enlargement of the concept of personal data contained in the Directive, in order to compulsorily comprise information related to legal persons, since there is no sense in such a distinction, because both kinds of data are processed in the same way and for the same purposes. It is worth noting that the European Court of Human Rights in the *Société Colas Est* Case (European Court of Human Rights 2002) extended to legal persons the protection of one sphere

[39] Austria and Switzerland have adopted the same approach (see Bygrave and Schartum 2009, 168).

of privacy ensured by Article 8 of the European Convention on Human Rights (De Hert and Gutwirth 2009).[40] The e-Privacy Directive (2002/58/EC), adopted a similar position, extending some of its provisions to legal persons (Article 29 Working Party 2007, 23).[41]

Finally, and equally important, is the "identifiability" of the data subject. None of the selected member states has incorporated reasonableness as a limit for the identification of the data subject. As discussed in the first section of this chapter, it is important to establish clear criteria whether a data subject is identifiable or not. A criterion that fulfils the objectives of the expression 'means likely reasonably to be used' contained in Recital 26 of Directive 95/46/EC is the one that uses the amount of time, expenses and labour to verify whether an individual can be identified from a specific processing of data and whether such data can be considered as personal data. In the same sense, a definition of anonymous data, taking into account the reasonableness of the identification of the data subject, should also be incorporated.

Therefore, my contribution for the discussions on the future of the Directive would be to broaden the concept of personal data to cover both legal and deceased persons and to incorporate clear criteria for the identification of the data subject, taking into account the time, expense and labour involved in this process, and the adoption of a definition of anonymous data following these same parameters (European Data Protection Supervisor 2011, 13).

13.5 Conclusion

In this contribution I analysed the concept of personal data as it is provided for by directive 95/46/EC and the interpretation given by the Article 29 Working Party and by the European Data Protection Supervisor in this regard.

Apart from that, I have considered the need to better specify the concept of anonymous data and the consequences it can have on the application of data protection. Regarding this issue, a "solution" was proposed and goes in a similar direction of the one adopted by Directive 95/46/EC, but putting more emphasis on the amount of time, expense and labour required to make a data identifiable.

After analysing the legal framework at the EU level, I have concentrated on the different experiences of three selected EU member states (France, Italy and UK).

[40] "*The Court has even gone so far as to recognise privacy protection to firms and business activities, which is non-mandatory feature of data protection regulation (which optionally allows Members States to recognise data protection rights not only to natural persons but also to legal persons).*"

[41] "*Some provisions of the e-privacy Directive 2002/58/EC extend to legal persons. Article 1 thereof provides that* '2. The provisions of this Directive particularise and complement Directive 94/46/EC for the purposes mentioned in paragraph 1. Moreover, they provide for protection of the legitimate interests of subscribers who are legal persons.' *Accordingly, Articles 12 and 13 extend the application of some provisions concerning directories of subscribers and unsolicited communication also to legal persons.*"

I have shown that there are considerable disparities in the way each member state incorporated the definition of personal data.

Considering this scenario, I have proposed some changes in the current EU concept of personal data, in order to stimulate not only further harmonisation within the EU area, but also (and perhaps, more importantly) to promote a better integration of markets which use information as an important raw material for their activities, without reducing the level of protection of individuals.

Acknowledgement I would like to acknowledge and thank Mike Wiesmeier for his valuable proof-reading assistance which helped to make the text much more readable. However, any mistake and lack of clarity remains entirely my fault.

References

Article 29 Working Party. 2007. Opinion 4/2007 on the concept of personal data. http://ec.europa.eu/justice_home/fsj/privacy/docs/wpdocs/2007/wp136_en.pdf. Accessed 2 Jan 2011.

Article 29 Working Party. 2008. Opinion 1/2008 on data protection issues related to search engines. http://ec.europa.eu/justice/policies/privacy/docs/wpdocs/2008/wp148_en.pdf. Accessed 2 Jan 2011.

Bensoussan, Alain. 2008. *Informatiques et libertes*. Paris: Éditions Francis Lefebvre.

Bianca, Cesare Massimo, Francesco Donato Busnelli. 2007. *La Protezione dei Dati Personali*. Tomo II. Milano: CEDAM.

Bygrave, Lee A., and Dag Wiese Schartum. 2009. Consent, proportionality and collective power. In *Reinventing data protection?* ed. Serge Gutwirth et al, 157–174. Springer.

Buttarelli, Giovanni. 2009. Speaking points of the Assistant European Data Protection Supervisor on the Council Working Group on e-Justice and interconnection of insolvency registers. http://www.edps.europa.eu/EDPSWEB/webdav/shared/Documents/EDPS/Publications/Speeches/2009/09-07-15_eJustice_insolvency_EN.pdf. Accessed 4 Jan 2011.

Carey, Peter. 2000. *Data protection in the UK*. London: Blackstone Press.

Carey, Peter. 2004. *Data protection: A practical guide to UK and EU law*. 2nd ed. Oxford: Oxford University Press.

Clayton, Richard. 2008. The Phorm 'Webwise' system, http://www.cl.cam.ac.uk/~rnc1/080518-phorm. pdf. Accessed 7 Nov 2010.

Committee of Ministers of the Council of Europe. 1997. Recommendation No. R (97) 5E on the Protection of Medical Data. https://wcd.coe.int/wcd/com.instranet.InstraServlet?command=com.instranet.CmdBlobGet&InstranetImage=564487&SecMode=1&DocId=560582&Usage=2. Accessed 4 Jan 2011.

De Hert, Paul and Gutwirth, Serge. 2009. Data protection in the case law of Strasbourg and Luxemburg: Constitutionalisation in action. In *Reinventing data protection?*, ed. Serge Gutwirth et al., 3–44. Springer.

Doneda, Danilo and Viola deAzevedo Cunha, Mario. 2010. Data protection as a trade resource in Mercosur in The Law of Mercosur, ed. Marcílio Toscano Franca Filho et al., 365–386. Oxford: Hart.

England and Wales Court of Appeal. 2003. Durant case. http://www.hmcourts-service.gov.uk/judgmentsfiles/j2136/durant-v-fsa.htm. Accessed 5 Dec 2010.

European Commission. 2010. Draft Communication from the Commission to the European Parliament, the Council, the Economic and Social Committee and the Committee of the Regions [COM(2010) 609 final]. http://ec.europa.eu/justice/news/consulting_public/0006/com_2010_609_en.pdf. Accessed 3 Jan 2011.

European Court of Human Rights. 2002. Société Colas Est *v.* France case. Application n 37971/97. http://cmiskp.echr.coe.int/tkp197/view.asp?item=1&portal=hbkm&action=html&highlight=37971/97&sessionid=64275468&skin=hudoc-en. Accessed 4 Jan 2011.

European Court of Justice. 2003. Lindqvist case (C-101). http://curia.europa.eu/jurisp/cgi-bin/gettext.pl?lang=en&num=79968893C19010101&doc=T&ouvert=T&seance=ARRET. Accessed 5 Jan 2011.

European Data Protection Supervisor. 2008. Opinion of 20 May 2008 on the proposal for a Regulation of the European Parliament and of the Council on European Statistics (COM(2007) 625 final). http://www.edps.europa.eu/EDPSWEB/webdav/site/mySite/shared/Documents/Consultation/Opinions/2008/08-05-20_Statistics_EN.pdf. Accessed 25 Jan 2010.

European Data Protection Supervisor. 2007. Opinion of 5 September 2007 on the proposal for a Regulation of the European Parliament and of the Council on Community statistics on public health and health and safety at work (COM(2007) 46 final). http://www.edps.europa.eu/EDPSWEB/webdav/site/mySite/shared/Documents/Consultation/Opinions/2007/07-09-05_Statistics_health_data_EN.pdf. Accessed 4 Jan 2011.

European Data Protection Supervisor. 2011. Opinion of 18 January 2011 on the Communication from the Commission to the European Parliament, the Council, the Economic and Social Committee and the Committee of the Regions—"A comprehensive approach on personal data protection in the European Union". http://www.edps.europa.eu/EDPSWEB/webdav/site/mySite/shared/Documents/Consultation/Opinions/2011/11-01-14_Personal_Data_Protection_EN.pdf. Accessed 9 Jan 2011.

European Parliament. 2010. SWIFT: MEPs to Vote on Backing or Sacking EU/US Data Sharing Deal, http://www.europarl.europa.eu/news/public/story_page/019-68537-039-02-07-902-20100205STO68536-2010-08-02-2010/default_en.htm. Accessed 19 Feb 2010.

French Act. 1978. n 78–17 on data processing, data files and individual liberties, http://www.cnil.fr/fileadmin/documents/en/Act78–17VA.pdf. Accessed 1 Aug 2011.

Garante per la protezione dei dati personali, Relazione. 2004. L'attuazione del Codice nel quadro della Costituzione per l'Europa. http://www.garanteprivacy.it/garante/document?ID=1093820. Accessed 15 Dec 2010.

Garante per la protezione dei dati personali Provvedimento del 25 luglio. 2007. http://www.garanteprivacy.it/garante/doc.jsp?ID=1434791. Accessed 22 Dec 2010.

Gediel, José Antônio Peres and Corrê a, Adriana Espíndola. 2008. Proteção jurídica de dados pessoais: A intimidade sitiada entre o Estado e o Mercado. *Revista da Faculdade de Direito—UFPR* 47: 141–153.

Information Commissioner's Office. 2001. Data Protection Act 1998. Legal guidance, http://www.ico.gov.uk/upload/documents/library/data_protection/detailed_specialist_guides/data_protection_act_legal_guidance.pdf. Accessed 4 Jan 2011.

Information Commissioner's Office. 2007. Data protection technical guidance determining what is personal data. http://www.ico.gov.uk/upload/documents/library/data_protection/detailed_specialist_guides/personal_data_flowchart_v1_with_preface001.pdf. Accessed 5 Dec 2010.

Information Commissioner's Office. 2008. What is personal data?—A quick reference guide. http://www.ico.gov.uk/upload/documents/library/data_protection/detailed_specialist_guides/160408_v1.0_determining_what_is_personal_data_-_quick_reference_guide.pdf. Accessed 5 Dec 2010.

Information Commissioner's Office. 2009. The guide to data protection. http://www.ico.gov.uk/upload/documents/library/data_protection/practical_application/the_guide_to_data_protection.pdf. Accessed 4 Jan 2011.

Kuner, Christopher. 2003. *European data protection law and online business*. Oxford: Oxford University Press.

Kuner, Christopher. 2007. *European data protection law—corporate compliance and regulation*. Oxford: Oxford University Press.

Lacoste, Jean-Marc. 2008. Pour une pleine et entière reconnaissance du droit à la protection des données à caractère personnel. Dissertation, Université deToulouse.

Laffaire, Marie-Laure. 2005. *Protection des données à caractere personnel*. Paris: Éditions d'organisation.

Mallet-Poujol, Nathalie. 2006. Protection de la vie privée et des données personnelles (Unofficial translation by the author). Legamedia, Février 2006, http://www.educnet.education.fr/chrgt/guideViePrivee.pdf. Accessed 1 Aug 2011.

Murray, Thomas H. 1997. Genetic exceptionalism and 'Future diaries': Is genetic information different from other medical information? In *genetic secrets: Protecting privacy and confidentiality in the genetic era*, ed. Mark A. Rothstein, 60–76. New Heaven: Yale University Press.

Nouwt, Sjaak. 2009. Towards a common European approach to data protection: A critical analysis of data protection perspectives of the Council of Europe and the European Union. In *Reinventing data protection?*, ed. Serge Gutwirth et al., 275–292. Springer.

Nugter, A. C. M. 1990. *Transborder flow of personal data within the EC: A comparative analysis of the privacy statutes of the Federal Republic of Germany, France, the United Kingdom and the Netherlands and their impact on the private sector.* Deventer: Kluwer Law and Taxation.

Ohm, Paul. 2009. Broken promises of privacy: Responding to the surprising failure of anonymization. University of Colorado Law School Legal Studies Research Paper No. 09–12, http://ssrn.com/abstract=1450006. Accessed 7 Nov 2010.

Reding, Viviane. 2011. The upcoming data protection reform for the European Union. *International Data Privacy Law* 1 (1): 3–5.

Sarmento e Castro, Catarina. 2005. *Direito da informàtica, privacidade e dados pessoais*. Coimbra: Almedina.

Sweeney, Latanya. 2000. Foundations of Privacy Protection from a Computer Science Perspective, http://dataprivacylab.org/projects/disclosurecontrol/paper1.pdf. Accessed 22 Feb 2011.

Tribunale di Roma, Sent. 2000. http://www.ictlex.net/?p=784. Accessed 1 Aug 2011.

Viola de Azevedo Cunha, Mario et al. 2010. La re-identificazione dei dati anonimi e il trattamento dei dati personali per ulteriori finalità: sfide alla privacy. *Ciberspazio e Diritto* 11 (4): 641–658.

Walden, Ian. 2002. Anonymising personal data. *International Journal of Law and Information Technology* 10 (2): 224–237.

Chapter 14
Towards a European eID Regulatory Framework

Challenges in Constructing a Legal Framework for the Protection and Management of Electronic Identities

Norberto Nuno Gomes de Andrade

14.1 Introduction

The difficulties, barriers and challenges in implementing a regulatory framework for a pan-European electronic identity (eID)[1] have been analyzed before in a number of studies. Deliverables pertaining to research projects funded by the European Union (EU), as well as study reports prepared for the European Commission in the areas of eID and eGovernment,[2] have focused on the legal complexities that currently hinder the realization of a pan-European eID scheme. In this respect, researchers and scholars have devoted more attention to legal barriers than to possible legal solutions. This paper attempts to fill this gap, and also to contribute to research on both these analytical dimensions. The article first summarizes the main legal obstacles and challenges to the implementation of a pan-European eID scheme and then suggests a conceptual framework of principles to address these challenges and overcome the obstacles. In summary, this paper contributes to the ongoing debate on the benefits of a regulatory framework for an electronic identity scheme for Europe by presenting a number of legal proposals that could facilitate the realization of such a scheme.

The views expressed in this article are purely those of the author and may not in any circumstances be regarded as stating an official position of the European Commission.

---

[1] See Appendix Terminology for the definition of the most relevant concepts and terms regarding electronic identity (eID) and electronic identity management (IDM) systems.
[2] This has been the case of studies done in the ambit of research initiatives such as the ones led by the Porvoo e-ID Group, Stork, MODINIS, and the IDABC program, as well as studies such as the European Commission (2005), prepared by the eGovernment subgroup of the eEurope Advisory Group.

N. N. G. de Andrade (✉)
Institute for Prospective Technological Studies (IPTS), European Commission,
Joint Research Center (JRC), Seville, Spain,
e-mail: norberto.andrade@ec.europa.eu

S. Gutwirth et al. (eds.), *European Data Protection: In Good Health?*,
DOI 10.1007/978-94-007-2903-2_14,

The article is structured as follows. Section 14.2 describes the relevance of eID for the general development of the information society. I will assess the importance of electronic identity for administration (public), business (private) and, above all, citizens. I will also highlight the role of eID as a key enabler of the economy.

Section 14.3 identifies the various legal gaps and barriers in the current EU legal framework that are hindering the creation of a fully-fledged pan-European eID. I will examine the following issues: the legal blurriness of EU competences in the field of eID; the divergence (and, sometimes, incompatibility) of approaches pursued by different Member States in the regulation of eID; the lack of a harmonized EU legal taxonomy in this area; and the uncertainties about the legal treatment and control of identity-related data used in eID transactions. This examination clearly shows that appropriate regulation regarding eID at European level is still lacking, as the current EU law does not provide a specific legal framework for eID. At the moment, legal regulation of eID is composed of principles, rules, and concepts "borrowed" from different EU legal instruments and national laws that could be better articulated to address the current state of legal fragmentation.

Section 14.4 presents a number of legal proposals which aim to embed electronic identity into the EU regulatory framework. A series of new principles that should underpin a future eID legal scheme are elaborated: the principles of user-centricity, anonymity and pseudonimity and the principle of multiple identities, identity portability, un-linkability and negotiation, among others.

14.1.1 Nota Bene

Before moving on, one important remark regarding the focus and scope of this paper must be made. This paper is devoted to the legal aspects of eID.[3] Hence, I will be looking at the main barriers [4] to the construction of a pan-European electronic identity scheme and the possible solutions from a strictly juridical point of view. Nevertheless, technological and organizational aspects of eID will also be taken into consideration

---

[3] It is also important to bear in mind that the scope of this article is limited to the management of the digital identities of individuals or natural persons. I am fully aware that issues concerning the management of online identities for entities or objects (namely through RFID tags) are growing in importance, but these are outside the scope of this paper.

[4] The analysis of the "specific barriers", or better, the analysis of the legal gaps which derive from particular legal instruments in EU law vis-à-vis the need to effectively and comprehensively regulate eID—namely from the three most relevant European directives in such area (the Data Protection, the eSignatures, and the Services directives)—go beyond the scope of this article. Nevertheless, and just for cataloguing purposes, one could mention the shortcomings of the current identifiability model of the data protection legal framework and the need to regulate the processing of certain instances of non-personal data as legal gaps of the data protection directive regarding the need to regulate eID. For further details, see (Andrade 2011a). In terms of specific issues missing from the eSignature directive that need to be solved in order to attain a successful implementation of a pan-European eID scheme, one could mention the lack of issuance procedures and the lack of a definition concerning the content and verification of eID. In this sense, see (Myhr 2008).

(European Commission 2003).[5] In fact, the technical and infrastructural elements of eID contribute directly to the formulation of the legal solutions proposed here. As we shall see later on, many of the new legal principles proposed are in fact derived from technological design principles, having already been tested in numerous research projects and technical prototypes. I will thus present a set of legal principles with a strong technical ascendancy.

In view of the intricate relationship between legal and technical aspects that this article will establish and address, I consider that the main challenge to European eID is not only technological but also legal. It is important to note that the technology[6] necessary to enable an interoperable eID across Europe already exists (Modinis-IDM-Consortium 2006, 7).[7] What is missing, in reality, is legal interoperability. It is the lack of legal harmonization that most inhibits cross border deployment of services based on electronic identity. Having said this, the article will focus on the legal framework that must be constructed in order to accompany and enforce the existing technological answers,[8] transposing some of the latter into operating full-fledged legal principles. In brief, the scope of the article is to identify the legal gaps and propose a number of principles that, ideally, could form the basis of a common EU legal framework for the protection and management of digital identities.

14.2 Relevance of eID

This section emphasizes the increasing socio-economic relevance and importance of electronic identities and examines how eID has been targeted by international organizations and by EU political agendas, declarations, action plans, and research funded projects.

---

[5] In reality, the need for a balanced mix between law and technology is not new. This alliance has been widely advocated under the label of "privacy by design." In this regard, the European Commission noted in 2003 that "...the use of appropriate technological measures is an essential complement to legal means and should be an integral part in any efforts to achieve a sufficient level of privacy protection." In the context of eID and taking into account the need to achieve a sufficient level of identity protection, I believe that technology should also contribute to an "identity by design."

[6] Microsoft, Shibboleth, Liberty Alliance, Passel, Sxip and other technology companies and consortia have devoted efforts to building digital IDM systems and tools.

[7] In effect, as the Modinis Interim Report observed: "A commonly heard remark is that for any given technical difficulty in the IDM sector the problem is not the unavailability of technical solutions, but rather an overabundance of possible solutions. Overlooking legal, cultural and socio-political perspectives, from a strictly technical point of view most hurdles to interoperate IDM systems would be fairly easy to overcome". One may therefore conclude that the most difficult obstacles posed to the creation of a pan-European eID are not technical, but are derived from the different legal approaches and socio-political sensitivities of EU Member States.

[8] In other words, the article does not focus directly on interoperable technical mechanisms and infrastructures enabling EU citizens to identify and authenticate themselves. The article, instead, focuses primarily on the legal framework that must be put into place in order to allow identification and authentication procedures to be carried out.

There is undoubtedly an increasing need today for identification and identity management. The development of ubiquitous networks of electronic communications and the general trends of globalization and increasing human mobility give rise to the need to ascertain "who is who" on the internet, in the street, in the usage of services, and in commercial transactions. Large investments made by governments[9] and companies are becoming essential for the provision of eGovernment services and interaction with the public administration—and also for the provision of commercial services and the deployment of marketing strategies—which aim to learn as much as possible about a potential customer's needs, habits, and preferences.

eID also brings various societal and economic benefits to European citizens. The ability to interact and transact remotely with various institutions and entities allows users to benefit from the provision of a wider number of services, most of which were previously only available through a physical visit. Moreover, eID based services will also increase the efficiency and convenience of use. Citizens will be able to access these services at any point of the day (24/7 availability) and from any geographical location (e.g., services that can be accessed through a mobile phone). The increased availability, efficiency, and convenience brought by services that rely on eID will also result in monetary gains for the users and have a positive impact on the environment.

As a result, electronic identity has become a key driver for the growth of the EU economy and the completion of the Single Digital Market. eID constitutes not only a fundamental enabler for the deployment of cross-border services within the EU27, but also an indispensable element for the increase of entrepreneurial activities in Europe. As observed in the Digital Agenda, "[e]lectronic identity (eID) technologies and authentication services are essential for transactions on the internet both in the private and public sectors" (European Commission 2010b, 11).[10]

In view of this, "it is clear that divergent rules with respect to legal recognition of eID and electronic signatures create barriers to the use of electronic communications and electronic commerce, and hinder the free movement of goods and services in the internal market" (Myhr 2008, 77). Therefore, the lack of a harmonized regulatory framework may not only create privacy and security issues affecting the construction of trustworthy online environments but may also compromise the development and the productivity of the increasingly interconnected and globalized economy in which we live, hampering the ability of entities to provide users with suitable services and applications.

Thus, interoperable electronic identities, at the European level, have been deemed essential for achieving the freedom of establishment and circulation of goods, capital

[9] Many EU Member States such as Germany have in the recent times deployed large scale eID projects (see Graux et al. 2009, 120), many of which are presently underway.

[10] Such strategic document envisages, moreover, specific and concrete actions in the field of eID. This is the case of Key Action 16, according to which the Commission will "[p]ropose by 2012 a Council and Parliament Decision to ensure mutual recognition of e-identification and e-authentication across the EU based on online 'authentication services' to be offered in all Member States (which may use the most appropriate official citizen documents—issued by the public or the private sector)".

Table 14.1 List of research and practical implementation projects devoted to eID and interoperability

STORK	https://www.eid-stork.eu/
CROBIES	http://ec.europa.eu/information_society/policy/esignature/crobies_study/index_en.htm
PRIME	https://www.prime-project.eu/
PrimeLife	http://www.primelife.eu/
Modinis IDM	https://www.cosic.esat.kuleuven.be/modinis-idm/twiki/bin/view.cgi/Main/WebHome
TURBINE	http://www.turbine-project.eu/
BEST	http://www.best-nw.eu/
PICOS	http://www.picos-project.eu/
ABC4Trust	https://abc4trust.eu/
SEMIRAMIS	http://ec.europa.eu/information_society/apps/projects/factsheet/index.cfm?project_ref=250453
FIDIS	http://www.fidis.net/

and services (Leenes et al. 2009). eID is also considered to be indispensable for the completion of the digital internal market, reducing administrative burden throughout Europe and allowing the EU-zone as a whole to attain a better competitive position (Leenes et al. 2009)

Consequently, the relevance of eID and the need for interoperable eIDs has been recognized in EU agendas and strategies (European Commission 2010c), action plans (European Commission 2010a),[11] declarations,[12] communications (European Commission 2010d),[13] studies (European Commission 2007),[14] and programs.[15]

The EU has also financed and supported a vast amount of research and practical implementation projects focusing on electronic identity and interoperability (see Table 14.1).

In addition to these projects, there are many other international networks and research centers in Europe carrying out important projects in this area, such as the PETWEB II[16] and the Porvoo Group.[17] Though they entail different approaches, methods, case-analysis and technologies, all these research initiatives have contributed to the development of generalized frameworks for trust and privacy-protective identity management (IDM) systems across Europe.

[11] In such Action Plan, the Commission has proposed a European Strategy on IDM to be attained by 2012, which includes legislative proposals on criminalization of identity theft and on electronic identity (eID) and secure authentication systems.

[12] Such as the Manchester Ministerial Declaration (2005) and the Lisbon Ministerial Declaration (2007).

[13] Such as the recent Communication from the European Commission (2010d).

[14] Namely the following studies: Commission, "Signposts Towards e-Government 2010."

[15] Such as the Stockholm Program, which defines the framework for EU police and customs operation, rescue services, criminal and civil law cooperation, asylum, migration and visa policy for the period 2010–2014.

[16] http://petweb2.projects.nislab.no/index.php/Main_Page

[17] http://www.vaestorekisterikeskus.fi/vrk/fineid/home.nsf/pages/6F4EF70B48806C41C225708B-004A2BE5

Furthermore, the need to develop an eID operational framework also stems from EU legal texts and instruments. Several single market initiatives and legal frameworks presuppose and rely on cross-border interactions between administrations, businesses, and citizens across Europe. Thus, the need to deploy a pan-European eID scheme also derives from EU-enacted legislation itself.[18]

Nevertheless, despite the various political declarations and initiatives in this area, the plethora of research projects, the proliferation of IDM systems and the wide array of advanced eID technologies, the creation of an encompassing, interoperable, pan-European eID scheme has not yet been accomplished. The fundamental reason for this, other than the organizational and technical challenges to interoperability that need to be addressed, is the presence of legal gaps and barriers in the EU legal framework. The main legal gaps and obstacles that hinder the creation of a full-fledged pan European eID are identified in Sect. 14.3.

14.3 Legal and Technical Barriers

This section describes the main barriers (encompassing both technical and legal difficulties) to the creation of a pan-European identity management infrastructure, which would allow existing national IDM systems to interoperate. In the analysis and description of these obstacles, I shall examine what one could call the 'general' barriers to a pan-European eID, that is, the obstacles that are not necessarily attached to any specific piece of legislation.

Although the article is mainly focused on legal barriers, I shall start with a fundamental technical barrier that is the Internet's lack of a proper identity infrastructure. As explained in the PRIME research project White paper

> The internet, by design, lacks unified provisions for identifying who communicates with whom; it lacks a well-designed identity infrastructure (Leenes et al. 2008, 1).[19] Instead, technology designers, enterprises, governments and individuals have over time developed a bricolage of isolated, incompatible, partial solutions to meet their needs in communications and transactions. The overall result of these unguided developments is that enterprises and governments cannot easily identify their communication partners at the individual level (Leenes et al. 2008, 1).

In certain contexts, the lack of an Internet identity infrastructure may not constitute a problem, promoting for instance freedom of expression (allowing people to freely express their ideas and opinions anonymously or through pseudonyms in online forums, for instance). In other contexts, the lack of an Internet identity infrastructure

[18] This is the case of the Directive on Services in the Internal Market (2006/123/EC), which article 8 constitutes an example of the necessity of interoperable eID, stating that "[. . .] all procedures and formalities relating to access to a service activity and to the exercise thereof may be easily completed, at a distance and by electronic means [. . .]."

[19] In effect, "[t]he Internet has an ID infrastructure often identifying only the endpoint of a communication: IP addresses. These are often unreliable to identify users" (Leenes et al. 2008, 1).

may hinder individuals, forcing them to 'over-identify' themselves, and disclose more personal data than is strictly necessary. Unlike real-world transactions, which can often be conducted in an anonymous fashion (by paying with cash without leaving any identity traces, for example), most online dealings require excessive disclosure of identifying data (this normally happens with online shopping, where detailed personal data is usually required to perform the transaction). At a more systemic level, the absence of an Internet identity layer also hampers commercial transactions and official government interactions, which rely on the proper identification of customers and citizens to provide their services.

14.4 The Diversity of Technical and Legal Approaches to eID, the Proliferation of Identity Management Systems and the Emergence of New Actors

One of the major factors blocking the development of interoperable IDM systems across Europe is the diversity (and, often, incompatibility) of technical and mainly legal approaches to the protection and management of electronic identities by EU Member States. As observed in previous studies and surveys (Graux et al. 2009, 106; Leenes et al. 2009, 25), EU Member States take different approaches to eID management systems, varying from the use of specific Public Key Infrastructures (PKI) and the inclusion of eID in non-electronic identity tokens (such as identity cards, driver licenses) to reliance on electronic signatures and two-factor authentication systems.

In addition to the variety of technical approaches, there is also a legal diversity of regulatory options and rationales. In this respect, while some EU Member States have developed national eID cards (such as Austria and Germany, among many others), others do not have an operational national identity card scheme (United Kingdom and Ireland). Furthermore, EU Member States also differ regarding the choice or not of unique identifiers, with some countries using national identification numbers for a wide variety of purposes and contexts, while others use several identification numbers with each one serving a single purpose within a specific context. It is worth noting that the use of unique personal identification numbers for multiple purposes and contexts has been considered unconstitutional in a number of countries (such as Germany, Hungary, and Portugal, among others).[20]

[20] This does not necessarily mean that unique identification numbers cannot be used in these countries, but that their use should be restricted to a specific context. In this way, countries tend to decree the use of separate sectoral identifiers (namely for tax and social security purposes). The use of sector based identifiers is, in effect, finding increasing adoption, partly as a consequence of the above mentioned constitutional restrictions.

Due to divergent legal regulation and organization in EU Member States, there is a proliferation of different IDM systems,[21] which render the eID process more and more complex. Furthermore, new actors and institutions are emerging in the data processing and eID fields.

We have thus surpassed the simple phase of having the same entity acting as both identity certifier and service provider. Today, there is a tendency to separate identity providers from service providers. Identity providers, on the one hand, act as trusted third parties, authenticating a user's identity. These entities, in addition, store user account and profile information. Service providers, also called 'relying parties,' on the other hand, accept assertions or claims about users' identities from the identity providers in order to offer them their services. Under a user-centric identity system, for instance, "[u]sers are allowed to choose identity providers independently of service providers and do not need to provide personal information to service providers in order to receive services" (OECD 2009, 17). In this model, users not only select what information to disclose when dealing with service providers, they also use several identity providers as well. They thus avoid storing all their information in one place (OECD 2009, 17).

We are thus confronted with an increasingly complex scenario, encompassing a wide set of actors such as identity holders, identity providers, registration authorities, and authenticating authorities.[22] Hence, in a typical eID management system, identity-related data is not simply sent or provided by a subject to a controller; rather the data is, in the process, authenticated by a third party. This new actor corroborates the authenticity of the citizen's/customer's identity, and then gives the trusted information to the public or private entity providing the service. We thus have identity providers and relying third parties. It is important to note that in these cases there is no explicit legal framework (Graux et al. 2009, 119).

In addition, and given the wide variety of technical and legal approaches followed by Member States, a fully-functional pan-European eID needs to articulate flows of data between eID holders, receiving parties, and certifying authorities from different countries. This can be quite a challenge (not only technically but also legally) when the receiving party has to handle eIDs from several certifying authorities, based in different countries and following different eID schemes, and the same challenge applies to certifying authorities, which "will have to relate to many receiving parties in different countries if they want eID holders to be able to make generic use of their eIDs" (Myhr 2008, 81). It is thus perfectly possible and probable that a relying party is situated in a different Member State from the one that has assigned the electronic identity. In these cases, the relying party will need to verify the eID at the authentication party in another Member State. Hence, cross border flows of eID

[21] Four main models of IDM systems can be identified within the massive proliferation of eID systems: the "siloed", the centralized, the federated and the "user-centric" IDM systems. For a detailed explanation of each of them, see OECD (2009, 16–17).

[22] One should bear in mind, though, that, in some circumstances, these different actors can coincide in the same entity. For example, an identity provider can also be an authentication authority, and a registration authority might also be an identity provider.

information can take place between the eID holder and the relying party, as well as between the relying party and the authenticating authority.

Another problem likely to emerge from this increasingly complex scenario is related to compliance with the Data Protection Directive rules. These require unambiguous consent from the data subject (the identity holder, also denominated the claimant), which may become complex when the data is not provided by the claimant directly (in an online form, for instance), or when data cannot be obtained from a certificate presented by the claimant (when taken from a certificate on a smart card inserted into a reader the claimant uses in the interaction) (Leenes et al. 2009, 32). This is the case "when the service provider (relying party) needs to obtain additional data, such as (certified) attributes and these can be, or even have to be obtained, from other sources than the user." (Leenes et al. 2009, 32).

As noted by specific eID research programs, these new generations of IDM systems "do not provide adequate safeguards for personal data and give individuals a limited control over their personal data" (Leenes et al. 2008).

The increase in different IDM systems and models poses also problems of accountability and transparency for how they are managed and operated, namely in terms of ascertaining responsibilities in case of an incident. The dilution of accountability and transparency of these systems will mainly affect the citizens and the consumers. Given the myriad of different digital identification systems and techniques, the registration and transfer processes for identity data will probably be less transparent. As a consequence, citizens and consumers will certainly have more difficulties in making informed choices as to which IDM systems to use.

14.4.1 EU Legal Competences

The problem of the distribution of competences between the EU and its Member States regarding a potential legislative action in the field of electronic identity is at the root of the increasingly diverse legal and regulatory approaches pursued by EU Member States.

Any proposal for EU legal intervention and regulation in the field of eID must analyze two important elements: competence and legal basis.

Firstly, an EU Institution adopting a legislative act in the area of eID must have the competence or the legal power to do so. Secondly, the legislative act (a Directive, for instance) must have a legal basis,[23] and reference must normally be made in the recitals to the concrete enabling power, generally to be found in the Treaty itself.[24]

[23] The basic principle underpinning legal basis was expressed in Case 45/86, Commission v. Council (Generalized Tariff Preferences) where the ECJ expressed the opinion that: "the choice of a legal basis for a measure may not depend simply on an institution's conviction as to the objective pursued but must be based on objective factors which are amenable to judicial review."

[24] In the case of delegated legislation, those references are located in an enabling legislative act.

In this manner, the main task is to find a way to legally anchor an eventual eID regulatory initiative to EU Law (both through Treaties and EU secondary legislation), that is, to identify specific area of EU competence and to specify a legal basis for a regulation regarding the implementation of a European eID system.

The relevant Treaty provisions concerning the issue of competences can be found in articles 2–6 of the Treaty on the Functioning of the European Union (TFEU). Three different categories of competence can be identified: exclusive, shared or complementary, and supporting or supplementary (Craig 2008, 8).[25] A brief survey of the different areas and categories of competence immediately confronts one with the considerable difficulty of assigning an eID regulatory initiative to a specific area of competence. This has to do with the fact that the regulation of (personal) identity covers a very wide field, cutting across a broad range of different EU areas and policies. Looking, on the one hand, at the distribution of competences between the Union and the Member States and, on the other hand, at regulating eID at the EU level, it is easy to see that the latter may involve different categories of competence at the same time (such as shared competences and competence to support, co-ordinate, or supplement) or different areas within the same category of competence. Therefore boundary problems may arise between the categories of competence to support and shared competences when inserting eID into the EU legal framework. For example, eID could come under the internal market, which is shared power, or it could be regarded as falling within administrative co-operation, where only supporting action is allowed. Furthermore, the regulation of eID may also affect distinct areas within the same category of competence, such as the internal market, consumer protection, and the area of freedom, security, and justice (among others).

Thus, the EU does not seem to have a direct mandate to regulate eID. Furthermore, regarding the distribution of competences in eID between the EU and Member States, it is worth mentioning paragraph 7 of article 8 of the DPD:

> Member States shall determine the conditions under which a national identification number or any other identifier of general application may be processed.

In other words, the requirements for processing these identifiers are to be defined by the Member States.

Moving from the topic of competences to the issue of legal basis, a legal disposition that could be invoked to sustain an EU legal regulation of eID is Article

[25] In more detail, such three categories are the following: Exclusive competence, according to which only the EU can legislate and adopt legally binding acts, the Member States being able to do so only if empowered by the EU or for the implementation of EU acts; Shared competence, which constitutes a 'general residual category,' (Craig 2008, 8), as it provides that the EU shall share competence with Member States where the Treaties confer on it a competence which does not relate to the areas referred in articles 3 and 6 TFEU (such dispositions deal, respectively, with the category of exclusive competence and with the competence according to which the EU is restricted to taking action to support, co-ordinate, or supplement the action of the Member States); Competence to support, co-ordinate, or supplement, which allows the EU to take action to support, co-ordinate or supplement the actions of the Member States, without thereby superseding their competence in these areas, and without entailing harmonization of Member State law (article 2(5) TFEU).

77(3) TFEU.[26] This article, contrary to the former EU Treaty, now allows for the adoption of measures and provisions on identity cards. Despite this innovation, the possibility of adopting such measures is still somewhat restricted, and requires a special legislative procedure (unanimity in the Council and a merely consultative role for the European Parliament). Furthermore, article 77 TFEU comes under the heading of border checks and immigration policies, and deals with identity cards. For these reasons, article 77 does not seem to be a suitable legal basis for eID, which encompasses electronic communications, covering a much wider spectrum of EU policies and areas. Nevertheless, Article 77(3) TFEU stands as a very important first step in legally framing identity in the EU Treaty, placing eID within the EU legal framework.

14.4.2 Control over Personal Data

The issue of control over personal data is not new, but it is intensified by the emergence of different IDM technical models for processing personal data.

Despite not being new, this issue is certainly exacerbated by the massive deployment of eID systems. This is particularly the case when personal data is re-used outside of the context in which it was initially granted, which, in principle, contravenes the provisions of the Data Protection Directive. Another related problem is the disclosure of more information than is actually needed for the purpose of the application. These situations contravene the provisions and the principles of the above mentioned Directive, namely the principles of fair collection and proportionality.

Depending upon the architectural model for the IDM system chosen, identity information may be stored in a myriad of different places and entities. In the case of Siloed IDM systems, identity information is stored in separate service provider accounts; in centralized IDM systems, however, it is stored in one main account. In addition, while in federated systems, the identity information is kept in separate accounts and in different locations by different service providers; in user-centric systems, identity information is stored by identity providers chosen by the user. These last two systems, despite their advantages over the former ones, offer no way of safeguarding data after it has been shared (OECD 2009, 18). In federated systems, users have little input into the business-partner agreements, and lose track of their data once it has been shared amidst the federation members. In user-centric systems, there is instead the risk of concentration in the market for identity providers, which would then undermine users' control over their own information.

[26] Article 77(3) TFEU: "If action by the Union should prove necessary to facilitate the exercise of the right referred to in Article 20(2)(a), and if the Treaties have not provided the necessary powers, the Council, acting in accordance with a special legislative procedure, may adopt provisions concerning passports, identity cards, residence permits, or any other such document. The Council shall act unanimously after consulting the European Parliament."

14.4.3 Lack of Common Taxonomy

The lack of a suitable, homogenous, unambiguous, and consistent terminology applied to the eID field has been identified by a series of studies and project deliverables.[27]

A legal taxonomy for eID[28] is not only lacking at the level of European legislation but also at the national level. The eID Interoperability for PEGS Analysis and Assessment Report interestingly noted that, in the countries surveyed, there is no legal definition of the concept of identity, and more importantly, of how an identity can be established in an electronic environment (Graux et al. 2009, 118). Austria comes closest to a legal definition in its eGovernment Act:

> Unique identity: designation of a specific person by means of one or more features enabling that data subject to be unmistakably distinguished from all other data subjects (Graux et al. 2009, 118).

Despite the general absence of regulatory frameworks detailing and defining what elements legally constitute an entity's identity, what authentication is and what specific requirements it entails, IDM systems do exist and operate. This is so because technology has stepped in and moved forward, regardless of law. The absence of law and legislation has not prevented technology from being developed, implemented, and applied in the field of eID.

An example of 'technology implementing law,' namely with regards to complying with the requirement for user consent, can be found in Italy where personal data is actually encrypted and cannot be accessed directly without the user's consent (Graux et al. 2009, 128). In this way, technology reinforces the principle of user control over personal data in electronic authentication processes.

As a result, technology seems to be providing the values of certainty and predictability in the regulation of relationships that law should provide. This point is well illustrated by the PEGS study, which remarks on the absence of legislation applicable to authentication processes and the role of PKI signature technology as an entity authentication mechanism.

> The mains reason for this is that, even if the legal framework does not strictly speaking address all relevant issues, the technology/technique behind PKI-based electronic signatures can still offer a large degree of certainty with regard to the entity using an electronic signature (especially when qualified certificates or qualified signatures are used), so that the use of electronic signatures is de facto an adequate tool for authentication, even if the legal basis for it is non-existent (Graux et al. 2009, 119).

As such, most of the current eIDM systems are working not on a 'legal basis,' but on a de facto 'technical basis.' There is thus a need to reintroduce law in this area in a way

---

[27] This is the case of the Modinis-IDM-Consortium (2006) Modinis Deliverable: D.3.9 IDM Issue Interim Report II1. In addition, the Modinis project developed a specific Terminology Paper (Modinis-IDM-Consortium 2005).

[28] See Appendix Terminology for an overview of the terminology use in the field of eID.

that assumes its regulatory functions accompanied by technology, and not replaced by it. It is exactly in this context, in order to re-articulate the relationship between law and technology that I will propose the principle of technological assistance.

14.4.4 Legal Barriers and Challenges: Conclusions

As a conclusion to our brief analysis of the legal barriers and challenges to a European eID—and reinforcing what has already been stated in similar studies—it is evident that an explicit legal framework for eID does not exist. As Myhr observed, "[e]ven though existing laws that regulate a paper-based environment and physical ID-cards to a large extent can also be applied to electronic communication and the use of eIDs, an appropriate regulation regarding eID on a European level is lacking" (Myhr 2008, 77) Furthermore, the application of the current EU legal framework (namely of the Data Protection, eSignatures and Services Directives) to eID is not sufficient to cover all the aspects involved in the protection and management of electronic identities. What could be described as the current legal framework applicable to eID is deeply fragmented, borrowing some elements from the Privacy Directive, the eSignatures Directive, national regulatory approaches and legislation, and others from technically-implemented solutions. In brief, there is no global view or overview of what is to be regulated and how.

14.5 Legal Solutions

As van Rooy and Bus observe, Europe needs a legal framework that "[e]nsures interoperability for trustworthy authentication across service domains of Member State public authorities, business and citizens" (van Rooy and Bus 2010, 403), allowing for "EU-wide trustworthy service provisioning in domains such as e-government, e-health, e-commerce, finances and social networks, and hence should support the provisioning of multiple identity instances from government-accredited to commercially accepted, ranging from strong identification to anonymity (van Rooy and Bus 2010, 403)."

In order to render different national and regional IDM systems interoperable within the EU, there is not only a need for technical interoperability, but also a fundamental need for legal interoperability. This section attempts to contribute to the latter by providing a series of common principles that are currently lacking from EU law and that could be contemplated in order to foster the vision of a pan-European eID scheme.

From the 1980's onwards, various international arrangements have formulated a number of key principles for the protection of personal data. This is the case of the Guidelines on the Protection of Privacy and Transborder Flows of Personal Data, adopted by the Organization for Economic Cooperation and Development (OECD) in 1980, and the Convention for the Protection of Individuals with regard to Automatic Processing of Personal Data, adopted by the Council of Europe in 1981. In the 1990s, the EU's Data Protection Directive (DPD) made a substantial contribution to this legislative effort with a list of principles stipulating the conditions in which personal data should be processed. These initiatives have enshrined an extensive list of principles regarding data collection, storage, and processing. These principles include collection limitation, data quality, purpose specification, use limitation, security safeguards, openness, individual participation, and accountability.[29]

As an addition to these existing principles, this section presents a series of legal principles and rules that, added to the current EU legal framework, fill some of the gaps in EU law and contribute to a more comprehensive and specific regulation of eID. These principles could also be seen as the foundations for a new shared European eID regulatory framework.

Relying upon the work done by initiatives and studies carried out in this area,[30] I will present a conceptual legal framework that groups the most salient findings gathered in these studies, clustering them into a number of general principles and overall rules that, together, complement the existing data protection principles. The objective is thus to present a conceptual framework of principles and guidelines able to orient and frame further specific legal provisions needed in the area of protection and management of eIDs. Formulating legal principles from the new dynamics brought by IDM systems can also help us in testing new solutions for present and upcoming legal problems. Dumortier rightly notes that "[t]he field of privacy and identity management will be an important laboratory where we can experiment how the law will function in our future global information society" (Dumortier 2003, 69).

The principles presented here are derived from the overarching principle of user-centricity. Under the umbrella of such guiding principle, we will then find a group of key principles and a group of procedural principles. The key principles reflect the application of the fundamental values of individual autonomy to the management of one's electronic identity, allowing users to act through multiple identities,

---

[29] The basic principles are listed in article 6 of the Data Protection Directive (DPD), and include the requirements that personal data must be: (a) processed fairly and lawfully; (b) collected for specified, explicit and legitimate purposes, and not further processed in a way incompatible with those purposes. Further processing of data for historical, statistical or scientific purposes shall not be considered as incompatible provided that Member States provide appropriate safeguards; (c) adequate, relevant and not excessive in relation to the purposes for which they are collected and/or further processed; (d) accurate and, where necessary, kept up to date; every reasonable step must be taken to ensure that data which are inaccurate or incomplete, having regard to the purposes for which they were collected or for which they are further processed, are erased or rectified; (e) kept in a form which permits identification of data subjects for no longer than necessary for the purposes for which the data were collected or for which they are further processed. Member States shall lay down appropriate safeguards for personal data stored for longer periods for historical, statistical, or scientific use.

Apart from these basic principles, article 7 of the DPD delineates the conditions under which personal data may be processed, amidst which we stress the requisite that "the data subject has unambiguously given his consent".

[30] Such as the EU/EC programs, commissioned studies, action plans, agendas, and research projects promoted in the eID area and mentioned in sect. 2.

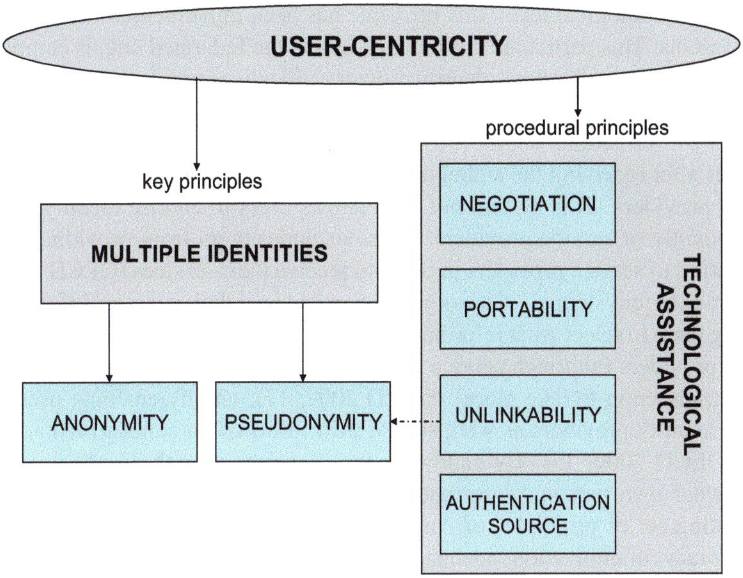

Fig. 14.1 The eID Legal Framework Principles

pseudonyms, or otherwise anonymously. The procedural principles operate at a more technical level, allowing users to keep their multiple identities separate (principle of unlinkability) and under their effective control (principles of negotiation, portability, and authentication source principle). These procedural principles, moreover, are derived from the principle of technological assistance, which underlines the important complementary role of technology in regulating eID (see Fig. 14.1).

In addition, it is important to note that the principles proposed here need to be complemented and implemented with concrete rules,[31] schemes, policy initiatives, and technological infrastructures in order to implement a fully-operational eID legal framework.

14.6 Principle of User-centricity

In order to "create the knowledge society by empowering the individual," (Reflection group on the Future of the EU 2030, 2010, 43) an eID legal framework should give the individual control over their own identity information. While respecting the interests of enterprises and society, the legal framework should place the individual at the core of the IDM system.

[31] In terms of concrete proposals for the achievement of a pan-European eID scheme, Thomas Myhr presents two concrete action proposals that the European Commission could take into consideration in order to achieve cross-border interoperability: (i) setting up requirements for Validation Authorities and self-declaratory schemes and (ii) setting up a quality classification system, where different national security levels can be mapped against neutral requirements adopted by the European Commission. See Myhr (2008).

At the technological level, this principle has been implemented in 'user-centric' IDM systems. This particular IDM model, unlike the federated one, is composed of service providers and various identity providers. Identity providers, in this model, act as trusted third parties and are in charge of authenticating users, storing user accounts and profile information. Service providers, also called 'relying parties', perform their activities after receiving the authenticated identity claims about their users from the identity providers. This system not only allows users to choose identity providers independently of service providers, it also excludes them from providing personal information to service providers in order to receive their services (OECD 2009, 17). The user-centric system gives users greater control over their personal information by enabling them to select what information they want to disclose when transacting with service providers (although service providers may still require certain information for the transaction to take place) (OECD 2009, 17); and by enabling users to use various identity providers as well, so that their information is not stored at just one place (OECD 2009, 17). By endowing the data subject with an effective control over his/her own personal information, the principle of user-centricity reinforces the existing set of principles of data protection, i.e., specification, fair collection and accuracy, minimization, adequacy, and proportionality, contributing also to the effective enforcement of a 'right to be forgotten' (European Commission 2010e, 8).[32]

It is important to stress that the principle of user-centricity, which protects users' interests in the control and management of their personal data, should be articulated with the interests of other relevant actors, namely governments and the private sector. Governments may also have a legitimate interest in accessing and sharing personal data. Be it for preventing terrorist actions, fighting cybercrime, or taxation purposes, the governments may be entitled to have access to users' personal data (Rundle 2006).[33] This is, in fact, one of the greatest challenges of building a coherent and operational eID legal framework: to conciliate the interests of individual citizens with those of the private sector and governments.

14.6.1 Principle of Multiple Identities

As Jones and Martin observed, "[t]he issue of what we consider to be the identity of a person has become increasingly complex as we have made ever greater use of

---

[32] That is, "the right of individuals to have their data no longer processed and deleted when they are no longer needed for legitimate purposes".

[33] As examples of governments' legitimate interest in accessing and sharing personal data, Mary Rundle lists the following: "For example, in fighting cybercrime, governments want authority to require Internet service providers to hand over subscriber information, among other data. To facilitate travel, governments have agreed to certain standards for a global system of electronic identity information. For taxation of international e-commerce, OECD members are seeking reliable ways to identify taxpayers. To counter the financing of terrorists or other criminals, governments seek to ensure that information on originators of wire transfer is available".

the facilities and services that have been made available by developing technologies and the Internet. In the past, people normally had one identity, while in the current environment it is acceptable to maintain separate 'identities' for different aspects of our online interactions" (Jones and Martin 2010, 1).

Hence, any given person can have different partial identities which they use in different contexts. In the offline world, an individual person can be a citizen of a specific country, an employee or an employer of a given company, a mother and/or a daughter in her family context, etc. In this way, and

> ... as individuals take on many different roles in the course of their life, different set of characteristics, corresponding to these different roles, are used to represent their identity. Each of these 'partial identities' includes both inherited 'timeless' characteristics (such as nationality, gender, etc) and characteristics that they have acquired during their life (such as diplomas, competences, etc.), or that they have been assigned or issued to fulfil this role (such as a position, some sort of authority, etc.) (Nabeth 2009, 38).

In the online world, and in addition to the different partial identities of the "physical world", an individual may have different accounts on various social networking sites (or within the same one), or he/she may hold different avatars in online games and virtual realities. An individual may also use pseudonyms for other kinds of interactions and present his/her civil identity for certain business transactions. In the digital world, a person may reveal and register selected information about his/her identity (disclosing certain attributes and not others) to a wide array of different institutions and service providers. These entities will then, based upon that information, assemble the (digital) identity of that person which can then vary quite considerably from one institution to another. In this manner, "[d]igital identities (and identifiers) can be constructed and issued by different organizations like the telephone company, the Internet provider, a social networking site, or an insurance company" (Leenes et al. 2009, 15).

Unlike the physical world and face-to-face interaction, where it is hard to avoid the disclosure of certain identity features (namely the physical and observable ones), in the digital world it is possible to reveal specific identity attributes while concealing others. It is even possible to create new attributes and features of ourselves, crafting and maintaining one or many new identities.

This new world of possibilities carries, nevertheless, problems and risks. The first problem is that citizens will tend to accumulate many "digital personae." As it will be difficult to keep track of what each of these digital personae has done online, the privacy of that "multifaceted" person will become more difficult to protect. The second problem relates to the loss of control over information concerning those partial identities once they are released. As observed elsewhere, "[u]nlike goods, data cannot be reclaimed without the possibility that a copy is left behind in several possible places" (Leenes et al. 2008, 9).

In this way, the principle of multiple identities should ensure that IDM systems provide its users with the necessary tools to keep their multiple identities under control, even after data is disclosed.[34] In this way, the principle of multiple identities also reinforces the principle of data minimization, as more user control over data disclosure (dispersed throughout its various *digital personae*) will lead to less disclosure of personal data.

The principle of multiple identities also aims to address the risks of using the same digital identity in the online world. As Poullet observes, "[i]t is clear that, most often, the same identification method or access key is used in different databases with as a result that our identity can be cross-referenced more easily" (Poullet 2010, 11). Taking into account that certain countries store the national registration number in all governmental databases, this "increases the possibility of cross-referencing the information and thus, enhances the power of the state (...) vis-à-vis the citizen (Poullet 2010, 11)." From this point of view, the principle of multiple identities contributes to the prevention of identity cross-referencing, thus equilibrating the balance of power between the state and the citizen. The principle of multiple identities has already been contemplated and developed at the technological level. The PRIME project, in providing privacy-enhancing identity management tools for individuals, conceived the PRIME Console as an instrument to manage users' personal data. Among its various features, the PRIME Console—as the interface to the user's IDM systems—would allow users to create partial identities (pseudonyms) and to associate personal data to these identities (Leenes et al. 2008, 5). Another example of a technical implementation of the principle of multiple identities (and of the principle of unlinkability, as we shall see next) can be found in the TURBINE project.[35] This research program planned to enable an individual "to create different 'pseudo-identities' for different applications with the same fingerprint, whilst ensuring that these different identities (and hence the related personal data) cannot be linked to each other."[36]

14.7 Principle of Anonymity and Pseudonymity

As a general principle, identity systems should facilitate anonymity and pseudonymity. They should also provide detailed rules regulating the use of anonymous and pseudonymous data. Thus, an IDM legal framework should explicitly regulate the cases in which people have the right to conceal their identity data (anonymization) or to present a different identity (pseudonymization), and the circumstances under which their identities can be unveiled. In this way, IDM systems should by default al-

[34] The PRIME research project, in its technical proposals and prototypes for privacy-identity management tools, envisaged three central means of controlling multiple partial identities: tracking one's data trail, support for rights enforcement and policy enforcement. See Ibid.

[35] The TURBINE project aims to develop innovative digital identity solutions, combining the secure, automatic user identification thanks to electronic fingerprint authentication; and reliable protection of biometric data through advanced cryptography technology. For further information, see http://www.turbine-project.eu/

[36] Ibid.

low for anonymous and pseudonymous interactions. This would be the case for most commercial transactions. Commercial service providers only need to know a limited number of specific attributes of a given client (such as age, address and payment information) to be able to successfully transact with them. For this kind of transaction, customers and citizens could interact through anonymous or pseudonymous identities. The principles of anonymity and pseudonymity, in this sense, are clearly related to the existing principle of data minimization. Exceptions to these principles would be established for certain and specific interactions with the public administration, in which it would be necessary to identify and/or authenticate the civil identity of a citizen (as a tax payer, a pension or benefits receiver). Apart from this exception, the principles of anonymity and pseudonymity applied to IDM systems acknowledge a known truth in today's commercial transactions: it is not the identity of the user that matters but rather a specific attribute. Once again, technology is one step ahead of law, as the privacy and identity management tools conceived by the PRIME research project duly document:

> ... anonymous, or pseudonymous interactions are the default within PRIME ... PRIME supports different forms of pseudonymous with different characteristics with respect to linkability (Leenes et al. 2008, 8).

The principle of pseudonym, once applied and embedded in IDM systems, would entail—for instance—the creation of transaction pseudonyms for customers.[37]

However, it is important to bear in mind that the principles of anonymity and pseudonymity are not absolute and should have their limits explicitly defined. Therefore, the principle of anonymity and pseudonymity should not prevent strictly and legally contemplated possibilities and mechanisms of revealing users' civil identities when the latter have breached their legal obligations or duties (Leenes et al. 2008, 11).[38]

The introduction of the principles of anonymity and pseudonymity should encompass both the regulation of the cases in which anonymous and pseudonymous identities are permitted, and the circumstances in which these identities can be revealed.

14.7.1 Principle of Unlinkability

In today's world, online service providers—on the one hand—tend to exchange information regarding users' habits, tastes, and preferences in order to address potential

[37] As remarked in the PRIME project White paper: "If I know your name, I can try to get data about you through all sort of channels, which is much more difficult if I only know your transaction pseudonym ghT55897" (Ibid).

[38] There are mechanisms to reveal the identity of users when warranted and under strict conditions. As a concrete proposal, it is suggested that "[o]ne of these conditions would be the use of a trusted third party that is contractually bound to reveal the civil identity of the user under certain circumstances."

customers with tailored-made products, services and offers. Users, on the other hand, can have a legitimate interest in remaining unidentified to some service providers and identified to others. Users should have the freedom to make a choice. To help them do so, the principle of multiple identities and the principle of pseudonymity have been proposed. However, in order to effectively implement these principles, a further principle should be put forward: the principle of unlinkability. It is not enough to be able to create and maintain multiple identities and pseudonyms, it is also necessary to keep them apart from each other, that is, unlinkable.

Unlinkability is necessary in the context of 'pseudonymization.' The different pseudonyms used by an individual should be isolated from each other to prevent 'full' (or 'exact') identities to be linked to these partial ones and to prevent one partial identity (in the form of a pseudonym) to be associated and clustered with another partial identity. Thus, the principle of unlinkability prevents both *de-pseudonymization* and *de-anonymization* of data (Ohm 2009),[39] that is, their re-identification.

The principle of unlinkability should thus secure the same degree of protection to different pseudonyms and to anonymized information. Otherwise, "[l]inking identities that do not share the same degree of anonymity, or that contain different sets of attributes may allow others to overcome pseudonyms and discover the user's identity" (OECD 2009, 14).

The concern about the risk of possible linkage between different identity representations has already been addressed by technology designers. For example, the PRIME project conceived the creation of multiple private credentials from a single master certificate. These credentials, which could correspond to different pseudonyms belonging to the same person, would not be linkable to each other or to the master certificate from which they are derived. Another 'technical' implementation of the principle of unlinkability can be found in the Austrian sourcePin, which works as an 'obfuscated identifier' (Graux et al. 2009, 115). This number is never used to directly authenticate the user in eGovernment applications; it is used instead to generate sector-specific personal identification numbers (PINs). The unlinkability principle comes into play through the use of cryptographic one-way functions, according to which "sector-specific identifiers are calculated so that the citizen is uniquely identified in one sector, but identifiers in different sectors cannot be lawfully cross-related."[40]

Touching upon a number of proposals advanced here (and as a way to recapitulate the principles presented so far), Dumortier argues that:

> Future solutions will have to give data subjects maximum possibilities to control and steer the use of their personal data. They should be flexible enough to offer possibilities for the data subject to reveal only the identification data that are necessary for particular circumstances.

---

[39] *De-anonymization* of data is becoming a recurrent phenomenon, posing new risks to privacy.

[40] In also observing the principle of unlinkability, the same study points out that the Czech republic plans to implement a similar system to the Austrian one, "based on the introduction of a 'basic personal identifier', which will be used to derive a set of personal identifiers for specific contexts, so that each individual will be identified by a different identifier in each context" (Ibid.), avoiding thus for different eIDs to be cross-related and linked.

> Anonymous use of network services should be guaranteed where it is reasonably admissible. If unconditional anonymity—whereby the identity of the user is irreversibly lost—is not feasible, privacy-protecting schemes for conditional anonymity have to be established. Consequently the use of multiple 'virtual identities' will have to be regulated (Dumortier 2003, 69).

To sum up, the principle of unlinkability should orient IDM systems to considerably reduce the risk of cross-referencing between the different kinds of pseudonyms and multiple identities used by the same person.

14.7.2 Principle of Negotiation

The principle of negotiation aims to introduce a greater degree of flexibility in the current regulatory model of data protection. The implementation of this principle would allow users to negotiate the terms and conditions of disclosure of their identity information with service providers as a prior step to the already contemplated legal possibilities of accessing, correcting, and deleting personal data. This would also strengthen the requisite consent, which today is deprived of any real meaning and force. In fact, today, users have to comply with the demands of service providers if they want to gain access to their services. There is a clear imbalance between the bargaining positions of these two actors. The user has to provide the data asked for and has no choice but to accept the privacy conditions stipulated by the service provider.[41] As a counterbalance to this currently pursued 'take it or leave it' approach, which undermines the idea of user consent, the principle of negotiation would endow users with more control over the processing of their own personal identity data. It derives from the principle of user-centricity and aims to reinforce and go beyond consent as a requirement for the lawful processing of personal data. The principle of negotiation thus serves to help the coming generation of IDM systems to empower users with tools that allow them to negotiate the conditions of the protection and management of their identities with service and identity providers. The PRIME project has already experimented with this idea. As stated in its White Paper:

> PRIME replaces the 'take it or leave it' approach to privacy policies by a system of policy negotiation. Both parties can express different kinds of policies relating to authorizations, data handling, and preferences. The user is assisted by the PRIME Console which helps in setting personal preferences and requirements, in converting preferences from machine readable form to human readable form and vice versa, and in automatically negotiating the user's preferences with the other party.[42]

The principle of negotiation entails that users express their preferences and negotiate the terms of their identity data disclosure with service providers.

---

[41] See (Leenes et al. 2008, 3).

[42] Ibid., 7.

14.7.3 Principle of Portability

This is a principle that is not derived from a privacy 'raison d'être', but from a strict and specific identity rationale. Privacy, seen from a more classical and negative perspective as a right to opacity or to seclusion, deals mostly with the concealment of certain private aspects from public knowledge and the protection of disclosed information from the public sphere. Identity, on the other hand, deals with the transmission of information to the public sphere, namely with its correct expression and representation to the public eye. According to this view, an important principle related to the protection and management of one's identity is the possibility of carrying one's identity information with oneself, that is, the principle of portability. This principle underlines the fact that preventing someone from taking his/her constructed identity information to another place constitutes an unjustified hindrance to the protection and management of one's identity.

The principle of portability is particularly relevant for reputations associated to eIDs, that is, for valuations and ratings of someone's identity attributes or skills expressed within a given online community or network. The construction of reputations in the online world is a growing trend. It is increasingly common for citizens and users to acquire reputations in the form of financial credibility, work recommendations issued by colleagues or other skills rating made by peers. However, and despite the development of these reputation circles, it is difficult—in the online world—to transfer reputations from one context to another. The move from one social network to another usually implies the need to build one's reputation from scratch. It is even more difficult to transfer one's reputation without revealing one's identity (be it the civil or a pseudonymous one). As noted in the PRIME project, "[t]ransferring reputations from one context to the next, without linkability of the underlying partial identities, is a feature that will prove valuable in online interactions."[43] Technology, once again, anticipates law, as PRIME proposes a technical system to handle this kind of reputation transfer through the issue of anonymous credentials. Here we have an interesting combination of the principles of portability and anonymity.

In summary, the principle of portability argues that online identities (including their reputations) should be inherently portable and not irremediably anchored to any given service or identity provider. Taking into account that the current data protection model is overly privacy-oriented (Andrade 2011b), this principle is innovative. The existing data protection model "only" allows for the right to access, correct and delete private information because, from a privacy point of view (as a seclusion instrument of opacity), it does not make much sense to talk about the right to move private information from one place to another. However, and as mentioned before, a right to portability makes sense in terms of an identity rationale. From an identity management point of view, it is crucial to have the possibility to carry our identity information from one service provider (e.g., a social network) to another.

[43] Ibid., 10.

14.7.4 The Authentication Source Principle

This principle derives from EU Member States' national legislations (namely from National Registers Acts, eGovernment Acts, and other pieces of national and regional legislation). According to a study on eID interoperability, "this principle implies that for each given attribute (piece of identity data), one and only one source is considered to be authentic, i.e., correct" (Graux et al. 2009, 112). Other sources for that attribute are dispensable.

As observed in the mentioned study, this principle "is relevant from a cross border interoperability perspective, because a consistent application of the authentic source principle means that a single correct source exists for each bit of information, which can facilitate the access and exchange of this information (Graux et al. 2009, 81)."[44]

This principle serves to help users manage and protect their digital identity, preventing them from having to provide the same information time and time again, ensuring that there is only one place in which information needs to be updated or corrected (Graux et al. 2009, 112). Thus, this principle reinforces the existing principle of data accuracy.

14.7.5 Principle of Technological Assistance

Law and legal solutions can only go so far. This is the case, for example, in the legal impossibility for the majority of EU Member States to allow (national) identity numbers to be used outside the Member State itself, along with the legal impossibility to establish a unique identifier to be used across every EU Member State. As the idea and project of a pan-European eID can only be implemented if citizens from one European country are able to use their eIDs to access services in a different EU country, Member States need to have some form of identifier when other EU national citizens make use of their services. This is the point where technical solutions must be devised and implemented. Given the legal impossibilities mentioned above, technology is the solution. In this way, and taking into account that one of the most problematic issues in cross-border IDM systems is the need for Member States to have some form of identifier when a foreign citizen makes use of their services, a "possibility to mediate this issue may be to use a one-way transformation function that unequivocally transforms a foreign ID number into one that may be locally stored" (Leenes et al. 2009, 32).

This example demonstrates that law can (and should) be complemented by technology so that they both form part of the regulatory framework. In other words, technology will fill the natural limits of law and assist the latter in enforcing its rules and dispositions.

---

[44] For more information on which countries surveyed in the PEGS study subscribed to an authentication source principle and to what extent that this principle has impacted their identity management policies, see (Graux et al. 2009, 81–84).

Several steps have already been taken in this direction. Article 29 Data Protection Working Party (1999), in Recommendation 1/99, explicitly stated that software and hardware industry products should provide the necessary tools to comply with EU data protection rules. This statement is an important manifestation of the principle of technological assistance. Other important steps taken on the implementation of this principle can be found in the support and development of Privacy Enhancing Technologies (PETs) and the "Privacy by Design" approach, as well as in the increasing trend of imposing liability on terminal equipment manufacturers and information system designers by Data Protection Authorities.

The principle of technological assistance may, for example, lead to the imposition of technical standards on terminal equipment manufacturers in order to ensure compliance in terms of digital identities protection and management. It may also lead to the construction of new and fully fledged rights.[45]

14.8 Conclusion

In its "Europe 2020" Strategy, the Commission alerts us to the need to overcome "the fragmentation that currently blocks the flow of online content and access for consumers and companies" (European Commission 2010c, 19) within the envisaged digital single market. Often in the offline world today, business and citizens still need to deal with 27 different legal systems for the same transaction (European Commission 2010c, 18). As this article has attempted to demonstrate, there is no specific legal framework for eID. The protection and management of electronic identities is currently regulated by a patchwork of different pieces of EU and national legislation, along with implemented technological initiatives. Many solutions and innovations, both at the technical and legal levels, have been developed by Member States and introduced into their national regulations. As an example, and going beyond the applicability of their generic data protection regulations, a number of Member States have subjected some or all unique identifiers used in their administrations to additional protection mechanisms (Graux et al. 2009, 115).[46]

Nevertheless, the existing legal and technological solutions, current EU and national laws, along with the present technical arrangements seem insufficient to cover the limitations of the current and fragmented EU legal framework for the eID area.

This article, contributing to the discussion on the need for a shared eID legal framework for the EU, has suggested a number of new legal principles that take into account the new dynamics of and demands for the protection and management of electronic identities.

[45] In this context, see Poullet's construction of a "new privacy right: the right to a privacy compliant terminal with a transparent and mastered functioning by its users", in (Poullet 2010, 27). Such right, as heavily based on technological components and technical requisites embedded into terminal equipments, constitutes what I would call a derivation of the principle of technological assistance.

[46] Member States have also implicitly introduced in their legislation the already alluded authentic source principle.

The principles listed in this article constitute the backbone of an eID legal framework that puts users at the center and empowers them with the means, both legally and technically designed, to remain anonymous or to use pseudonyms, to manage multiple identities, to keep them separate and irretraceable, to negotiate the terms of their identity management preferences, to carry and freely move their identity information, among other possibilities. Furthermore, the listed principles would contribute to an even stronger protection of users' privacy, strengthening trust, confidence and security in the online world of electronic communications and transactions.

More than technology, we need law. We need a shared encompassing legal framework, which guarantees that electronic identities can unobtrusively travel across different EU Member States, enabling access to services and transactions. The list of new principles described in this article aims to orient and contribute to this endeavor.

Acknowledgments Thanks to Ioannis Maghiros for very helpful comments.

Appendix: Terminology

This annex provides a general overview of the most relevant concepts, terms, and notions regarding electronic identity (eID) and electronic identity management systems (eIDM).[47] It lays down the terminological grounds on which the legal analysis provided in the article is based.

The processing of electronic identities involves a wide array of technical terms that must be clarified in order to understand what the creation of a pan-European eID entails and implies. In fact, in order to discuss the creation of a European electronic identity and the legal challenges to such an endeavor, we need first to understand what electronic identity is. In order to comprehend the notion of **electronic identity**, we also need to understand other related and important concepts and processes, such as **attributes**, **credentials**, **identification**, **authorization,** and **partial identities**.

Starting with the basics, we should first distinguish between an entity and a quality. Any specific entity (a human being, for instance) has a number of qualities or attributes. The sum of these attributes make up one's identity (namely one's exact identity).[48] The notion of **"attribute"** is of utmost importance because, depending on the context or on the attribute in question, it can refer to a **"full identity"** (when it is used to unequivocally identify a given individual) or to a **"partial identity"** (when it refers to an identity characteristic of a given person without revealing his/her full

---

[47] This section relies upon various studies that have provided detailed "glossary-type" definitions of the various terms and notions employed in the area of eID. This is the case of the FIDIS project, the MODINIS, PrimeLife, STORK and specific studies, such as Pfitzmann and Hansen (2010).

[48] In order to distinguish the concept of exact identity from the one of partial identity, I shall also use the term 'full identity'.

or entire identity, (Pfitzmann and Hansen 2010, 31)[49] that is, without identifying him/her in absolute terms).[50]

Another important term is **'identifier'**. A unique identifier can be defined as "an attribute or a set of attributes of an entity which uniquely identifies the entity within a certain context" (Graux et al. 2009, 113).[51] Two classes of identifiers can be distinguished which are primary digital identifiers, which are directly connected to a person (name, address, mobile phone number, password, or electronic signature) and secondary digital identifiers, which are not directly connected to an individual (cookies, IP addresses, or RFID tag numbers).

Also relevant is the notion of **identity claims**, which is intimately connected with **credentials**. In the offline world, claims that an individual is of certain age or lives at a given address are certified by third parties, namely by the State when it issues certificates supporting these claims (e.g., passport, ID card, or driver's license). In the online world, there are entities specifically designated for the certification of identity claims. "[O]nline certifiers can, by means of cryptographic techniques (security tokens), vouch for certain claims in a secure manner that cannot be tampered with" (Leenes et al. 2008, 8). While paper-ID aims to identify physically present individuals, electronic ID provides credentials to enable citizens to remotely identify themselves. While conventional ID functions on the basis of personal appearance and paper-based proof of identity (certificates, identity cards, showing one's signature or photograph), eID is based upon more complex processes and mechanisms.

Such processes of identity recognition are developed and carried out by **identity management** (IDM) **systems** . The overall objective of eIDM systems is to associate information with people, enabling transactions between different parties in an ecosystem of mutual confidence and trust. IDM, at a more general level, can be defined as "[s]ystems and processes that manage and control who has access to resources, and what each user is entitled to do with those resources, in compliance with the organization's policies (Leenes et al. 2008, 1). On the administrators' side, IDM systems allow organizations, businesses, companies, and institutions to grant, control, and manage user access to information, applications, and services over a wide range of network services. This access is conducted through authentication methods (passwords, digital certificates, hardware or software tokens) and authorization

[49] The distinction between full and partial identity I here propose presents a different *nuance* from the one advanced by Pfitzmann and Hansen regarding complete and partial identities: "A partial identity is a subset of attribute values of a complete identity, where a complete identity is the union of all attribute values of all identities of this person", in (Pfitzmann and Hansen 2010, 31). While for these authors, partial identities may encompass attributes through which a person can be identified; I define partial identities as covering those attributes that do not necessarily identify a given person, classifying the ones that do as full identities. In sum, the difference between full and partial identities has to with identifiability, equating to the difference between information that relates to an identified or identifiable person, and information that does not.

[50] As we have seen, this specific characteristic of the processing of eIDs enables the use of multiple identities by the same individual.

[51] Though numbers (such as national register numbers, VAT numbers, certificate numbers, etc) are the most common (and, in fact, the default) form of unique identifier, "any sufficiently unique set of attributes pertaining to a specific entity can serve the exact same purpose" (Graux et al. 2009, 113).

rights. On the users' side, IDM systems provide (or should provide) them with the necessary tools to manage their identities and control the use of their personal data. IDM systems can widely vary in terms of applications requiring different degrees of identification, access control, and credentials.

The functioning of IDM systems involves two main processes or components which are **identification** and **authentication**.

While the purpose of identification is to "link a stream of data with a person," (Myhr 2008, 77) the process of authentication can be defined as "the corroboration of the claimed identity of an entity or of a set of its observed attributes" (Graux et al. 2009, 113). In this respect, a distinction can be made between an authentication process that determines one's exact identity and an authentication process that determines one's specific quality or attribute (partial identity). In the latter situation, a given application authenticates the entity only to verify whether he or she has a specific required quality (such as being an adult, being a resident of a given region, city, etc).[52] The process is thus carried out without revealing or knowing who exactly the person is. "The application determines the entity's status, not his/her identity (Graux et al. 2009, 113)." In the other situation, the application authenticates one person by determining his/her exact identity. Here, authentication processes sufficient information to distinguish and select one individual from all others, one specific person out of all mankind.

In other words, the authentication process corresponds to the verification of the authenticity of an identity. Authentication must effectively prove that a person has indeed the identity that he/she claims to have. In this way, the authentication process requires elements/instruments such as identity cards, passports, or a key (proving to a technical infrastructure the right to access). In brief, authentication is the process of associating and permitting a specific identity or set of identity-related credentials to access specific services.

The authentication phase thus requires the presentation of a **"credential"**, i.e., "data that is used to authenticate the claimed digital identity or attributes of a person (OECD 2007, 12). Examples of digital credentials include an electronic signature, a password, a verified bank card number, a digital certificate, or a biometric template (OECD 2009, 6). Several actors can be identified in the authentication process of electronic identities. Within the eGovernment area, and as explained in one of the deliverables of the STORK project:

> the eID process generally comprises five roles, which will be present in most Member States' eID models. First of all, there is an (1) authority that registers the citizen that wants to obtain an eID. This authority is related to the (2) organization that provides an electronic token and the credentials (hence, the eID) that can be used in eGovernment authentication. In addition, the process of authentication comprises the role of (3) an authority that authenticates the token that is used by the citizen. Next to the authenticating party, there is (4) a relying party that depends on this electronic authentication for the purpose of interaction or transaction, e.g. in the eGovernment service. Of course, there is also (5) an entity that claims a particular identity (e.g., the citizen or a delegate) (Leenes et al. 2009, 25–26).

---

[52] (Graux et al. 2009, 113) As we shall see, it is based on this type of authentication that I will argue in favor of a principle of multiple identities.

In a European context, the concept of **interoperability** is of paramount importance. Electronic identities will have little value for free movement of persons, goods, services and capital, and the stated objectives of constructing a fully operational single digital market, if they are not recognizable outside national borders and across different EU Member States. Interoperability is generally defined as "the ability of a system or a product to work with other systems or products without special effort on the part of the user, covering both the holder of the eID and the counterparty on the receiving end of electronic communication" (Myhr 2008, 77). It has both technical and legal/organizational dimensions.

A pan-European eID can be roughly defined as an "eID issued to persons, mainly natural persons but also legal persons (enterprises, etc.), which can be used in cross-border transactions, and is accepted by all states within the EU (Myhr 2008, 77)." A pan-European eID is closely connected to the notion of interoperability, which "mainly comprises the possibility of a citizen from one country to use the authentication system from this country to have access to an application in another country" (Leenes et al. 2009, 15).[53]

To conclude, and in line with previously mentioned proposals for an eID terminology (Pfitzmann and Hansen 2010), the term eIdentity is used in this paper to indicate a set of personal information and data relevant to a human's identity when stored and transmitted via electronic systems, including but not limited to computer networks (that is, digitized). Taking into account that, in the offline world, an identity is established from an extensive set of attributes associated with an individual (e.g., name, height, birth date, employer, home address, passport number), it is relevant to note that, in the online world, an individual identity can be established by combining both real world and digital attributes (OECD 2009, 6) (such as passwords or biometrics[54]). Electronic identities are thus identities that are constructed out of the various identity-attributes related to a given person (which together compile his/her identity information), processed electronically by technically supported IDM systems, and that are then recognized by public and private entities (such as national governments and private companies) (Leenes et al. 2009, 16).[55]

[53] Typical use cases of an interoperable eID, which are currently being developed by Stork, "are when a citizen of country X can use the electronic identity and authentication scheme of his or her home country for a license application, or when a student from country Y can register for a scholarship in country X with her home authentication scheme, without a need to register herself in country Y" (Leenes et al. 2009, 16).

[54] "Biometrics are measurable biological and behavioral characteristics and can be used for strong online authentication. A number of types of biometrics can be digitized and used for automated recognition. Subject to technical, legal, and other considerations, biometrics that might be suitable for IDM use include fingerprinting, facial recognition, voice recognition, finger and palm veins", (OECD 2009, 7).

[55] From a more technological perspective, the technical solution most commonly used in electronic communication identifying the person/holder of eID is PKI (public key infrastructure), which uses a pair of 'keys': a public key used for signing an electronic document and a private key linked to a certificate and used by the receiver to validate the signature. In this way, PKI can be used to detect

References

Andrade, Norberto Nuno Gomes de. 2011a. Data protection, privacy and identity: Distinguishing concepts and articulating rights. In *Privacy and identity management for life: 6th Ifip Wg 9.2, 9.6/11.7, 11.4, 11.6/Primelife International Summer School, Helsingborg, Sweden, August 2–6, 2010, revised selected papers*, eds. S. Fischer-Hübner, P. Duquenoy, M. Hansen, R. Leenes and G. Zhang, 90–107. Berlin: Springer.

Andrade, Norberto Nuno Gomes de. 2011b. The right to privacy and the right to identity in the age of ubiquitous computing: friends or foes? A proposal towards a legal articulation. In *Personal data privacy and protection in a surveillance era: Technologies and practices*, eds. C. Akrivopoulou and A. Psygkas, 19–43. Hershey: Information Science Publishing.

Article 29 Data Protection Working Party. 1999. Recommendation 1/99 on invisible and automatic processing of personal data on the internet performed by software and hardware.

Craig, Paul. 2008. The treaty of Lisbon, process, architecture and substance. *European law review* 33 (2): 137–66.

Dumortier, Jos. 2003. Legal considerations with regard to privacy protection and identity management in the information society. *112e rapport annuel, hochschüle für technik und architektur biel, tilt* 15: 66–69.

European Commission. 2003. First report on the implementation of the data protection directive (95/46/EC). Brussels.

European Commission. 2005. Signposts towards eGovernment 2010.

European Commission. 2007. A roadmap for a pan-European eIDM framework by 2010—V.1.0.

European Commission. 2010a. Delivering an area of freedom, security, and justice for Europe's citizens: Action plan implementing the Stockholm programme. Brussels.

European Commission. 2010b. A digital agenda for Europe.

European Commission. 2010c. Europe 2020: A strategy for smart, sustainable and inclusive growth. Brussels.

European Commission. 2010d. Towards interoperability for European public services.

European Commission. 2010e. A comprehensive approach on personal data protection in the European union. In *European Commission*. Brussels.

Graux, Hans, Jarkko Majava, and Eric Meyvis. 2009. eID interoperability for PEGS—update of country profiles—analysis & assessment report.

Jones, Andy, and T. Martin. 2010. Digital forensics and the issues of identity *Information security technical report* 1–5.

Leenes, Ronald, Jan Schallaböck, and Marit Hansen. 2008. Prime (privacy and identity management for Europe) white paper.

Leenes, Ronald, Bart Priem, Carla van de Wiel, and Karolina Owczynik. 2009. Stork—towards pan-European recognition of electronic IDs (eIDs)—D2.2—report on legal interoperability.

Lisbon Ministerial Declaration, eGovernment Conference. 2007. Reaping the Benefits of eGovernment of the PortuguesePresidency of the European Council and of the European Commission. Accessed 19 Sept 2007, Lisbon, Portugal.

Manchester Ministerial Declaration, eGovernment Conference. 2005. Transforming Public Services of the United Kingdom Presidency of the European Council and of the European Commission. Accessed 24 Nov 2005, Manchester, United Kingdom.

Modinis-IDM-Consortium. 2005. Modinis Study on identity management in eGovernment. Common terminological framework for interoperable electronic identity management—Consultation Paper V.2.01.

---

if a document has been changed without authorization after it was sent. In addition, eIDs "may be stored on smart cards or other devices but may also be received from a central authority during an authentication process" (Leenes et al. 2009, 16).

Modinis-IDM-Consortium. 2006. Modinis study on identity management in eGovernment, identity management issue interim report II1.

Myhr, Thomas. 2008. Legal and organizational challenges and solutions for achieving a pan-European electronic ID solution or I am 621216–1318, but I am also 161262–43774. Do you know who I am? *Information security technical report* 13 (2): 76–82.

Nabeth, Thierry. 2009. Identity of Identity. In *The future of identity in the information society: Challenges and opportunities*, eds. Kai Rannenberg, Denis Royer and André Deuker, 19–69. Berlin: Springer.

OECD. 2007. OECD recommendation on electronic authentication and OECD guidance for electronic authentication.

OECD. 2009. The role of digital identity management in the internet economy: A primer for policy makers.

Ohm, Paul. 2009. Broken promises of privacy: responding to the surprising failure of anonymization. *University of Colorado law legal studies research paper no. 09–12* (2009).

Pfitzmann, Andreas, and Marit Hansen. 2010. A terminology for talking about privacy by data minimization: Anonymity, unlinkability, undetectability, unobservability, pseudonymity, and identity management (version V0.34).

Poullet, Yves. 2010. About the e-privacy directive: towards a third generation of data protection legislation? In *Data protection in a profiled world*, eds. S. Gutwirth, Y. Poullet and P. de Hert, 3–30. Dordrecht: Springer.

Reflection group on the Future of the EU 2030. 2010. Project Europe 2030. Challenges and opportunities—a report to the European council by the reflection group on the future of the EU 2030.

van Rooy, Dirk, and Jacques Bus. 2010. Trust and privacy in the future internet—a research perspective. *IDIS—identity in the information society* 3 (2): 397–404.

Rundle, Mary. 2006. International personal data protection and digital identity management tools. *Berkman Center research publication no. 2006–06.*

Chapter 15
From the Protection of Data to the Protection of Individuals: Extending the Application of Non-discrimination Principles

Daniel Le Métayer and Julien Le Clainche

15.1 Introduction

The unprecedented deployment of information and communication technologies has made possible the development of myriads of new services but it has also given rise to a variety of threats to individual rights that must be taken very seriously:

- Data protection rights: the extensive use of data collection and knowledge inference techniques[1] undermines the effectiveness of existing data protection regulations.
- Privacy: the facilities offered by the internet to publish and have access to information lead to more and more privacy infringements.
- Non-discrimination: automatic data processing techniques can be applied to huge amounts of available information to build (individual and group) profiles which can be used to treat people differently, which makes it easier to commit large scale, discreet discriminations.

Simultaneously, the evolution of the technology has also increased the interactions between these three types of rights. For example, there is no doubt that misuses of personal data can adversely affect privacy and self-development (resulting in the unwanted disclosure of personal data to third parties, in identity theft, harassment through email or phone calls, etc.), or lead to a loss of choices or opportunities

This work was funded by an INRIA postdoctoral position.

[1] We use the term "inference" here to denote the derivation of new knowledge on the basis of available data. This new knowledge may typically involve facts (e.g. a taxi driver's address inferred from the GPS data provided by his cab) or predictions (such as the likely destination of a vehicle on the basis of previous journeys).

D. Le Métayer (✉)
Institut National de Recherche en Informatique et Automatique (INRIA), Grenoble Rhône-Alpes Research Center, 655 avenue de l'Europe, 38334 Montbonnot, France
e-mail: Daniel.Le-Metayer@inria.fr

S. Gutwirth et al. (eds.), *European Data Protection: In Good Health?*,
DOI 10.1007/978-94-007-2903-2_15, © Springer Science+Business Media B.V. 2012

(e.g. enabling a recruiter to obtain information over the internet about political opinions or religious beliefs of a candidate and to use this information against him). As a matter of fact, privacy breaches and discriminations based on data processing are probably the two most frequent and the most serious types of consequences of misuses of personal data.

In this chapter, we focus on one of these interactions, the relation between personal data protection and anti-discrimination regulations, and argue that an extended application of the latter can help strengthening the former. We first review the increasing need for data protection before identifying the weaknesses of existing regulations and their consequences (Sect. 15.2). We then turn to anti-discrimination regulations and make a comparison with personal data protection considering both the types of data concerned and their respective *modus operandi* (Sect. 15.3). From this comparison, we make proposals for a stronger synergy between data protection and anti-discrimination regulations (Sect. 15.4) and draw some conclusions (Sect. 15.5). As far as legal systems are concerned, we focus on European regulations in this chapter, with examples mostly drawn from the French legislation and jurisprudence.

15.2 Data Protection: Increasing Needs, Decreasing Effectiveness

As Simon Nora and Alain Minc emphasised already in 1978 in their report on the computerisation of society, "*this growing overlap between computers and telecommunications, which we will call "telematics", will not form an additional network, but another kind of network (...) It will transform our cultural model (...) it constitutes a common factor enabling and accelerating all other technologies. Especially insofar as it radically alters the processing and storage of information, it will change the nervous system of organisations and of society as a whole.* [Telematics], *unlike electricity, does not carry an inert current, but information, i.e. power*" (Nora and Minc 1978). Associating information with power naturally leads to a major issue which is the potential means of control of this power and the establishment of adequate counter-powers to keep a balance between entities which can collect, produce and have access to information and individuals who do not have the same abilities or can be the targets of such collections or treatments of information.

Looking at it more closely, information actually confers two different, yet complementary, types of power: the *power of knowledge* and the *power of action*.[2] As a first approximation, the collection of information can be associated with the power of knowledge when the use of information seems more related to the power of action. Obviously, personal information is the first type of information which confers power on individuals. Personal data regulations therefore constitute a significant part of the necessary counter-powers. From the legal point of view, the European regulation on

[2] The power of action is a translation of the Latin maxim attributed to Bacon (1597). For more substantial developments, see Stehr (2000).

personal data protection is based on *a priori* procedures (typically notifications and authorisation requests): no matter whether any individual suffers any actual loss or harm, the failure to complete prior formalities, even without malicious intentions, is sufficient to constitute a criminal offence.[3] We can thus argue that, to some extent, personal data protection was originally intended to control the power of knowledge. In contrast, privacy protection and anti-discrimination regulations both relate more directly to the control over the power of action: an offence is established only if an individual has actually suffered from a privacy breach[4] or a detrimental decision is made unlawfully on the grounds of a discriminatory criterion.[5] These differences of approaches can be well justified by historical and theoretical reasons and could lead to complementary means to protect individual rights. We argue however that the *a priori* procedures which form the basis of data protection regulations are weakened by the recent technological and legal evolutions (Sect. 15.2.1) and this weakening has in turn an impact in terms of privacy and discrimination (Sect. 15.2.2).

15.2.1 A Priori Checks: A too High and too Low Barrier

Under European regulation, the completion of prior formalities by the data controller is one of the conditions for personal data processing to be lawful.[6] These formalities however do not necessarily lead to thorough checks by the national data protection authorities. For example, in France the notification procedure[7] does not impose any verification from the French data protection authority (CNIL), which has only to record it and issue a receipt. In contrast, the authorisation procedure under Art. 25 of the French data protection law[8] does require more extensive checks as the CNIL has to provide a reasoned decision. In practice, the CNIL may prohibit the processing, authorise it, or issue an authorisation with reservations, which amounts to authorising the processing if specific modifications or additional guarantees are implemented.

---

[3] Art. 24 of European Directive 95/46/EC: "*The Member States shall adopt suitable measures to ensure the full implementation of the provisions of this Directive and shall in particular lay down the sanctions to be imposed in case of infringement of the provisions adopted pursuant to this Directive*".

[4] Art. 9, § 1, French Civil Code: "*Everyone has the right to privacy*". See also Directive 2009/136/EC of 25 November 2009 amending Directive 2002/22/EC on universal service and users' rights relating to electronic communications networks and services, Directive 2002/58/EC concerning the processing of personal data and the protection of privacy in the electronic communications sector and Regulation (EC) No 2006/2004 on cooperation between national authorities responsible for the enforcement of consumer protection laws.

[5] Art. L. 225–1 (and the following) of the French Penal Code. See also, Council Directive 2000/43/EC of 29 June 2000 implementing the principle of equal treatment between persons irrespective of racial or ethnic origin and Council Directive 2000/78/EC of 27 November 2000 establishing a general framework for equal treatment in employment and occupation.

[6] Art. 18 and following of European Directive 95/46/EC.

[7] Art. 23 and 24 of French law 78–17 of 6 Jan. 1978.

[8] No mention will be made here of data processing for security purposes on behalf of the state, under Art. 26 and Art. 27 of the law of 6 January 1978.

However, the *a priori* control procedures have been weakened by the transposition of Directive 95/46/EC into French law, leading to a revision of the data protection law of 6 Jan. 1978 (6 August 2004). In fact, not only have notifications become the common, by default, procedure, but the appointment of a "personal data protection official"[9] releases organisations from any obligation of notification. This weakening of *a priori* controls has been offset by an increased emphasis on *a posteriori* checks, at least with respect to personal data processing in the private sector.[10] This evolution is justified by the unprecedented proliferation of data processing in the new information society and the practical impossibility to submit all these treatments to *a priori* checks.

It is already the case today with the internet, but the phenomenon will take new proportions with "ubiquitous computing"[11] or "ambient intelligence" (RFID chips, sensors, the "internet of things", etc.): information and communication technologies will make it more and more easy to collect vast amounts of personal data automatically and without notice from the individuals. The impact of these technologies is even amplified by the increasingly sophisticated knowledge inference and data mining techniques which make it possible to produce new personal data and accurate profiles, or to de-anonymise data, using ever larger volumes of available information. It should be pointed out that the origins of this evolution are not exclusively technical but also social, since many internet users deliberately contribute to populating this gigantic database.[12] Another consequence of the development of knowledge inference techniques is that the frontier between anonymous data and identifying data tends to blur and to evolve: data which can be considered as anonymous at a given time in a given context can become identifying later on because new, seemingly unrelated data has been released, generated or forwarded to a third party, giving rise to the possibility of "re-identification" (see Ohm 2010; Narayanan and Shmatikov 2010). Several authors have already pointed out that, far from being a panacea, anonymisation should rather be viewed with extreme caution.[13] Actually, as stressed

---

[9] Art. 18 of European Directive 95/46/EC and Art. 22 III of French law 78–17 of 6 Jan. 1978.

[10] When the law of 6 January 1978 was amended by the law of 6 August 2004, the formalities preceding the constitution of government data processing were slimmed down considerably, while the powers of the CNIL to carry out *a posteriori* verifications are not binding on the state. For further information on the powers of the CNIL with respect to public-sector data records, see Le Clainche (2005).

[11] "Ubiquitous" computing refers to the integration into the human environment (e.g. within objects, clothes and even, in extreme cases, implanted under the skin) of a variety of small computing devices (sensors, actuators, etc.) with the capacity to spontaneously collect data, communicate and perform simple computations.

[12] Although such behaviour often results from the lack of awareness of the subjects and their ignorance of the risks of de-anonymisation and undesired use of the disclosed data.

[13] Paul Ohm (2010): *"These scientists have demonstrated they can often 'reidentify' or 'deanonymize' individuals hidden in anonymized data with astonishing ease. By understanding this research, we will realize we have made a mistake, labored beneath a fundamental misunderstanding, which has assured us much less privacy than we have assumed. This mistake pervades nearly every information privacy law, regulation, and debate, yet regulators and legal scholars have paid it scant attention".*

by Serge Gutwirth and Mireille Hildebrandt (Gutwirth and Hildebrandt 2010), the legal status of the profiles themselves is another striking illustration of the limitations of European data protection regulation: one could argue that group profiles built from anonymised data fall outside the scope of Directive 95/46/EC, and are instead ruled by intellectual property laws, thus offering protection to those building these profiles rather than to the individuals, even when these profiles may be used to run procedures or take decisions (unilaterally, possibly unfairly, and generally without any requirement to provide motivations) affecting them.

To summarise, we have argued in this section that *a priori* checks, even though they could represent in theory a very strong protection, are no longer effective enough, and become more and more both a too high barrier (considering the huge amount of data flows in the digital society) and a too low barrier (because certain types of data which can have an impact on our individual life, such as anonymous profiles, can still escape their application field). In the next section, we study the consequences of these limitations in terms of privacy and discrimination.

15.2.2 Impact in Terms of Privacy Breaches and Discrimination

As argued in the introduction, the increased inadequacy of the *a priori* controls which form the basis of data protection regulations can lead to misuses of personal data with strong impact in terms of privacy and discrimination. As an illustration, the teacher grading website "note2be.com" was prosecuted for two offences: failing to obtain prior consent for processing personal data and privacy breach. The French data protection authority and the judge took the view that consent was necessary, but they came to different conclusions with respect to the alleged privacy breach: the CNIL considered the disclosure of the workplace address as a privacy breach, while the judge held the opposite view.[14] Another recent case illustrates the potential risks in terms of discrimination: in a public report, the non-profit organisation "SOS Racisme" claimed that ethno-racial data records were a tool for discrimination (Thomas 2009) and criticised Air France for processing ethnic data records on his cabin personnel to meet customers' requests.[15] More generally, the development of profiling techniques which are typically based on the analysis of personal data (even if the data may be subsequently "anonymised" in a more or less robust way) has the effect of increasing the differences of treatments between individuals, both in the private sector (e.g. services offered or prices set on the basis of profiles) and in the public sector (e.g. monitoring for security purpose). As a matter of fact, the first reason for elaborating profiles is often to be able to provide personalised services, which in many situations can be perfectly legitimate[16] but can also imperceptibly

---

[14] For legal study of these rulings, see Lepage (2008).

[15] Customers favouring personnel from a particular ethnic origin.

[16] Actually, personalization has always been a common business practice, and the point made here is obviously not to object to personalization in general or even to stigmatise it.

turn into various forms of discriminations. This widespread use of profiling and the associated risks, especially as regards discrimination, have already been studied and denounced by a number of lawyers (Gutwirth and Hildebrandt 2008; Gutwirth and Hildebrandt 2010; Hildebrandt 2009; Zarsky 2002) and sociologists (Lyon 2003).

15.3 Non-discrimination: Protecting Individuals

Even though they have different origins and objectives, and they are governed by different bodies of rules, the rights for personal data protection and non-discrimination interact in different ways and it can be helpful to review their differences and similarities before drawing some lessons and providing suggestions to improve their effectiveness. Considering the scope of this study, namely the new risks posed by data processing technologies, we distinguish two complementary aspects of data protection and non-discrimination rights: the types of data which are covered by the protections (Sect. 15.3.1) and the types of controls which are provided on these data (Sect. 15.3.2).

15.3.1 Similar Types of Data

Let us first consider the types of data covered by non-discrimination regulations. Article L. 225–1 of the French Penal Code prohibits the use of certain types of data in specific contexts. These categories of data and contexts are extensively enumerated in the law:

- Art. 225–1 (excerpt): "*Discrimination comprises any distinction applied between natural persons by reason of their origin, sex, family situation, state of health, handicap, sexual morals, political opinions, union activities, or for being a member or not a member (or supposed to), of a given ethnic group, nation, race or religion*".
- Art. 225–2: "*Discrimination defined by article 225–1, committed against a natural or legal person, is punished by two years' imprisonment and a fine of € 30,000 when it consists:*

  1. *of the refusal to supply goods or services;*
  2. *of obstructing the normal exercise of any given economic activity;*
  3. *of the refusal to hire, the sanction or the dismissal a person;*
  4. *of subjecting the supply of goods or services to a condition based on one of the factors referred to under article 225–1;*
  5. *of subjecting an offer of employment to a condition based on one of the factors referred to under article 225–1*".

As for the European Directive 2000/43/EC "implementing the principle of equal treatment between persons irrespective of racial or ethnic origin", its scope includes "conditions for access to employment", "social protection", "social advantages", "education" and "access to and supply of goods and services which are available to the public, including housing". As far as the European Convention on Human Rights is concerned, its Art. 14 states that "The enjoyment of the rights and freedoms set forth in this Convention shall be secured without discrimination on any ground such as sex, race, colour, language, religion, political or other opinion, national or social origin, association with a national minority, property, birth or other status." Its scope is thus larger than that of the Directive, the expression "other status" leaving the door open to a non-limitative list of potential grounds.

Interestingly, the new Art. 8 § 1 of the European directive 95/46/EC sets out a list of sensitive data ("*special categories of data*") which, by default, may not be collected or processed. These sensitive data include any data "*revealing racial or ethnic origin, political opinions, religious or philosophical beliefs, trade-union membership, and the processing of data concerning health or sex life*". Beyond these types of data that are clearly common to these areas of law, information about pregnancy or disabilities, which are considered as discriminatory, are also related to health and can therefore be considered as sensitive data in the sense of the data protection directive. The same can be said about sexual preferences which are considered both as sensitive and discriminatory data.

As far as differences are concerned, one may notice that gender and age are considered as discriminatory factors but not as sensitive data in the sense of the data protection Directive. On the other hand, the data protection Directive states that "*Processing of data relating to offences, criminal convictions or security measures may be carried out only under the control of official authority, or if suitable specific safeguards are provided under national law*" and "*Member States may provide that data relating to administrative sanctions or judgments in civil cases shall also be processed under the control of official authority.*" Offences and criminal convictions are thus considered as sensitive data in the sense of the data protection Directive but not as discriminatory factors. This observation naturally leads to the following question: are these residual differences really justified by the respective goals of these two types of regulations or should they rather be seen as the fortuitous result of different histories. First, the fact that certain types of data can be seen as potential sources of discriminations without necessarily being considered as sensitive in the sense of data protection regulations seems quite reasonable: for example, it may be considered unfair to treat people differently based on information such as the gender but this information, even if undoubtedly personal (attached to a person), can hardly be considered as sensitive. The other type of difference (sensitive data not considered in anti-discrimination regulations) may be more questionable though: for example, wouldn't it be unfair to use information about an offence concerning a person who has already been sanctioned in court and would thus have to suffer a "double punishment" (unless, of course, this information can have an objective value in the context where it is used, for example the fact that a person has been convicted for armed robbery in the context of the recruitment of a bank security agent). Indeed, it is the case that the specific status granted to sensitive data in European data protection regulation is justified by the risks that could result from the processing of such data, which should also lead to ban the use of such data as discriminatory (i.e. selection) criteria.

More generally, one can argue that the close proximity between the types of data covered by data protection and non-discrimination rights stems from their common objectives to ensure fairness and to re-establish some kind of balance, or at the very least to reduce the imbalance of powers, between the individuals and those who may threaten their rights. Paradoxically, this shared reference to sensitive data also introduces difficulties in the implementation of data files for the assessment of discrimination practices: although Directive 95/46/EC provides exceptions to the ban on collecting sensitive data, the legality of such processing and the conditions under which it is permitted are still under discussion in some European countries (Ringelheim 2010).

15.3.2 A Priori Versus A Posteriori Controls

As stated in Sect. 15.2.1, the first measures for data protection in Europe are the formalities to be followed by the data controllers before embarking on the collection or treatment of personal data. The application modes of anti-discrimination laws are rather different. These laws prohibit the use of certain discriminatory criteria for specific purposes, but it would be difficult, if not impossible, to require that all actions falling under these purposes (e.g. service provision or hiring) go through an administrative procedure to confirm that they are not prohibited by law. Indeed, one can hardly conceive a system, other than in the context of a totalitarian regime, in which all actions which could potentially fall under anti-discrimination laws should be declared beforehand in order to confirm that they are lawful. For this basic reason, anti-discrimination regulations put more emphasis on the compensations for damages than on *a priori* verifications. This practical reason is reinforced by the civil law origin of anti-discrimination regulations in countries like France (even though they have since found their way into criminal law as well).

In conclusion, one should notice that the differences identified here between data protection laws and anti-discrimination laws are diminishing over time: as suggested in Sect. 15.2.1, the implementation of data protection laws evolves towards stronger emphasis on *a posteriori* checks, this shift on emphasis being justified by the growing difficulty to control data collection, which makes it necessary to be extremely vigilant on the use made of the data.

15.4 Towards a Synergy Between Data Protection and Anti-discrimination Regulations

In order to address to the issues raised by the technological developments and the new threats to individual rights that they make possible, it can be helpful to distinguish two very different types of data collection:

1. The collection of data as part of formal procedures with clearly identified parties or in the course of clearly identified events, recognised as such by the subjects (e.g. when submitting a file, filling a questionnaire, using a smart card or providing one's fingerprint to get access to a building).
2. The apparently insignificant and almost continuous collection of data that will become more and more common in the digital society (digital audit trails, audio and video recordings, etc.). This collection may be more or less perceived or suspected by the subject (which does not mean that he is necessarily aware of the potential risks concerning the subsequent use of the data or its divulgation to third parties), or remain completely invisible and unsuspected, the frontier between the two situations depending on the quality of the information provided by the controller and the level of awareness of the subject. Another worrying phenomenon—which could in fact be considered as a processing as well as a collection—is the automatic generation of new knowledge using data mining and knowledge inference techniques. In this kind of situation, the subject may ignore not only the process but also the generated knowledge itself, even if this knowledge concerns him (e.g. his preferences, the probability that he could accept a given offer or the risks that he could develop a given disease) and could be used to take actions affecting him (e.g. not offering him a job or an insurance contract or adjusting the price of a service up to the level he would be prepared to pay).

The regulations on personal data protection were originally designed to address the first type of situation. Efforts are made to adapt them to the complex issues raised by the second type of data collection but they tend to be increasingly ineffective in these situations. The main cause of this ineffectiveness is their underlying philosophy of *a priori* and procedural controls. The digital world is based on the circulation and processing of data, and data collection is no longer a one-off event but a commonplace process, that will even become a permanent phenomenon with the advent of ubiquitous computing. Furthermore, the boundaries between personal and non-personal data are more and more blurring,[17] as well as the frontiers between the private and public domains, and also the differences between data collection and data processing.[18] In view of these developments, *a priori* checks and procedures are too rigid or simply impossible to implement. As a result, requirements which may once have represented genuine protections are becoming purely formal obligations, leaving individuals more and more helpless to protect their personal data in the digital world. Just to take an example, on the internet the requirement for prior consent generally turns into the mindless acceptance of users eager to gain access to a website or a service and who hardly take the time to read the question, not to mention the privacy policy of the site.

In order to better address the second type of situation mentioned above, we believe that two types of evolutions are necessary:

[17] The questions raised about the status of group profiles illustrates this difficulty (see the discussion above).

[18] As an example, does knowledge inference fall under data collection, data processing, or both?

1. The first one is to put greater emphasis on the protection of the subjects against data misuse, which would involve more stringent *a posteriori* checks and the integration within the scope of the law of all types of discriminatory processing, i.e. all processing resulting in significant differences of treatment between individuals whenever such differences are not justified by objective grounds that are not solely based on the interests of the data collector (e.g. cost effectiveness or revenue maximisation[19]).
2. The second one is to assess the data processing by the yardstick of its potential harm to individuals, which suggests relying more on civil law than on criminal law and applying a thorough "risks versus benefits" analysis to evaluate the legitimacy of the data processing.

As regards potential harm to individuals, one may observe that most infringements to personal data protection regulations result either in privacy breaches[20] (excessive disclosure of information, intrusive actions such as phone calls or emails, etc.) or in various forms of discriminations (in the common sense of the term, even if those discriminations are not necessarily considered as such in the legal sense and sanctioned by existing anti-discrimination regulations) such as losses of chances to get access to certain offers, benefits or services (job, insurance, etc.) or to get such access under reasonable conditions.[21] This observation, combined with the convergence sketched in the previous sections, leads us to call for the establishment of stronger connections between personal data protection regulations and these two other categories of rights, in particular the right to be protected against unfair discriminations. Anti-discrimination laws also present significant advantages to cope with the continuous information flows which characterise the new digital society:

- There are more flexible as they are not based on *a priori* procedures and administrative formalities.
- Being rooted in civil law, they put emphasis on compensations for damages.

In addition, in certain countries like France, anti-discrimination laws explicitly provide for collective legal proceedings[22] (akin to the American "class actions") which

[19] On this subject, reference could be made to the detailed analysis by Zarsky (2002).

[20] The Dataloss db group maintains a database of data breaches with statistics about the types of data, breaches and business concerned: http://datalossdb.org/latest_incidents.

[21] As an illustration, the CNIL has conducted in 2009 an investigation on the STIC, a large national police database of recorded offences. According to its annual report, this database contains a lot of erroneous or obsolete records because 80% of the decisions to close an investigation for lack of evidence are not forwarded by the courts. This situation is especially alarming considering that the STIC can be used in the administrative enquiries required in the recruitment process of certain categories of professions, which, according to the CNIL, concerns about one million people in France.

[22] Art. 1263–1 of the French Civil Procedure Code: "*Associations regularly reported since at least five years and intending, by their constitutions, to fight against discriminations may bring an action in court*".

may, to a certain extent, tend to restore the balance of powers between the organisations in position to collect and process personal data or to apply differentiating treatments and the individuals who may suffer from such treatments.

It must be recalled, however, that, under substantive law, the protection against discriminations is restricted to very specific types of information (sex, handicap, race, religion, etc.) and purposes (recruitment, supply of services, etc.) which are comprehensively enumerated in the law. The preciseness of this definition contributes to the effectiveness of the law because it makes it possible to apply clear criteria, but it is also a strong limitation, especially in the perspective suggested here to apply anti-discrimination laws to all kinds of unfair differences of treatments based on the processing of personal data. Indeed, this generalisation would make it necessary to lift the current restrictions on the legal definition of discrimination. But such expansion of the scope of anti-discrimination regulations should of course be handled with great care to maintain the effectiveness of the protection.

Another possible instrument to establish stronger connections between personal data protection and anti-discrimination regulations is Art. 15 of European directive 95/46/EC, which applies to decisions producing legal effects on individuals or significantly affecting them.[23] One might think that this article could be applied to cases of discriminations (such as the refusal to supply a service or its supply on very disadvantageous terms) based on personal data processing. To make this provision genuinely effective however, it would be useful to clarify and broaden its scope, in particular to ensure that a processing involving insignificant or purely formal human interventions would not systematically fall outside its scope (Bygrave 2001). Another prerequisite for its application is that the individuals concerned are actually in a position to exercise their rights, the first condition being to be aware or informed of the fact that a decision is based on automatic data processing, for instance the use of a profile (Hildebrandt 2009). As already argued by Lee Bygrave (Bygrave 2001), even if the principles underlying this article are sound and highly commendable, much has still to be done to make it truly effective. To make things worse, the scope of this article may be further reduced in the transposition of the Directive by member states. For example, in France, it is limited to decision with legal consequences: Art. 10 of the law 78–17 states that "*No legal ruling involving the appraisal of an individual's conduct can be taken on the grounds of any automated processing of personal data intended to assess certain aspects of his or her character. No other decision producing legal effects on an individual can be taken solely on the grounds of any automated processing of data intended to establish the profile of that individual or assess certain aspects of his or her character*" without reference to decisions which "*significantly affect him*", as stated in the Directive. One option to implement the approach suggested here would thus be to take the opportunity of the future revision

[23] Art. 15 paragraph 1 of European Directive 95/46/EC states that: "*Member States shall grant the right to every person not to be subject to a decision which produces legal effects concerning him or significantly affects him and which is based solely on automated processing of data intended to evaluate certain personal aspects relating to him, such as his performance at work, creditworthiness, reliability, conduct, etc*".

of the European directive 95/46/EC to reinforce, to revivify Art. 15 and clarify its scope so that it could be applied to all cases of discriminations based on personal data processing.

Two complementary measures are needed to make this approach truly realistic. The first one is to strengthen the means of the national data protection authorities to implement *a posteriori* controls which are sufficient to dissuade data controllers from misusing data. These means are first in terms of funding and manpower, but they should also include enhanced legal support with respect to the accountability of the data controllers[24] and technical solutions enabling more secure and effective verifications (Le Métayer et al. 2009; Le Métayer and Monteleone 2009).

The second complementary measure concerns the possibilities for individuals to get real compensations in the event of unlawful use of personal data. Again, this possibility is a prerequisite to ensure that *a posteriori* controls can effectively have a deterrence effect towards data controllers.[25] One desirable development in this respect would be to make it possible to victims of misuses of personal data to resort to collective legal proceedings ("class actions") as they already can do it for specific kinds of discriminations in certain European countries.

From an institutional viewpoint, we can notice that the former French high commission against discriminations and for equality (HALDE)[26] and the French data protection authority (CNIL) have executed a partnership agreement in March 2006

[24] Accountability should involve a requirement of transparency to ensure, for example, that data controllers cannot resort to intellectual property right protection law to object to the disclosure to the national authority of the algorithms used to process the data (including, for example, profiling algorithms).

[25] Many examples in the past have shown the deterrence effect of class actions and their benefits for consumers. Regarding personal data, the recent loss by Sony of a huge amount of personal information (77 million names, addresses, email addresses, birthdates, passwords and logins, profile data, purchase history and possibly credit cards according to the Dataloss web site http://datalossdb.org/about) illustrates the difference in terms of means of defense between European and American consumers: 55 purported class-action complaints have been filed in the United States against Sony, which places the company in a difficult position (Sony is now seeking coverage of the damages by its insurers). In Europe, national data protection authorities conduct their own investigations but, whatever their conclusions will be, they will have very little means of pressure against a worldwide company like Sony and consumers would have to file complaints on an individual basis. As far as the deterrence effect is concerned, it is still too early to fully assess it in this case, but a number of measures have already been taken by Sony, based on a combination of technical, organizational and legal means (see Sony identity theft protection program: http://blog.us.playstation.com/2011/05/05/sony-offering-free-allclear-id-plus-identity-theft-protection-in-the-united-states-through-debix-inc/).

[26] The constitutional law 2008–724 of 23 July 2008 on modernising the institutions of the Fifth Republic (the current constitution in France) in France's official gazette J.O.R.F 171 of 24 July 2008, p.11890, plans to merge HALDE within a new authority called the "Defender of Rights". For more information on this "Defender of Rights", which was put in place in 2011, see « Loi organique no 2011–333 du 29 mars 2011 relative au Défenseur des droits, JORF n° 0075 du 30 mars 2011 », p. 5497 and « Loi organique n° 2011–334 du 29 mars 2011 relative au défenseur des droits, JORF no 0075 du 30 mars 2011 », p. 5504.

on the grounds that the "*legal competencies of both authorities may prove complementary in many cases, as discriminatory practices are in fact likely to be based on processing personal data, whether or not computerised*".[27] This agreement provides for the exchange of information including the disclosure by one authority of information required for the other take actions.[28] It also includes provisions for the organisation of joint inspections, staff training and communications.

15.5 Conclusion

To sum up, starting from the observation that it is increasingly difficult to effectively control *a priori* all data collections or the production of new knowledge on individuals, we have argued that a possible option is to strengthen *a posteriori* controls on the use of personal data and to ensure that the victims of data misuses can get compensations which are significant enough to represent a deterrence for data controllers. We have also argued that the consequences of such misuses of personal data often take the form of unfair discriminations and this trend is likely to increase with the generalisation of the use of profiles. For this reason, we advocate the establishment of stronger connections between anti-discrimination and data protection laws, in particular to ensure that any data processing resulting in unfair differences of treatments between individuals is prohibited and is subject to effective compensations and sanctions.

Needless to say, the evolution suggested here is by no means a final statement or answer to the very complex issues addressed in this chapter. As discussed in Sect. 15.4 and by Schreurs et al. (2008), the scope and conditions of application of current anti-discrimination laws are too narrow for them to provide sufficient protections in the context of automated processing. One of the most challenging tasks for the future will be the extension of the definition of discrimination criteria to ensure that the scope of non-discrimination regulations covers all (or most) situations where computers could be used to introduce unfair differences of treatments between people. But where to place the red line between acceptable business practices and unfair discriminations is a very delicate (and political) matter.

It should be clear also that the evolution advocated here is meant to provide complementary means to protect individual rights and should not lead to weaken existing protections, including *a priori* controls when these controls are possible. The shift of emphasis from *a priori* to *a posteriori* checks should apply essentially to situations in the second category mentioned above (the unobtrusive and almost

[27] HALDE ruling no. 2006–45 of 13 March 2006 and CNIL ruling no. 2006–077 of 21 March 2006. The agreement is available from the HALDE website: http://www.halde.fr/IMG/pdf/Convention_CNIL.pdf.

[28] The CNIL has also executed an agreement with the French Directorate General for Competition, Consumer Affairs and Prevention of Fraud (DGCCRF). This agreement is intended to encourage the exchange of information between the two authorities in order to reinforce their control measures.

continuous collection of apparently insignificant data) and must not undermine notification obligations, authorisation requests or the requirement for consent for the first category (the collection of data as part of formal procedures with clearly identified parties or in the course of clearly identified events recognised as such by the individuals concerned) where they remain appropriate. It is also clear that one of the primary purposes and *raison d'être* of personal data regulations is to protect a model of democratic society (Rouvroy and Poullet 2009), and this objective must in no way be jeopardised by the evolutions suggested here. In particular, it is necessary to maintain the principle of an absolute barrier, a personal data sanctuary, ensuring that in certain situations or for certain types of data, because the common interest is at stake, the subject's consent is not a sufficient condition to make personal data processing legitimate and that prior authorisation from the data protection authority is expressly required.

References

Bacon, Francis. 1597. *Meditationes Sacrae.*

Bygrave, Lee. 2001. Minding the machine: Art 15 of the EC Data Protection Directive and automated profiling. *Computer Law and Security Report* 17:17–24.

Gutwirth, Serge, and Mireille Hildebrandt, eds. 2008. *Profiling the European citizen: Cross-disciplinary perspectives.* Springer Verlag.

Gutwirth, Serge, and Mireille Hildebrandt. 2010. Some caveats on profiling. In *Data protection in a profiled world,* ed. Serge Gutwirth, Yves Poullet and Paul de Hert, 31–41. Springer Verlag.

Hildebrandt, Mireille. 2009. Who is profiling who? Invisible visibility. In *Reinventing data protection,* ed. Serge Gutwirth et al., 239–252. Springer Verlag.

Le Clainche, Julien. 2005. Pouvoirs *a posteriori* de la CNIL: les risques de l'excès de prudence [CNIL's authority to conduct *a posteriori* verifications: the risks of being over-cautious]. *Revue Lamy Droit de l'Immatériel* 11:43–47.

Le Métayer, Daniel, Shara Monteleone, and Joël Moret-Bailly. 2009. Les ressources du droit alliées aux moyens de la technologie: application à la protection des données personnelles [Combining the resources of law and the resources of technology: application to personal data protection]. *Revue Lamy Droit de l'Immatériel* 51:65–82.

Le Métayer, Daniel, and Shara Monteleone. 2009. Automated consent through privacy agents: Legal requirements and technical architecture.*The Computer Law and Security Review* 25 (2): 136–144.

Lepage, Agathe. 2008. Les professeurs notés sur Internet [Teachers graded on the internet]. *Communications Commerce Electronique* 4:58.

Lyon, David, ed. 2003. *Surveillance as social sorting—Privacy risk and digital discrimination.* Routledge.

Narayanan, Arvind, and Vitaly Shmatikov. 2010. Privacy and security: Myths and fallacies of personally identifiable information. *Communications of the ACM* 53 (6): 24–26.

Nora, Simon, and Alain Minc. 1978. *L'informatisation de la société.* Documentation française.

Ohm, Paul. 2010. Broken promises of privacy: Responding to the surprising failure of anonymization. *UCLA Law Review* 57:1701.

Ringelheim, Julie. 2010. Recueil de données, catégories ethniques et mesure des discriminations: un débat européen [Data collection, ethnic categories and discrimination assessment: a European debate]. *Revue trimestrielle des droits de l'homme* 21 (82): 269–314.

Rouvroy, Antoinette, and Yves Poullet. 2009. The right to informational self-determination and the value of self-development: Reassessing the importance of privacy for democracy. In *Reinventing data protection,* ed. Serge Gutwirth et al., 45–76. Springer Verlag.

Schreurs, Wim, Mireille Hildebrandt, Els Kindt, and Michaël Vanfleteren. 2008. Cogitas, Ergo Sum: The role of protection law and non-discrimination law in group profiling in the private sector. In *Profiling the European citizen: Cross-disciplinary perspectives,* ed. Mireille Hildebrandt and Serge Gutwirth, 241–270. Springer Verlag.

Stehr, Nico. 2000. Le savoir en tant que pouvoir d'action [Knowledge as power of action]. *Sociologie et société* 32 (1): 157–170.

Thomas, Samuel. 2009. *Le fichage ethno-racial: un outil de discrimination.* [Ethno-racial data records: a tool for discrimination]. SOS Racisme.

Zarsky, Tal. 2002. Mine your own business! Making the case for the implications of the data mining of personal information in the forum of public opinion. *Yale Journal of Law and Technology* 5 (4): 17–47.

Chapter 16
On the Principle of Privacy by Design and its Limits: Technology, Ethics and the Rule of Law

Ugo Pagallo

16.1 Introduction

In the first edition of *The Sciences of Artificial* (1969), Herbert A. Simon lamented the lack of research on "the science of design" which characterized the curricula of both professional schools and universities throughout three decades after the Second World War. In the phrasing of the Nobel laureate, the reason hinged on academic respectability, because "in terms of the prevailing norms, academic respectability calls for subject matter that is intellectually tough, analytic, formalizable, and teachable. In the past much, if not most, of what we knew about design and about artificial sciences was intellectually soft, intuitive, informal, and cook-booky" (Simon 1996, 112).

Thirty years later, in *Code and Other Laws of Cyberspace* (1999), Lawrence Lessig similarly stressed the lack of research on the impact of design on both social relationships and the functioning of legal systems, that is, how human behaviour may be shaped by the design of spaces, places and artefacts (*op. cit.*, pp. 91–92).

Thenceforth, the scenario has dramatically changed. Not only, according to Simon, an academically respectable "science of design" has emerged since the mid 1970s, when the *Design Research Centre* was founded at Carnegie Mellon University (the institute became the "Engineering Design Research Centre" in 1985). Significantly, over the last 10 years, legal scholars and social scientists have increasingly focused on the ethical and political implications of employing design mechanisms to determine people's behaviour through the shaping of products, processes, and Information & Communication Technology (ICT)-interfaces and platforms.

On one hand, let me mention work on the regulatory aspects of technology in such fields as universal usability (Shneiderman 2000); informed consent (Friedman et al. 2002); crime control and architecture (Katyal 2002, 2003); social justice (Borning et al. 2004); allegedly perfect self-enforcement technologies on the internet (Zittrain 2007); and design-based instruments for implementing social policies (Yeung 2007).

Ugo Pagallo (✉)
Law School, University of Torino, via s. Ottavio 54, 10124 Torino, Italy
e-mail: ugo.pagallo@unito.it

S. Gutwirth et al. (eds.), *European Data Protection: In Good Health?*,
DOI 10.1007/978-94-007-2903-2_16,

On the other hand, following seminal work on the ethics of design (Friedman 1986; Mitcham 1995; Whitbeck 1996), and privacy (Agre 1997), it is noteworthy that scholars have examined data protection issues raised by the design of ICT, by the means of value-sensitive design (Friedman and Kahn 2003; Friedman et al. 2006), legal ontologies (Abou-Tair and Berlik 2006; Mitre et al. 2006; Lioudakis et al. 2007), projects on platforms for privacy preferences (P3P), (Jutla and Zhang 2005; Cranor et al. 2008; Reay et al. 2009) and PeCAN platforms (Jutla et al. 2006; Jutla 2010), down to the topology of complex social networks (Pagallo 2007). In addition, the idea of incorporating data protection safeguards in ICT was the subject matter of both "Privacy by Design. The Definitive Workshop" organized in Madrid in November 2009 (Cavoukian 2010), and the "Intelligent Privacy Management Symposium" held at Stanford University, CA., on 22–24 March 2010 (the program is online at http://research.it.us.edu.au/magic/privacy2010/schedule.html).

Although the idea of embedding privacy safeguards in information systems and other types of technology is not new, e.g., recital 46 and Article 17 of the European Union (EU) directive 46 from 1995 (D-46/95/EC), privacy commissioners have been particularly active in recent times. For example, in the document on "The Future of Privacy" from the 1 December 2009, the European authorities on data protection, that is, the EU Working Party Article 29 D-95/46/EC have frankly admitted that a new legal framework is needed and, more particularly, it "has to include a provision translating the currently punctual requirements into *a broader and consistent principle of privacy by design*. This principle should be binding for technology designers and producers as well as for data controllers who have to decide on the acquisition and use of ICT" (WP29 2009). Among the examples of how the new principle can contribute to better data protection, the WP29 recommends what Norman Potter presented in his 1968 book *What is a Designer* (Potter 2002) as "environmental design" (i.e. designing spaces) and "product design" (i.e. forging objects).

As an illustration of the first kind of design, think about people's anonymity and the challenge of protecting people's privacy in public (Nissenbaum 1998). While the use of, say, CCTVs proliferates and seems unstoppable, the European authorities on data protection propose to design video surveillance in public transportation systems, in such a way that faces of individuals cannot be recognizable (WP29 2009).

Similarly, when making personal data anonymous is considered a priority, matters of design also concern how we organize data processes and product design. A typical instance is given by the WP29's example on the processing of patient names in hospitals via information systems, where patient names should be kept separated from data on medical treatments or health status. Likewise, in accordance with the principle of controllability and confidentiality of the data to be processed, biometric identifiers "should be stored in devices under control of the data subjects (i.e. smart cards) rather than in external data bases" (WP29 2009).

(In the third section of the paper, I address another kind of design that Norman Potter calls communication design. A good example is given by the user friendliness of ICT interfaces and public complaints against Facebook's data protection policies. Whether or not we buy this form of privacy by design, the social network announced on 26 May 2010, to have "drastically simplified and improved its privacy controls" which previously amounted to 170 different options under 50 data protection-related settings...)

Meanwhile, the Ontario's Privacy Commissioner, Ann Cavoukian, has developed the formula "privacy by design" she invented in the late 1990s, so as to cope with the "ever-growing and systemic effects" of both ICT and large-scale networked data systems (Cavoukian 2009). After more than 10 years of efforts and increasing success, the Commissioner organized the aforementioned "definitive workshop" on the principle of privacy by design in November 2009. On that occasion, Cavoukian summed up the idea of handling today's data protection issues, according to seven principles:

1. We have to view data protection in proactive rather than reactive terms, making privacy by design preventive and not simply remedial;
2. Personal data should be automatically protected in every IT system as its default position;
3. Data protection should accordingly be embedded into design;
4. The full functionality of the principle which follows from (2) and (3) allows a positive-sum or win-win game, making trade-offs unnecessary (e.g. privacy vs. security);
5. A cradle-to-grave, start-to-finish, or end-to-end lifecycle protection ensures that privacy safeguards are at work even before a single bit of information has been collected;
6. No matter the technology or business practices involved, the design project should make data protection mechanisms visible and transparent to both IT users and providers;
7. Finally, the principle "requires architects and operators to keep the interests of the individual uppermost by offering such measures as strong privacy defaults, appropriate notice, and empowering user-friendly options" (Cavoukian 2010). In other words, privacy by design requires an individual-focused respect for user privacy.

In the light of these seven tenets, I admit that the principle of privacy by design looks particularly promising in such different fields as data protection in CCTV systems, biometrics, social networks, smart environments, data loss prevention and more. The principle may in fact represent a turning point in how we address most of the challenges in data protection due to the development of cloud computing, the internet of things, or the semantic Web (Kim et al. 2002; Jutla and Xu 2004; Breuker et al. 2009), by strengthening people's habeas data and allowing us to prevent the risk of hampering economic growth due to alleged privacy reasons. Moreover, the principle shows an effective way to solve some of the extra-territorial legal effects and jurisdictional issues created by digital technology, since privacy assurance can become a default mode of operation for both private companies and public institutions in transnational law (Pagallo 2008).

However, this success entails its own risks, such as current misapprehensions in today's debate and divergent interpretations of the principle among commissioners and scholars. Whereas some propose a version of the principle of privacy "as" design, that is, making most legal provisions on data protection preventive and automatic, it is far from clear what type of design mechanism the WP29 is referring to, when claiming that privacy by design "should be binding" for data controllers, technology designers and producers (WP29 2009). Should the aim be to integrate compliance with regulatory frameworks through design policies or, conversely, should the aim be to prevent harm-generating behaviour from occurring?

In order to offer a hopefully comprehensive view of these issues, this chapter is presented in three sections.

First, I examine the idea of making all the legal provisions on data protection automatic, according to points (ii), (iii), and (v) of Cavoukian's scheme (2010). As shown by 10 years of efforts on the development of platforms for privacy preferences, "the P3P specification is not yet mature enough in terms of element definitions to handle many legal subtleties cleanly" (Jutla 2010). Far from being mere subtleties, however, the first section of the chapter aims to show that such legal hurdles to the "end-to-end lifecycle" of data protection regard some of the most important notions of the legal framework, that is, highly context-dependent normative concepts like data controller, security measure or, even, personal data.

Secondly, these difficulties emphasize the ethical issues of design and the strong moral responsibilities behind the use of alleged perfect self-enforcement technologies. Whereas individual preferences play a crucial role in determining levels of access and control over information in digital environments, people's behaviour would unilaterally be determined on the basis of automatic techniques rather than by choices of the relevant political institutions (Lessig 2004). In the name of individual autonomy, this is why I propose to frame the ethical issues of design and its modalities, by adopting a stricter version of the principle (Pagallo 2009).

Thirdly, such a stricter version of privacy by design is examined in connection with the democratic rule of law and the principle that individuals have to have a say in the decisions affecting them. As suggested by the European Data Protection Supervisor (EDPS), Peter Hustinx, in the Opinion from 25 July 2007 (2007/C 255/01), the challenge of protecting personal data "will be to find practical solutions" through typical transnational measures such as "the use of binding corporate rules by multinational companies" and "international agreements on jurisdiction" (*op. cit.*, § 44). Analogously, efforts should aim at "promoting private enforcement of data protection principles through self-regulation and competition" (*op. cit.*, § 65), while "accepted standards such as the OECD-guidelines for data protection (1980) and UN-Guidelines could be used as basis" (*op. cit.*, § 44).

To conclude, privacy by design should encourage people to change their conduct (e.g. user-friendly interfaces), or limit the effects of harmful behaviour (e.g. security measures) by strengthening people's rights and broadening the range of their choices. There is, indeed, "respect for user privacy" (Cavoukian 2010), when averting both the risks of paternalistic drifts and further conflicts of values in the realm of privacy by design. Rather than a "cradle-to-grave lifecycle" of automatic protection, let us reinforce the pre-existing individual autonomy (Pagallo 2011a).

16.2 Technology and its Limits

I mentioned some of the different ways the scholars have addressed points (ii), (iii) and (v) of Cavoukian's scheme, so that personal data should automatically be protected in every IT system as its default position and even before a bit of information has been collected. Leaving aside value sensitive design-approaches, P3P and PeCAN platforms, let me focus on current efforts in Artificial Intelligence (AI) & Law and, more specifically, in legal ontologies, so as to stress the first limit of the principle of privacy by design, that is, current state-of-the-art in technology.

Legal ontologies is the field of AI that aims to model concepts traditionally employed by lawyers through the formalization of norms, rights and duties, in fields like criminal law, administrative law, civil law, etc. (Breuker et al. 2009; Casanovas et al. 2010). The objective is that even a machine should comprehend and process this very information, by preliminarily distinguishing between the part of the ontology containing all the relevant concepts of the problem domain through the use of taxonomies (e.g. ontological *requirements*), and the ontology which includes both the set of rules and restraints that belong to that problem domain (e.g. ontological *constraints)*. An expert system should thus process the information in compliance with regulatory legal frameworks through the conceptualization of classes, relations, properties and instances pertaining to that given problem domain of data protection. Following what has been said about the ongoing project on the "Neurona Ontology" developed by Pompeu Casanovas and his research team in Barcelona, Spain, the goal is to implement new technological advances in managing personal data and provide organizations and citizens "with better guarantees of proper access, storage, management and sharing of files" (Casellas et al. 2010). By programming the software of the system to comply with regulatory frameworks of data protection, it is feasible to help company officers and citizens "who may have little or no legal knowledge whatsoever."

In technical terms, we should pay attention to the bottom-up approach that starts from legal concepts defined by scholars. A traditional top-down approach works well for the *topmost level*, where the representation instruments are at the disposal of the ontology-builders and the basic conceptual primitives such as relation, role, qualia, processes, etc., are precisely defined. However, a lot of issues arise when the *core ontology* level is taken into account, because the amount of information involved in the project of making data protection safeguards automatic is hardly compressible. Simply put, data protection regulations not only include "top normative concepts" such as notions of validity, obligation, prohibition, and the like. These rules present also highly context-dependent normative concepts like personal data, security measures, or data controllers. In order to grasp some of the difficulties of embedding data protection safeguards in a software program, simply reflect on three facts:

1. In the aforementioned document on "The Future of Privacy", the EU WP29 warns that "Web 2.0 services and cloud computing are blurring the distinction between data controllers, processors and data subjects";

2. In the Opinion from the 1 February 2010, the EU WP29 insisted that "the concept of controller is a functional concept, intended to allocate responsibilities where the factual influence is, and thus based on a factual rather than a formal analysis. Therefore, determining control may sometimes require an in-depth and lengthy investigation" (doc. 00264/10/EN WP 169)
3. Finally, on 23 March 2010, the European Court of Justice declared that liability of online referencing service providers depends on "the actual terms on which the service is supplied." In other words, according to the judges in Luxembourg, it is necessary to determine "whether the role played by that service provider is neutral, in the sense that its conduct is merely technical, automatic and passive, pointing to a lack of knowledge or control of the data which it stores" (*Google v. Louis Vuitton case*, § 114 of the decision).

The difficulty to program the WP29's "factual influence" of the data controller or the ECJ's "actual terms" of the service provided on the internet, does not mean that projects on legal ontologies and privacy by design should be abandoned. On the contrary, these difficulties suggest a bottom-up rather than a top-down approach, in order to lawfully process growing amounts of personal data. By splitting the work into several tasks and assigning each to a working team, we should start from smaller parts and sub-solutions of the design project, to end up with "global answers" (Casellas et al. 2010). The evaluation phase consists in testing the internal consistency of the project and, according to Herbert Simon's "generator test-cycle," entails the decomposition of the complete design into functional components. The test generates alternatives and examines them against the set of *requirements* and *constraints*, so that "important indirect consequences will be noticed and weighed. Alternative decompositions correspond to different ways of dividing the responsibilities for the final design between generators and tests." (Simon 1996, 128)

Further criteria and empirical methods have been proposed: apart from functional efficiency, we should consider the robustness, reliability, and usability of design projects. Evaluation and verification of the design can additionally employ automated and regression-oriented tests, use of prototypes, internal checks among the design team, users tests in controlled environments, surveys, interviews and more (Flanagan et al. 2008). On this basis, we can quantify the growing amount of personal data processed in compliance with regulatory frameworks. This is the focus of the research on legal ontologies and the support of privacy preservation in location-based services (Mitre et al. 2006), the management of information systems (Abou-Tabir and Berlik 2006; Casellas et al. 2010), or middleware architectures for data protection (Lioudakis et al. 2007), each of which aims at integrating smaller parts and sub-solutions into the design. Remarkably, there are even cases where the conceptualization of classes, relations, properties and instances pertaining to a given problem domain, does not seem particularly complex, for example, the design of information systems for hospitals to ensure that patient names are kept separated from data on medical treatments or health status (WP29 2009).

However, by lawfully processing growing amounts of personal data, it does not follow that goals (ii), (iii) and (v) of Cavoukian's scheme, that is, making data protection automatic by design, are at hand. Besides the difficulty of formalizing highly context-dependent concepts such as data processor or data controller, designers must take into account that privacy is not a zero-sum game between multiple instances of access and control over information. Personal choices play indeed the main role when individuals modulate different levels of access and control, depending on the context and its circumstances (Nissenbaum 2004). Moreover, people may enjoy privacy in the midst of a crowd and without having total control over their personal data, whereas total control over that data does not necessarily entail any guarantee of privacy (Tavani 2007). Such constraints emphasize the first limit of the principle: in accordance with today's state-of-the-art, no expert system allows us to fully achieve goals (ii), (iii) and (v) of Cavoukian's principles of privacy by design. To the best of my knowledge, it is impossible to programme software so as to prevent, say, forms of harm-generating behaviour as simple as defamations, but leaving aside technical details, how about the desirability of such a project?

16.3 Ethical Constraints

Some of the most relevant problems concerning today's data protection hinge on the information revolution and the fact that no clear legal boundaries exist in digital environments. State-action is often ineffective due to the ubiquitous nature of information: while citizens of nation states are often affected by conduct that the state is unable to regulate (e.g. spamming), this situation may also lead to the illegitimate condition where a state claims to regulate extraterritorial conduct by imposing norms on individuals, who have no say in the decisions affecting them (Post 2002). According to the 2007 EDPS Opinion, "this system, a logical and necessary consequence of the territorial limitations of the European Union, will not provide full protection to the European data subject in a networked society where physical borders lose importance (. . .): the information on the Internet has an ubiquitous nature, but the jurisdiction of the European legislator is not ubiquitous" (Hustinx 2007).

The ineffectiveness of state-action depends on how ICT allows information to transcend traditional legal borders, questioning the notion of the law as made of commands enforced through physical sanctions. Spamming is again a good example for it is par excellence *transnational* and does not diminish despite harsh criminal laws (as the *CAN-SPAM Act* approved by the U.S. Congress in 2003). Since the mid 1990s, as a consequence, companies and big business have tried to find out a remedy for the apparent inefficacy of state-action in protecting their own rights. While lobbying national and international law-makers in the copyright field, some of the most relevant companies focused on how to enforce their (alleged) exclusivity rights through the development of self-enforcement technologies, for example, Digital Rights Management (DRM). By enabling right-holders to monitor and regulate the use of their own copyright protected works, companies would have prevented unsolvable problems involving the enforceability of national laws and conflicts of law at the international level.

However, whether or not DRM works and can be held to be fair, the aim of privacy by design, that is, to exert "automatic control" over personal information is even more debatable than the use of DRM technology for the protection and enforcement of digital copyright. Whereas Steve Jobs (2007) conceded in his *Thoughts on Music* that DRM-compliant systems raise severe problems of interoperability and, hence, antitrust-related challenges, the aim of privacy by design to automatically prevent harm-generating conducts from occurring looks problematic for three reasons.

First, we have evidence that "some technical artefacts bear directly and systematically on the realization, or suppression, of particular configurations of social, ethical, and political values" (Flanagan et al. 2008). Specific design choices may result in conflicts between values and, vice versa, conflicts between values may impact on the features of design. Consider the different features that privacy by design acquires, once data protection is grasped in terms of property rights or human dignity, of total control or contextual integrity, of restricted access or limited control over digital information. At the end of the day, should an artefact be designed in accordance with the opt-in model for users of electronic communication systems or, vice versa, according to the opt-out approach? Moreover, reflect upon the information system of hospitals which I mentioned in the previous section: should we privilege the efficacy and reliability of that information system in keeping patient names separated from data on medical treatments or health status? How about users, including doctors, who may find such mechanism too onerous?

Secondly, attention should be drawn to the difficulties of achieving such total control. Doubts are cast by "a rich body of scholarship concerning the theory and practice of 'traditional' rule-based regulation [that] bears witness to the impossibility of designing regulatory standards in the form of legal rules that will hit their target with perfect accuracy" (Yeung 2007). The worthy aim to prevent people's privacy infringements involves *strong moral responsibility* of both the designers and public authorities, in that use of self-enforcement technologies collapses "the public understanding of law with its application eliminating a useful interface between the law's terms and its application" (Zittrain 2007). As a response to the inefficacy of state-action in digital environments, the development of this type of technology risks to curtail freedom and individual autonomy severely, because people's behaviour would unilaterally be determined on the basis of technology, rather than by choices of the relevant political institutions. In the phrasing of Larry Lessig, "the controls over access to content will not be controls that are ratified by courts; the controls over access to content will be controls that are coded by programmers" (Lessig 2004).

Finally, there is the issue of security in balancing different levels of access and control via software: the expert system should not be capable to only balance personal preferences *and* matters of "property rights" (Spinello 2003), "trade-offs" (Nissenbaum 2004), or "integrity" (Grodzinsky and Tavani 2008), which often depend on contextual choices. In fact, design projects should be capable to evaluate this (sensitive) information safely, although experts warn that "the only truly secure system is one that is powered off, cast in a block of concrete and sealed in a lead-lined room with armed guards—and even then I have my doubts" (Garfinkel and Spafford 1997). Whereas the use of self-enforcement technologies may be compatible with the precautionary principle in the area of intellectual property rights (Weckert and Moor 2004; Clarke 2005), this does not seem to be the case of privacy by design. DRM's shortcomings principally impact on companies that employ such devices to protect their own copyright protected files: in the case of privacy, however, the use of alleged self-enforcement technologies would directly impact on everyone of us as "informational objects" (Floridi 2006).

Therefore, I suggest abandoning the idea of making data protection automatic by design, so as to prevent every harm-generating conduct from occurring.

Rather, we should focus on other mechanisms we may aim at through design, that is, both the aim to encourage the change of people's behaviour via user friendly interfaces and to decrease the impact of harm-generating conducts through "digital air-bags" as encryption and other security measures (von Ahn et al. 2008).

Let us examine projects on privacy by design, by considering today's "habeas data" in connection with the principle of the rule of law.

16.4 Habeas Data

As shown by the proceedings of the 2009 IVR 24th World Congress in Beijing, China, on *Global Harmony and the Rule of Law* (see http://www.ivr2009.com/), not only "harmony" is a very controversial concept of the millennial political tradition of China, but Western scholars are sparkly divided by the meaning of the "rule of law" as well (whether or not we conceive it as the German *Rechtsstaat*, the French *État de droit*, the Spanish *Estado de derecho*, the Italian *Stato di diritto*, and so forth). While the appeal of the formula historically derives from Plato's distinction between the "empire of the laws," that is, rule by law, and "empire of men," namely, rule under the will of men, it is unclear whether the view of the rule of law adopts a thin-procedural or a thick-substantive approach to distinguishing between rule of law, rule by law, etc. (Summers 1993). It is noteworthy that "despite a growing empirical literature, there remain serious doubts about the relationship, and often causal direction, between the rule of law and the ever-increasing list of goodies with which it is associated, including economic growth, poverty reduction, democratization, legal empowerment and human rights" (Peerenboom 2009).

In this context, it suffices to dwell on the traditional connection between the rule of law and the principle of *habeas corpus*, that is, individual protection against arbitrary (both public and private) action. Over the last two decades, several provisions on data protection, for example, Article 8 of the EU Charter of Fundamental Rights, have complemented the traditional version of the principle of *habeas corpus*, linked to the physical body, with a new kind of protection, that is *habeas data*, as an extension of that protection to the electronic body of each individual (Rodotà 2006). What is at stake with the principle of privacy by design accordingly concerns whether some kinds of "electronic protection" violate people's right to have a say in the decisions affecting them, that is, what the German Constitutional Court frames in terms of the individual "informational self-determination." As well known, the *Bundesverfassungsgericht* (BVerfG) has furthered the concept since its *Volkszählungs-Urteil* ("census decision") from 15 December 1983.

Furthermore, we have to ascertain whether protection of the electronic body via design policies may entail what Kant criticized as paternalism (Kant 1891). By adopting a sort of automatic *habeas data*, the threat is to impinge on the "property of the will" to rule over itself, so that, according to *Grounding for the Metaphysics of Morals*, the risk is to overprotect individuals against every harm and, even, against themselves. In the light of the panoply of projects and approaches in the field of data protection mentioned in this paper, it is crucial to preliminarily distinguish three aims of design (Yeung 2007), so as to take sides on the legal constraints of the principle:

1. Design may prevent harm-generating behaviour from occurring;
2. Design may aim to decrease the impact of harm-generating conducts;
3. Design may encourage the change of social behaviour.

Although design is not necessarily digital (Lessig 1999), the information revolution has induced a more sophisticated legal enforcement than, say, the installation of speed bumps in roads to reduce the velocity of cars. In the case of data protection, scholars should determine what kind of design mechanism is compatible with the tenets of the rule of law, in order to ensure the minimization and quality of the data, its controllability, transparency, and confidentiality, down to the user friendliness of information interfaces.

The first aim of design mechanism, that is, the prevention of harmful conducts thanks to the use of self-enforcement technologies, seems highly problematic in this context. Besides the technical and ethical reasons that make such a protection neither feasible nor desirable in the realm of *habeas data*, perfect automation of data protection mechanisms impinges on the individual right to the "informational self-determination"—that is, the *informationelle Selbstbestimmung* of the BVerfG—which includes the right to determine whether personal data can be collected and, eventually, transmitted to others; the right to determine how that data may be used and processed; the right to access that data and, where necessary, to keep it up to date; besides the right to delete that data and to refuse at any time to have the data processed. Since the enforcement and guarantee of most of these rights are beyond today's state-of-the-art in technology (see Sect. 16.2), it follows that an automatic *habeas data* would impose norms on subjects who have no say in the decisions affecting them (Lessig 2004; Zittrain 2007), thereby making people lose their capacity for making moral choices (Brownsword 2005). Instead of letting people determine autonomously levels of access and control over personal data, depending on personal choices and circumstances, the use of self-enforcement technologies seems incompatible with a basic tenet of the democratic rule of law—autonomy.

But, how about the remaining mechanisms of privacy by design, that is, when the aim is not to prevent certain actions from being chosen overall, but to merely inspire a different conduct by encouraging people to change their behaviour or decreasing the impact of harm-generating conducts? Are these aims compatible with the rule of law?

On one hand, design mechanisms closely regard point (vii) of Cavoukian's principles of privacy by design, that is, the individual-focused approach respectful of user privacy. The idea is well represented by current efforts on security measures, location-based services, friendly interfaces, P2P overlay platforms, default settings and more (Pagallo 2011b). In all the examples of *this* type of design mechanisms, it is arguably correct to stress that "privacy assurance must ideally become an organization's default mode of operation" (Cavoukian 2009). The aim to decrease the impact of harm-generating conducts, as air-bags do in cars, does not seem to impinge on individual autonomy and personal data, because ICT mechanisms as well as air-bags are designed to respect people's choices when they, say, drive cars or modulate different levels of privacy, according to the context. As an instance of "digital air-bags," consider "the power of defaults" (Kesan and Shah 2006), so that we can ensure that values of design are appropriate for novice users and, still, the system improves efficiency. Likewise, reflect on modifications to user interfaces by increasing, or reducing, the prominence of a default setting, so as to allow users to configure and use their software as they deem appropriate. Moreover, consider security measures, such as reCAPTCHA, that aim to prevent automated programs from abusing online services (von Ahn et al. 2008). The aim of such design projects that reduce the effects of harmful conducts fully respects the Kantian principle of autonomy because the only responsibility, both legal and moral, which is at stake with this type of design mechanism concerns the technical meticulousness of the project and its reliability, as it occurs with security measures for the information systems of an atomic plant or a hospital.

On the other hand, by encouraging the change of social behaviour, design projects suggest to assess the impact of design choices on people's conduct. This is the case of the free-riding phenomenon on P2P file-sharing networks, where most peers tend to use these systems to find information and download their favourite files without contributing to the performance of the system. Whilst this behaviour is triggered by many properties of P2P applications like anonymity and hard traceability of the nodes, designers have proposed ways to tackle the issue through incentives based on trust (e.g. reputation mechanisms), trade (e.g. services in return), or alternatively slowing down the connectivity of the user who does not help the process of file-sharing (Glorioso et al. 2010). The editorials in *The Economist* aside, some scholars have nevertheless stressed a threat of paternalism behind the very idea of encouraging the change of people's behaviour (Kruner 2003; Volkman 2003). After all, this type of design mechanism may represent a way of modelling social conduct so as to protect people against all forms of harm. This threat makes urgent a normative viewpoint such as information ethics (Floridi 2005), online privacy policies (Tavani 2007), ethics of design (Friedman 1986; Mitcham 1995; Whitbeck 1996; Flanagan et al. 2008), machine ethics (Moor 2006; McLaren 2006), and more, for we should previously test the goodness of data protection laws, in order to prove the goodness of our own design projects. Is there a way to ensure that privacy by design does not violate the anti-paternalistic stance of the rule of law by encouraging people to change their conduct? How about conflicts between values that necessarily reverberate on design choices? Is, say, Jeffrey Rosen right, when stressing the fear that "cultural differences will make thoughtful regulation difficult" in data protection? What does it mean for data protection that "the French will bare their breasts but not their salaries and mortgages, and the reverse is true in the U.S."? (As Rosen declares in Mills 2008.)

Although it is notoriously difficult to solve conflicts of values with their divergent interpretations, we might prevent most issues in today's *cul de sac* by embracing one of the several examples and design mechanisms put forward by the EU Working Party's document on "The Future of Privacy." Whether or not you agree that the European legal framework "is clearly and deeply flawed as an account of what informational protection is all about" (Volkman 2003), we need not sympathize with Brussels to follow the proposal that the principle of privacy by design should be implemented in accordance with a bottom-up rather than a top-down approach, that is, depending on individual autonomous choices via self-regulation and competition among private organizations (WP29 2009).

As a result, besides a stricter version of privacy by design as a way to decrease the "informational entropy" of the system through "digital air-bags," we find a further design mechanism compatible with the rule of law. When encouraging people to change their behaviour by the means of design, the overall goal should be to reinforce people's pre-existing autonomy, rather than building it from scratch. In the wording of the EU privacy commissioners, the principle should enable business and individuals to "take relevant security measures by themselves" (WP29 2009).

16.5 Conclusions

It is unlikely that privacy by design will offer the one-size-fits-all solution to the problems in the realm of data protection, although privacy by design is a good candidate for understanding how we have coped with privacy issues over the last few years. The principle may in fact be a turning point in how we address most of the challenges in data protection, by strengthening people's habeas data, without hampering economic growth for alleged privacy reasons. In different fields as data protection in CCTV systems, biometrics, social networks, smart environments, data loss prevention and more, projects are increasingly processing growing amounts of personal data in compliance with current normative frameworks, strengthened by the capacities of computers to draw upon the tools of AI and operations research.

Notwithstanding the merits, however, there are three reasons why we should be aware of the limits of privacy by design. These limits are especially relevant when the aim is to automatically protect personal data as the default position of every ICT system, even before a single bit of information has been collected, that is, points (ii), (iii) and (v) of Cavoukian's scheme on the principle (Cavoukian 2010). Let me sum up these limits.

First, I mentioned work on legal ontologies, value-sensitive design, P3P or PeCAN platforms, so as to show the limits of today's state-of-the-art in technology. Besides the difficulty of modelling highly context-dependent normative concepts as data controllers and the "neutrality" of the services provided on the internet, designers should take into account that privacy is not a zero-sum game but concerns personal choices on levels of access and control over information that often depend on the context. Making all the provisions of data protection automatic is simply out of reach.

The second limit involves the ethical constraints of the approach and the process of both law-making and legal enforcement. Not only conflicts between values do impact on the features of design but, vice versa, design choices may result in further conflicts between values. Since privacy may be conceived in terms of human dignity or property rights, of contextual integrity or total control, it follows that privacy by design acquires many different features. Moreover, self-enforcement technologies risk to curtail freedom and individual autonomy severely, because people's behaviour would be determined on the basis of design rather than by individual choices.

Finally, two tenets of the rule of law, that is, autonomy and anti-paternalism, stressed the legal constraints of privacy by design as a means to prevent harm-generating behaviour from occurring. By adopting a sort of automatic *habeas data*, the risk is to impinge on what Kant defined the "property of the will" to rule over itself and, two centuries later, the BVerfG presented as the individual right to "informational self-determination." Leaving aside the technical unfeasibility of goals (ii), (iii) and (v) of Cavoukian's model, it is undisputable that the more personal choices are wiped out by automation, the bigger the threat of modelling social conduct via design, that is, Kantian paternalism.

As a consequence, this chapter has proposed a stricter version of the principle of privacy by design which seems to be technically feasible, ethically sound and lawful. On one hand, in accordance with goals (i) and (vi) of Cavoukian's scheme (2010), privacy by design can legitimately aim to automatically reduce the impact of harm-generating behaviour, so that "privacy assurance must ideally become an organization's default mode of operation" (Cavoukian 2009). Besides values of design that are appropriate for novice users and, hence, procedural constraints for changing the setting of the interfaces on voluntary and fully informed basis, I mentioned security measures that aim to prevent automated programs from abusing online services: "digital air-bags" as friendly interfaces, P2P overlay platforms or default settings will not impinge on individual autonomy, no more than traditional air-bags affect how people drive. On the other hand, in connection with point (vii) of Cavoukian's model, privacy by design can legitimately aim to encourage the change of social behaviour if, and only if, the goal is to strengthen people's rights by widening the range of their choices. This is the version of the principle put forward by the example of both the WP29 and the European Data Protection Supervisor, when endorsing the enforcement of data protection through self regulation and competition (Hustinx 2007; WP29 2009), thus preventing claims of paternalism by fostering individual habeas data.

The result is a final convergence over the "full functionality" of the principle, that is, point (iv) of Cavoukian's scheme. A positive-sum or win-win game becomes possible by embedding data protection safeguards in technology with the aim to encourage people to change their conduct as well as to decrease the effects of harmful behaviour. As shown by current work on legal ontologies, middleware architectures for data protection, the management of information systems, and more, trade-offs such as privacy vs. business, privacy vs. security, privacy vs. copyright, are not always necessary. However, it is not only a matter of business and security—privacy by design concerns a basic tenet of the rule of law such as the principle of autonomy.

References

Abou-Tair, D. el Diehn I., and Stefan Berlik. 2006. An ontology-based approach for managing and maintaining privacy in information systems. *Lectures notes in computer science*, 4275: 983–994 (Berlin-Heidelberg: Springer).

Agre, Philip E. 1997. Introduction. In *Technology and privacy: The new landscape*, eds. Philip E. Agre and Mark Rotenberg, 1–28. Cambridge: The MIT Press.

von Ahn, Luis, Maurer, Benjamin, McMillen, Colin, Abraham, David, and Manuel Blum. 2008. reCAPTCHA: Human-based character recognition via web security measures. *Science* 321 (5895): 1465–1468.

Borning, Alan, Friedman, Batya, and Peter H. Kahn. 2004. Designing for human values in an urban simulation system: Value sensitive design and participatory design. *Proceedings of eighth biennial participatory design conference*, 64–67. Toronto: ACM Press. http://www.urbansim.org/pub/Research/ResearchPapers/vsd-and-participatory-design-2004.pdf. Accessed 23 Dec 2010

Breuker, Joost, Casanovas, Pompeu, Klein, Michel C.A., and Enrico Francesconi (eds.). 2009. *Law, ontologies and the semantic web*. Amsterdam: IOS Press.

Brownsword, Roger. 2005. Code, control, and choice: Why east is east and west is west. *Legal Studies* 25 (1): 1–21.

Casanovas, Pompeu, Pagallo, Ugo, Sartor, Giovanni, and Gianmaria Ajani (eds.). 2010. AI approaches to the complexity of legal systems. *Complex systems, the semantic web, ontologies, argumentation, and dialogue*. Berlin: Springer.

Casellas, Nuria, Torralba, Sergi, Nieto, Juan-Emilio, Meroño, Albert, Roig, Antoni, Reyes, Mario, and Pompeu Casanovas. 2010. The Neurona ontology: A data protection compliance ontology. Paper presented at the intelligent privacy management symposium, Stanford University, CA., USA. 22–24 March 2010.

Cavoukian, Ann. 2009. *Privacy by design*. Ottawa: IPC.

Cavoukian, Ann. 2010. Privacy by design: The definitive workshop. *Identity in the Information Society* 3 (2): 247–251.

Clarke, Steve. 2005. Future technologies, dystopic futures and the precautionary principle. *Ethics and Information Technology* 7 (4): 121–126.

Cranor, Lorrie F., Egelman, Serge, Sheng, Steve, McDonald, Aleecia M., and Abdur Chowdhury. 2008. P3P deployment on websites. *Electronic Commerce Research and Applications* 7 (3): 274–293.

Flanagan, Mary, Howe, Daniel C., and Helen Nissenbaum. 2008. Embodying values in technology: Theory and practice. In *Information technology and moral philosophy*, eds. Jeroen van den Hoven and John Weckert, 322–353. New York: Cambridge University Press.

Floridi, Luciano. 2005. Information ethics, its nature and scope. *Computers and Society* 36 (3): 21–36.

Floridi, Luciano. 2006. Four challenges for a theory of informational privacy. *Ethics and Information Technology* 8 (3): 109–119.

Friedman, Batya. 1986. Value-sensitive design. *Interactions* 3 (6): 17–23.

Friedman, Batya, Howe, Daniel C., and Edward Felten. 2002. Informed consent in the mozilla browser: Implementing value-sensitive design. *Proceedings of 35th annual hawaii international conference on system sciences* 247. IEEE Computer Society.

Friedman, Batya, and Peter H. Kahn Jr. 2003. Human values, ethics, and design. In: *The human-computer interaction handbook*, eds. Julie A. Jacko and Andrew Sear, 1177–1201. Mahwah: Lawrence Erlbaum Associates.

Friedman, Batya, Kahn, Peter H. Jr., and Alan Borning. 2006. Value sensitive design and information systems. In *Human-computer interaction in management information systems: Foundations*, eds. Ping Zhang and Dennis Galletta, 348–372. New York: Armonk.

Garfinkel, Simson, and Eugene Spafford. 1997. *Web security and commerce*. Sebastopol: O'Reilly.

Glorioso, Andrea, Pagallo, Ugo, and Giancarlo Ruffo. 2010. The social impact of P2P systems. In *Handbook of peer-to-peer networking*, eds. Xuemin Shen, Heather Yu, John Buford and Mursalin Akon, 47–70. Heidelberg: Springer.

Grodzinsky, Frances S. and Herman T. Tavani. 2008. Online file sharing: Resolving the tensions between privacy and property interest. In *Proceedings of ETHICOMP2008 "Living, Working and Learning Beyond Technology"*, eds. Terry W. Bynum, Maria Calzarossa, Ivo De Lotto and Simon Rogerson, 373–383. Mantova: Tipografia Commerciale.

Hustinx, Peter. 2007. Opinion of the European data protection supervisor on the communication from the commission to the European parliament and the council on the follow-up of the work program for better implementation of the data protection directive. *Official Journal of the European Union* 27 Oct. 2007, C 255: 1–12.

Jobs, Steve. 2007. Thoughts on music. http://www.apple.com/hotnews/thoughtsonmusic/. Accessed 20 April 2009.

Jutla, Dawn N., and Liming Xu. 2004. *Privacy agents and ontology for the semantic web. Americas conference on information systems*. New York City: CUSP.

Jutla, Dawn N., and Yanjun Zhang. 2005. Maturing E-privacy with P3P and context agents. In *Proceedings of IEEE international conference on E-Technology, E-Commerce and E-Service*, 536–541. Hong Kong.

Jutla, Dawn N., Bodorik, Peter, and Yanjun Zhan. 2006. PeCAN: An architecture for user privacy and profiles in electronic commerce contexts on the semantic web. *Information Systems* 31 (4–5): 295–320.

Jutla, Dawn N. 2010. Layering privacy on operating systems, social networks, and other platforms by design. *Identity in the Information Society* 3 (2): 319–341.

Kant, Immanuel. 1891. *Kant's principles of politics, including his essay on perpetual peace. A contribution to political science* (1795), (trans: Hastie W.). Edinburgh: Clark.

Katyal, Neal. 2002. Architecture as crime control. *Yale Law Journal* 111 (5): 1039–1139.

Katyal, Neal. 2003. Digital architecture as crime control. *Yale Law Journal* 112 (6): 101–129.

Kim, Anya, Hoffman, Lance J., and C. Dianne Martin. 2002. Building privacy into the semantic web: Ontology needed now. *Semantic web workshop 2002*. Honolulu, Hawaii. http://semanticweb2002.aifb.uni-karlsruhe.de/proceedings/Position/kim2.pdf. Accessed on 23 Dec 2011.

Kesan, Jay P. and Rajiv C. Shah. 2006. Setting software defaults: Perspectives from law, computer science and behavioural economics. *Notre Dame Law Review* 82:583–634.

Kuner, Christopher. 2003. *European data privacy law and online business*. Oxford: Oxford University Press.

Lessig, Lawrence. 1999. *Code and other laws of cyberspace*. New York: Basic Books.

Lessig, Lawrence. 2004. *Free culture: The nature and future of creativity*. New York: Penguin Press.

Lioudakis, Georgios, Koutsoloukasa, Eleftherios, Tselikasa, Nikolaos, Kapellakia, Sofia, Prezerakosa, Georg, Kaklamani, Dimitra and Iakovos Venieris. 2007. A middleware architecture for privacy protection. *The International Journal of Computer and Telecommunications Networking* 51 (16): 4679–4696.

McLaren, Bruce. 2006. Computational models of ethical reasoning: Challenges, initial steps, and future directions. *IEEE intelligent systems* 2006 (July/August): 29–37.

Mills, Elinor. 2008. To be anonymous or not to be, that is the privacy question: interview to jeffrey rosen. *News blog*. http://news.cnet.com/8301-10784_3-9889255-7.html. Accessed 15 Oct 2010.

Mitcham, Carl. 1995. Ethics into design. In *Discovering design*, eds. Richard Buchanan and Victor Margolin, 173–179. Chicago: University of Chicago Press.

Mitre, Hugo, Gonzàlez-Tablas, Ana Isabel, Ramos, Benjamin, and Arturo Ribagorda. 2006. A legal ontology to support privacy preservation in location-based services. *Lectures notes in computer science*, 4278: 1755–1764 (Berlin-Heidelberg: Springer).
Moor, James. 2006. The nature, importance, and difficulty of machine ethics. *IEEE intelligent systems* 21(4): 18–21.
Nissenbaum, Helen. 1998. Protecting privacy in an information age: The problem of privacy in public. *Law and Philosophy* 17 (5–6): 559–596.
Nissenbaum, Helen. 2004. Privacy as contextual integrity. *Washington Law Review* 79 (1): 119–158.
Pagallo, Ugo. 2007. Small world-paradigm and empirical research in legal ontologies: A topological approach. In *The multilanguage complexity of European law: Methodologies in comparison*, eds. Gianmaria Ajani, Ginevra Peruginelli, Giovanni Sartor and Daniela Tiscornia, 195–210. Florence: European Press Academic.
Pagallo, Ugo. 2008. *La tutela della privacy negli Stati Uniti d'America e in Europa: Modelli giuridici a confronto*. Milano: Giuffrè.
Pagallo, Ugo. 2009. Privacy e design. *Informatica e diritto* 1:123–134.
Pagallo, Ugo. 2011a. Designing data protection safeguards ethically. *Information* 2 (2): 247–265.
Pagallo, Ugo. 2011b. The trouble with digital copies: A short km phenomenology. In *Ethical issues and social dilemmas in knowledge management organizational innovation*, eds. Gonçalo J. Morais da Costa, 97–122. Hershey: IGI Global.
Peerenboom, Randy. 2009. The future of rule of law: The challenges and prospects for the field. *Hague Journal on the Rule of Law* 1 (1): 5–14.
Post, David G. 2002. Against "Against Cyberspace". *Berkeley Technology Law Journal* 17 (4): 1365–1383.
Potter, Norman. 2002. *What is a designer*. London: Hyphen Press.
Reay, Ian, Dick, Scott, and James Miller. 2009. A large-scale empirical study on P3P privacy policies: Stated actions vs. legal obligations. *ACM transactions on the web* 3(2): 1–34.
Rodotà, Stefano. 2006. The retention of electronic communication traffic data. *Revista d'Internet, dret i política* 3:53–60.
Shneiderman, Ben. 2000. Universal usability. *Communications of the ACM* 43 (3): 84–91.
Simon, Herbert A. 1996. *The sciences of the artificial*. Cambridge: The MIT Press.
Spinello, Richard A. 2003. The future of intellectual property. *Ethics and Information Technology* 5 (1): 1–16.
Summers, Robert S. 1993. A formal theory of rule of law. *Ratio Iuris* 6 (2): 127–142.
Tavani, Herman T. 2007. Philosophical theories of privacy: Implications for an adequate online privacy policy. *Metaphilosophy* 38 (1): 1–22.
Volkman, Richard. 2003. Privacy as life, liberty, property. *Ethics and Information Technology* 5 (4): 199–210.
Weckert, John and James Moor. 2004. Using the precautionary principle in nanotechnology policy making. *Asia Pacific Nanotechnology Forum News Journal* 3 (4): 12–14.
Whitbeck, Caroline. 1996. Ethics as design: Doing justice to moral problems. *Hastings Center Report* 26 (3): 9–16.
Working Party (WP) Article 29 D-95/46/EC. 2009. *The future of privacy*. 02356/09/EN–WP 168.
Yeung, Karen. 2007. Towards an understanding of regulation by design. In *Regulating technologies: Legal futures, regulatory frames and technological fixes*, eds. Roger Brownsword and Karen Yeung, 79–108. London: Hart Publishing.
Zittrain, Jonathan. 2007. Perfect enforcement on tomorrow's internet. In *Regulating technologies: Legal futures, regulatory frames and technological fixes*, eds. Roger Brownsword and Karen Yeung, 125–156. London: Hart Publishing.

Chapter 17
The Right to Forget, the Right to be Forgotten

Personal Reflections on the Fate of Personal Data in the Information Society

Ivan Szekely

As I am writing this essay, which is concerned not so much with codified law as with moral rights and values in a changing world, I am conscious of the fact that I am up against a stiff headwind, in a social climate where the prevalent trend in public discourse on recent history favors the public exposure of crimes and criminals, real and imaginary; where the political dialogue confuses the increasingly outworn problem of informers under a totalitarian regime with the issue of identifying with, or showing loyalty to, the previous government's ideology; where scheming historians mistake the unveiling of the previous regime's transgressions for probing into and disclosing people's private lives; where the millions of naïve Internet users take the claims made by the big IT corporations about the eternal life of information technology at face value; where "the code is the law"; and where not only the cohort of the technical intelligentsia (an interested and willing party), but also the social science elite of postmodern society (people dazzled by the chimera of a relentlessly changing market for attention on the one hand, and unable to comprehend the real forces of social interaction on the other) all seem to fall for the promise that everlasting storage of, and ready access to, all the information, at all times and in all places, actually paves the way to the redemption of mankind, and in any case new technology will solve social problems and make people happier.

In short, the present milieu does not provide favorable conditions for the tranquil contemplation of forgetting. But still, what is it that lends actuality to the phenomenon of forgetting? It is a transformation which can readily be defined, but it also has some very basic consequences, which the majority of people tend to overlook. In the

---

As the chapter's subtitle suggests, my intention has been to present my views in a format that is decidedly different from the standard style of academic writing. Amalgamating the languages of scientific and literary approaches, the chapter is meant to establish a common thread that runs through the separate topics of data protection literature—a leitmotif centered on the problematics of remembering and forgetting, if you will. As a result, readers will not find any numbered sections, bullet points, footnotes or end notes which would hinder the continuity of reading. There is, however, an annotated reference list at the end.

I. Szekely (✉)
Eotvos Karoly Policy Institute, Budapest, Hungary
e-mail: szekelyi@ceu.hu

S. Gutwirth et al. (eds.), *European Data Protection: In Good Health?*,
DOI 10.1007/978-94-007-2903-2_17,

course of human history, forgetting was the norm and remembering the exception. Now it seems to be the other way around: it is the act of forgetting, or the ability to forget, that is becoming the exception.

*

It would be easy to explain all this simply by reference to advances in computer technology, digital information processing, and cheap storage capacity. While all of this undeniably constitutes the technical basis for the phenomenon, the mere fact that this development started in the first place, and has continued unabated ever since, is the consequence not only of the selfless efforts of individuals who have made use of their constitutional right to carry out research, but also of a hitherto inconceivable concentration of interest and power, which has used this technology to effect a dramatic transformation in state and society, in the business and personal spheres, and in private and public relationships. If people find modern society too complex and incomprehensible, then they will find this "information society" all the more so, and I would even venture to say it has been deliberately designed to be so, in order to prevent its citizens from comprehending, or even wishing to comprehend, the social trends, including those that have a direct bearing on their own fate. We may call it a specific manifestation of Beck's risk society, where, under the permanently changing conditions, our experiences do not constitute solid enough ground for our decision-making; where the future is largely unpredictable and our decisions have unforeseeable consequences; and where, despite all this, we live out our short- and long-term plans.

Today, not even the world of Internet can escape the attention of social critics, nor is Web 2.0 exempt from critical reviews by political economists. However, I would once again like to call attention to the fact that since the laws are being written by the code writers (IT experts and their paymasters), these theoretical arguments never even register on the radar screen of the data processing monopolies and the power centers—for them, the only restraining force appears to be bad publicity and the wrath and withdrawal of the masses who (unwittingly) supply them with their data. And even if these ideas were to attract the attention of socially responsible legislators, no laws that could possibly result would have any chance of either practical implementation or adequate management of the actual problems, since from the moment of their inception they would be in conflict with the technologies and organizations controlling the IT world—and also the associated commercial, political, and ideological power structures.

It would be easy to claim that the current method of information storage, which holds out, at least according to its promoters, the promise of eternal life for our data, is merely a quantitative development in the evolutionary process that characterizes the history of mankind so far. However, just as genetic engineering is not merely a more effective version of our previous breeding selection practices, and just as war is not merely a more effective continuation of politics, the current scale and perspective of digital information storage cannot be regarded simply as a step forward in the improvement of efficiency.

What is memory? It is many things: a need; a luxury; a natural aptitude; a key aspect of culture; a tool for survival; a field of science; the ability to foretell the future; and many more. There is internal and external memory, short, intermediate and long-term memory, there is individual and collective memory, visual, auditive, and notional memory. There is factual memory and emotional memory; implicit memory and memory of principles, relations, responsibilities, friendships and loves. There are techniques to retain memories and there are institutions to do the same.

However, there is one aspect that has always been a prominent feature of memory: the principle of selectivity, the recurring act of assessment. Whenever we go through photographs or letters that have come to us through inheritance, we look through them and read into them; at first we decide to keep most of them, but when the next house move or redecoration is upon us, we start to throw some of them out—and not merely on account of a shortage of storage space. In the end we only retain the documents which are the most important for us, and also the most characteristic, with the greatest power to jog our memory. These are the ones that we bequeath to the next generation, who—at least up till now—repeat the same process all over again. There is a well-known counter-argument, claiming that if a document or photograph or any recorded information can survive the selection process, it will eventually become a priceless treasure for some researcher in the future, be that a private individual interested in family history or a professional student of the past. And it is true that their special ability to survive is *precisely* that makes them so precious, lending them a value that they would never have in a world where everything was kept forever. If we were to keep all the ruins from every historical period, all our worn-out cloths and chipped crockery, then we would not only overcrowd our physical world, but also lose our ability to distinguish between what matters and what does not. To preserve what is significant and discard the rest is what traditional archives are trying to do in their professional capacity: from time to time their representatives show up at the originating organizations and select the documents that should be archived, or preserved for future generations. They may make mistakes in trying to ascribe importance to some documents, in which case future researchers will be disappointed that others have not been preserved—but this is the very nature of long-term memory: all the documents of an archives are used by the readers, researchers or users in contexts and for purposes that are different from the original ones.

It is precisely this process, the acts of selecting, evaluating and deleting, which is missing from the promise and present state of eternal digital memory. Instead of the "extended arm" imagery applied to our brain and memory, which was how machines were visualized as the extensions of manpower in the not so distant past, here we are dealing with a totally incomprehensible system, or a complex of many systems, operated by robots for purposes completely unknown to us. The declared purposes, at the level of the promotional slogans, at least, are invariably about making our lives easier and introducing new services that are more affordable, or assisting us in managing our affairs more efficiently, or establishing e-democracy.

But before we, together with all those consumers living in the more developed societies—the masses of consumer idiots, to borrow the expression used by some of the more critical authors—decide to buy into the claim that the Internet and the ever-increasing information storage designed to last forever is in fact the charitable deed of a superior being, a free public service, which has only one purpose: to offer us, free of charge, more and more services in education and entertainment, we should note that we never hear about *e-dictatorship*, only about *e-democracy* (even though dictators and dictatorial organizations could lay their hands on our personal data preserved for eternity in IT-land just as easily as charitable organizations), which is somewhat analogous to diagrams of productivity figures which invariably move upwards on the charts displayed on walls, or airplanes rising proudly, or photogenic models raising their eyes optimistically above the horizon—we could continue with the list of the long-established clichés in the world of promotion. However, thanks to the dedicated efforts of advertisement psychologists, these clichés are actually working, to the extent that they permeate the thinking of large masses of people. And the crucial elements of this thinking are comfort, (seemingly) low price and high speed, as well as the absence of any need to pay attention and apply critical thinking. A certain amount of temporary attention is generated by the so-called "danger discourses," but these are never directed at the essential features and are didactically misconceived: warnings along the lines of "Honey, never make contact with strangers on the Internet!" usually elicit the opposite reaction from the target audience.

*

Nowadays, forgetting is an expensive business. Selection, evaluation and scrapping are all expensive and labor-intensive processes (in other words, they require much time and attention), so instead we keep the hundreds of photographs taken during a hike or a party without selection, along with the masses of e-mails and carelessly drawn sketches. By now even in this area the situation has been reversed: previously it was the memory part (the recording, the storage and dissemination of knowledge) that was costly and labor-intensive and the limits of the processes led to the natural decay of information, in a way constituting an ecological balance between current, semi-current and historical information.

But why should we want to forget? Even if we were to pretend that we had lost our mental and emotional premises, social axioms, a priori notions and our whole value system, we could still list a host of explanations for the advantages and necessity of forgetting in the capacity of an objective, external observer.

On the individual level, if we did not have the ability to forget, we would share the fate of the mnemonists described in well-documented case studies: Funes and Shereshevsky both experienced the incurable condition they suffered from, for which the outside world actually celebrated them, as a lifelong imprisonment—in other words, we would be unhappy. And even if the accumulative and unselected memories resided outside our operational memory, somewhere in a continuously accessible backup storage area, the implications would still be similar. On the one hand, we would be unable to make use of one of the most important elements of our personal informational autonomy, the freedom to decide whether to store or share information about ourselves, to control the fate of our data, and to determine what to share with whom, and for how long. As long as our backup storage is not a notebook kept in a securely locked drawer, but a supposedly private, electronic mail box or document storage facility accessible not only to the user but also to a circle of people of unknown size and dubious intentions, the beautiful idea of information self-determination remains an illusion.

On the other hand, if we were to preserve everything, we would become prisoners of our past. And what I am referring to here is not simply our acts, later re-evaluated and repented, nor our deviations of various grades of severity, nor our transgressions, but the freedom to control our lives and to develop our personalities. The early ideals of personal freedom had always included the possibility of starting life anew, and the chance to leave our personal history behind. "Go west!," the Americans urged us, although that advice can now be heard only in romantic Western movies, as the Wild West, the terra incognita, no longer exists.

Naturally, our mentioning this ideal should not be construed as a latent support for murderers and other villains in their attempts to put their guilty past behind them—and we only need to emphasize this in light of the common experience of human rights campaigners, who are regularly accused of being hell-bent on "defending criminals." The decision about the fate of these types of information clearly cannot be left to the persons concerned—at least for a certain period of time. Information related to crime and punishment (in practical terms: criminal personal data) is exempted from the main rules of information self-determination: it is regulated by special provisions in constitutional law, which permit the curtailing of the individual's information autonomy on principles invoking the interests of society at large. At the same time, however, after the expiration of a certain period whose length depends on the severity of the crime, the legal accountability for these officially sanctioned deviant acts will be ended, implying that the obligation to keep a record of them will be lifted and, as a special category of sensitive data, control over such records, at least in principle, will be returned to the data subjects. In modern democracies based on the rule of law, only war crimes are exempt from the statute of limitations, meaning that information about people found guilty in such crimes will never be private; any other acts condemned by society or prosecuted by law will sooner or later lose their relevance.

But as we are confronted with certain cases listed as illustrations by Viktor Mayer-Schönberger in his seminal book on the need for forgetting, we find that these are not isolated and individual incidents; instead, here we are dealing with an essential feature of a system incapable of forgetting. In his book we can read about an elderly and well-respected professional, who in the 1960s was found guilty of the offense of taking LSD. Thanks to the eternal memory of digital technology, he was still made to pay the price in his seventies, when a border official refused him entry after identifying him as a suspicious element.

It is not just individual cases of trauma caused to people later in life by digital memory that we see here; we can also discover the possibility of citizens' actions (occasionally supported by the law) to implant RFID tags or miniature radio transmitters under the skin of persons convicted of child molestation, so that their movements can be constantly monitored, and thus forever barring them from shedding the consequences of their acts; or the possibility of people's names ending up in DNA databases set up for the mandatory genetic identification of individuals judged suspicious or dangerous, and the impossibility of having one's name removed from such databases even in a clear case of mistaken identity.

*

The chances of forgetting are closely connected to the extent of surveillance, which provides source material for digital memory. But does any community have the right to spy on people, or to stigmatize them in any way, for the purpose of ensuring wider acceptance for its own norms? Living in the public eye, or in forced openness, is a well-known phenomenon in certain Protestant societies: the reason they have no curtains in their windows is not because this is how Big Brother set up its Panopticon, nor because the inhabitants of the house do not have the right to screen their private sphere from the outside world, but simply, because it is *not the proper thing to do*—the people inside the house must feel the public's eye on them at all times, so that they behave precisely as demanded by the community's morals and preferences.

However, the problem of individuals perpetrating serious offenses is different from the above-mentioned manifestations of ideological coercion or the enforcement of preferences in taste: in their cases we talk about real stigmatization. The question is whether the risk of the possible repetition of deviant acts justifies permanent limitations on the individuals' autonomy, their excommunication and placement in a virtual pillory—even when the statute of limitations, in the legal as well the moral sense, has already expired in relation to the concrete offences.

The word *risk* has attained key importance here, not only from the viewpoint of the individual child-molesters, but also regarding the new and supposedly modern—in my view, however, fundamentally flawed—concept of society. I refer to the so-called actuarial society, named after insurance mathematicians who weigh risks down to fractions of a percentage point, which confesses to principles that have served to justify the establishment of a surveillance society—or to use a less polite term, a voyeur society. According to this concept, it makes no sense to talk about normalcy and deviancy; it makes no sense to explore the individual and social motives behind the individuals' actions; it is enough to consider the statistical probability of any given offense. As to the question of what exactly qualifies as deviant, "bad" and, therefore, reprehensible behavior, it is up to our bosses and business and political elite to decide. And on the point of minimizing the statistical probability of the occurrence of criminal acts, this can be achieved by increasing the likelihood of being detected, which in turn is best served by keeping everyone under constant surveillance. This, and this alone, will prevent "criminals" from carrying out criminal acts—according to this ideology. Naturally, this will also be the only thing that will stop those who engage in the surveillance profession from committing crimes, such as abusing the specific knowledge they have acquired while spying on other people. Therefore they, too, must be placed under surveillance, along with those who spy on them, and so forth. All this is closely linked to society's approach to deviant behavior. The system even has its own name: it is called New Penology. The full-grown system, quite understandably, relies not only on real-time surveillance, but also on infinite digital memory, ever-growing personality profiles, and ever-more accurate probability calculation.

Instead of Panopticon, which suggests the existence of a central hub of surveillance (and is meant to generate in the subjects the feeling that they are under surveillance, thus enforcing their compliance to the prescribed norm), a better metaphor for the society described above would be Peripticon: this, too, is meant to keep society in check by giving its members the feeling that they are being watched, but here nobody knows who the observer is; from where and when surveillance is conducted; and who will make use of the results of the observation; when, and in what way.

In such a society—as in every surveillance society that never forgets—individuals will develop distorted personalities, as they can never behave as free persons (even if they are not aware of it at every moment); instead, their behavior will be shaped by expectations, pressures and opportunities. This is already different from the situation which the German Federal Constitutional Court described in its momentous ruling more than a quarter of a century ago; namely, that anyone who is left in the dark as to what information his communication partners hold about him has a limited freedom of decision-making—what we have now is no longer about communication partners, or any kind of partners, only an incorporeal, virtual environment of surveillance. The overall effect of such changes in individual behavior is likely to alter the fabric of society, and in my opinion, to alter it for the worse.

And as for the spies and the non-forgetters, the current situation offers them an even more dangerous weapon: one way to bring about a zero-tolerance society. The possibility of storing information on everyone, of retrieving and using it at any time against anybody, is the perfect means to detect and sanction the slightest deviation from the ideologically, politically or commercially preferred behavior.

*

The degrees of memory and forgetting are related to the degrees not only of surveillance but also of public transparency. Today we are inclined to think that public transparency is a binary phenomenon: information is accessible to either everyone or no one. Nevertheless, the historically or functionally emerging institutions of public transparency have not evolved along such logic. Public transparency has degrees, and that applies even to access to public information, even though this seemingly contradicts the existing legislation on freedom of information. Still, public transparency always had some kind of functional purpose specificity. In a court hearing, the control of publicity is meant to guarantee the honesty and fairness of the legal process; however, control in this case should be understood as control by the people affected and their immediate entourage, rather than unbridled Internet publicity for people who are totally unfamiliar with the context. The earlier practice, whereby the names of the offenders are published according to "local customs," served to exert a restraining influence on the local population, rather than to build the profile of the offender on a social networking site of immense proportions (where these data, incidentally, may be construed as special category sensitive data according to EU data protection norms). Nowadays, this kind of publicity has been removed from the umbrella of the original intentions and legitimacy.

We can see a similar kind of functional target specificity in the way people share various segments of the information they are privy to, with the various circles of people they are in contact with. There are things that we are only willing to share with our spouse, while we may limit other information to family circles, and there are types of information we would divulge at our workplace, in the supermarket, or while speaking at a political rally. We behave very similarly, when in our personal lives we share with the various circles of the outside world only some segments of our identity. We are teachers at school, valued customers in the supermarkets (and we are not asked to show any identification to prove it), patients with a medical history at our GP, buddies in our local pub, and occasionally stupid fools "who can't be more careful" on the subway. They each form segments of our identities, which partially overlap, yet in specific situations and circumstances we never reveal all of them together; not because we have "something to hide," but because in a specific life situation only one specific segment of our identity has a function and relevance—the others quite simply do not belong there. These partial identities, along with the information associated with them, also have a relevant temporal aspect; at a class reunion organized many years after college or during a get-together of former colleagues it would be anachronistic to revert to the old class or workplace hierarchy among the participants, even though telling old stories usually forms an essential part of such reunions.

However, this thoroughly wired world of Internet can easily connect these partial identities, and can even create new ones in the process. But it doesn't even stop there: it markets them and exploits them. There are whole computer technology systems designed specifically for the management of people's identities. These are developed to make it easier for the stronger party in information transactions (the administration or the service provider) to manage the user identities of its clients, actual or potential.

One of the few developments that may still offer a glimpse of hope in the present environment of information technology has been the emergence of user-centric identity management systems, of which the most advanced and the most suitable for systemized use are PRIME and its successor, PrimeLife. Such systems would be built into the "Internet" as an invisible intermediary layer; in other words, they would be built into the complete network services seen and used by us, automatically implementing most of the rules that deal with our data and identities. The software itself would execute all the relevant provisions of the data protection laws; it would carry out the data protection provisions agreed by the service provider and the client, while also implementing the individual user's preferences within the above framework. By selecting the appropriate settings of a PRIME-compatible application, one could have one's photographs automatically deleted from any social networking sites after the expiration of a specified period of time—as long as the social networking site itself is PRIME-compatible. It would even delete copies of these photographs forwarded to any other, PRIME-compatible websites. Similarly to the method employed by the indexing function of search-engines, an ideal, user-centric system of identity management would search the remotest corners of the Internet in order to perform its task, which in this case would be the deletion of photographs. (At this point, my students never fail to ask: What if someone has already copied the pictures onto a pen drive? Well, it's true that no PRIME system will ever be able to erase those photos from a pen drive, but as soon as anyone attempts to upload the pictures back on the Internet, the system will instantly delete them.) However, the broad scale implementation of such systems is not expected to happen soon, not only because of the technical snags, but also on account of the powerful opposition of the adversely interested monopolies of information management.

On the basis of what I have written so far in criticism of the information management monopolies, my approach may come across as slightly activist. While on the one hand I am not suggesting that all that I have written about the problems of forgetting is completely value-neutral, on the other hand I would like to add a cautious word of praise for one of those monopolies: Microsoft.

One of the most promising developments in the information technology of identity management in the past few years has been the concept of private credentials. Together with a number of related software applications, it has been developed by Stefan Brands, probably the most talented member of the new generation of cryptographers. These virtual certificates allow us to take our real-life partial-identities and transfer them to the online world, or even to use them for building new ones. We can use these private credentials to identify ourselves in various online situations, without having to worry that someone will connect our partial-identities and take advantage of them without our approval. These certificates can also be used to verify our various entitlements, as well as our age, qualifications and other personal information, without allowing the information seeker to connect the various databases that are being accessed in the process of authenticating the information. Naturally, the task of "forgetting" the data about the partial identities, i.e. the deletion of the information, will continue to be left to a system not unlike PRIME, but in this way we can be sure that information will be floating around in the promised "eternal" memory of Internet only in fragments, rather than in the form of complete profiles and biographies.

Well, three years ago Microsoft bought Brands' company 'Credentica,' together with all its registered patents. Information experts, as well as most people taking an interest in the deal, were all convinced that this would be the death of Brands' concept. The software giant would simply lock up the patents in its vaults and thus try to prevent the spread of the idea. Brands was the only one who continued to maintain that Microsoft's intentions were, indeed, sincere and the company was serious about incorporating the patents into its software packages; everyone else had their doubts. And then recently, the unexpected happened: Microsoft built the system of private credentials into some of its software packages, or rather, it made provisions to allow the incorporation of the system, and it even published its source code under the name U-Prove. We must not, however, delude ourselves. Microsoft was not starting to act on the basis of a sudden access of human kindness: it was still driven by business considerations; still, this move undoubtedly helped to considerably improve its image in the profession as well as among the organizations of civil society. Obviously, this was not the sole motivation behind the decision and the company's business strategy also called for this move, but still, it would not hurt, and it might even help, if the demand for information autonomy—including guarantees for self-determination in the various phases of remembering and forgetting—were to be met by the market also as a result of commercial pressure.

*

There is one thing we must not forget even in the optimistic, privacy-friendly technocratic milieu of user-centric identity management systems, namely the fundamental difference between human memory and computer memory.

Memory is like a dog that lies down where it pleases, Nooteboom writes. He was, of course, referring to human memory, using a metaphor based on the behavior of another creature, man's best friend, the dog, rather than, say, one's frequently freezing computer. In comparison to its digital counterpart, human memory is *imperfect*: it distorts, omits and selects; it seizes upon some memories and blows them up out of proportion, while relegating others to insignificance, not to mention the fact that it has a tendency to re-evaluate the past from time to time. We may experience déja vu or remember an imaginary memory; in a life-threatening situation, our whole life may flash before our eyes; we may invent incidents that allegedly took place in the past, and after repeating them a great many times we, too, may stop doubting their reality. We may use mnemonics, write diaries, take photographs and archive our e-mails; we may use memoirs or artworks to aid our memory; still, our memories will always be produced and interpreted inside our self. For this very reason, human memory is also *perfect*: it is perfectly human.

Confronting human memory with factual history can often have unwanted consequences. Shortly before my mother died, I decided to take her back to the scene of her childhood vacations—if I may relate a personal memory here. Throughout my childhood I often heard stories from her about a fabulous family estate set in a huge orchard, which also featured an uncle working as a physician in a mine and keeping exotic plants; various family friends, including a half-witted painter; lunches in the garden, labyrinth-like pergolas and a beautiful house. All this, of course, referred to the period before the Second World War, but I thought that perhaps the two of us should make an effort to revisit her past, or at least recover traces of it—and my mother concurred in this.

It was a mistake. After reaching our destination—a large village just outside Budapest, which has since then been elevated to town status—we found an elderly man of more or less the same age as my mother, who could still recall the garden and its owners and could give us directions. Driving on roads that were almost impassable for cars, we had a hard time finding it. Of course, what we found there was a mere shadow of all the things that had stood there back in the old days, but even so, it became quite clear to us that there had never been a fabulous estate here, with a huge orchard and a majestic house. We saw a small plot of land and the remains of a small house, located in a not particularly attractive setting. People who revisit the scenes of their childhood—kindergarten or elementary school—after many years, invariably find the place, which once meant the whole world to them, shockingly small and insignificant; something similar happened to us there. Sure, there were photos and surviving objects, but these only seemed to serve the purpose of fuelling our memory, rather than replacing it. We should never have gone there: then we would never have experienced the distress of witnessing the demise of this important element in individual and family mythology.

Individual memory, but also family and communal memory, consists of such elements, and it is these elements, with all their fallibilities, that after all constitute our culture. We need not take this burden—the burden of human remembrance—off the shoulders of our descendents.

*

Thirty. Thirty what? Years? Euros? The average weight of something? Of whom? Of what?—It is a well-known fact that a piece of data has no meaning by itself: it only has meaning in context. More precisely, data are always produced in a context: we might say, in the context of some information. When we place the data back into their original context, we can retrieve the information. But what happens if we record the data in one context and then interpret them in a different one? Well, in that case we shall get new information. This is what computerized data processing systems and computer networks routinely do. It is a welcome and promising prospect from the viewpoint of scientific/technological development, but has controversial and negative tendencies in the realms of human communities.

The term "function creep" is frequently used in the data protection jargon. As long as a computer system retains data recorded for a specific purpose in its memory, there will always be a great temptation to flout all rules of purpose specificity and make use of those data in a different context by exploiting the growing technological potentials of data analysis. It is in fact more than just a temptation: by now it has become part of the mainstream procedures, involving techniques which are being taught on information science courses all over the world. We build up data warehouses by accumulating personal data, which are no longer relevant and lack legality or legitimacy for data processing (in other words, they are designated for oblivion), in order to bring them back to life and make them suitable for analysis according to some uniform criteria. We employ data mining techniques to extract information from this sea of accumulated historical data—and to draw conclusions in addition to, or in some cases completely independent of, the original purpose, in the hope of discovering certain patterns, such as the signs of secret proclivities, the existence of which the data subjects themselves may be completely unaware.

Of course, there are techniques that take into account the interests of both the data management monopolies and the data subjects and, therefore, achieve the desirable balance between remembering and forgetting. For one thing, it is quite possible to build data warehouses and employ data mining techniques using data that have been stripped of all personal aspects. In such cases the data records are still individual: they contain the complete history of the clients, their gender, age, shopping habits, etc.—everything, short of their name, address and other data usable for personal identification. This will not prevent us from carrying out sophisticated data analysis, the type that we would do on personalized data; we can draw interesting conclusions on the basis of behavior patterns; also, if the identifying data (age, occupation, home address, etc.) of a person newly registered in the system are available, we stand a good chance of guessing that person's preferences and predilections—without remembering the personal identity of the old data subjects. There is only one thing that such a system is definitely not capable of doing: providing support for direct marketing campaigns based on former clients' behavior patterns, which are analyzed subsequently and then used to target those very same clients (incidentally, in clear violation of the law). And if there is still a need to carry out data analysis on personal data beyond the original purpose, we have the whole arsenal of PPDM (Privacy Preserving Data Mining) methods. By employing them, we can conceal the link between the data and the data subject, sometimes through the use of statistical manipulation and sometimes by adding what, in information theory, is known as "noise."

As these examples have shown, sometimes it is the new technology itself that offers solutions to compensate for the harmful side effects of new technology. However, the application of such methods is by no means widespread in practice.

Up to now, the divergent academic fields concerned with the study of memory have constituted the home ground of psychologists. Now it appears that this area will become the monopoly of IT professionals. And the majority of the information experts seem reluctant to come to the rescue of the weaker party: they do not exactly exert themselves to develop and operate systems that would serve the interests of data subjects.

Curiously enough, the social scientists have not yet shown much interest in sounding out the views, knowledge base and attitudes of IT professionals. The studies have focused on users, the young, the elderly, consumers, citizens and suchlike populations; the views of IT professionals have almost never been surveyed. Admittedly, there is some controversy about the definition of an IT professional nowadays, when so many people use computers in their daily work; yet, it is a meaningfully defined population. And if it is true that the code is the law, then the code makers are the de facto lawmakers—it is evident that their views, along with the views of their paymasters, have a crucial bearing on the direction the development of information system affecting and controlling our lives will take. All this will change thanks to some groundbreaking research which has been initiated in this neglected area: an international research project named BROAD that is now well under way. The analysis of the survey data has not yet been completed, so it would be too early to release the results, but the project's starting hypothesis is perhaps worth mentioning here: it is postulated that, at least in Central and Eastern Europe, the majority of IT professionals are socialized so as to serve the more powerful side in the field of information (the authorities, the business monopolies, the service providers—in other words, their paymasters); it is the latter's priorities that this majority has internalized, with only a small minority with markedly divergent values thinking differently—they are the ones that the authorities (mistakenly) call "hackers." The position of the majority is understandable: they receive their commissions, salaries and career opportunities from the stronger party, so why should they worry about the weaker side, the data subjects of the information management systems? Then there is also the typical thinking of technocrats, which makes no substantive difference between humans and inanimate objects, living creatures and abstract entities, no distinction between an RFID dog tag and a chip implanted under the skin of a human being.

IT professionals also play a crucial role in the development and popularization of the visions of eternal digital memory. And while the first popular utopias emerged from the works of early science fiction writers—we should remember H. G. Wells' "World Brain" from 1938—the iconic figures of the present are either IT professionals themselves, or users who represent the worldview of IT professionals. Sometimes even the scientific advisors, in some ways affected by their war experiences, seemed unable to resist the allures of popular utopia: as early as 1945, American journalists were writing about the memory extender machine, or memex, which would enable people to access and tap the knowledge base of earlier ages. Perhaps the best-documented person living today is Gordon Bell, a well-known figure among the designers of the early computers and networks, who in his old age is being sponsored by Microsoft to constantly wear a digital device that takes a digital picture every 30 seconds, capturing virtually every aspect of his life and then archiving it for digital eternity.

Today several information experts hold on to the naïve, utopian notion that the total information produced in the present as well as in the future will soon be orbiting around the Earth as some kind of a public utility service, thus forever defeating all limitations of space and time. The only problems that we are likely to have then will arise from the technical aspects of digital forgetting. Such a world would be controlled by engineers (IT professionals) and the ethical application of the information would be guaranteed by the wiring of the system and the wisdom of respectable old IT professionals. It would be quite interesting to conduct a discourse analysis on the social composition and motives of the groups that develop such a worldview; on the identification and motivation of the groups that are interested in disseminating and raising the popularity of such views; and on whether they have the capacity to look beyond the boundaries of these views and whether they have the willingness to listen to arguments and proposals that are outside the realm of such a paradigm.

*

Forgetting and forgiving. Somehow these are also related concepts. Mind you, I am fully conscious of the fact that proposing this idea would be beyond the pale in current public discourse, and that in the milieu of a transitional society the lessons of the past, or the overcoming of long-suppressed traumas, would always enjoy primacy over reconciliation and letting bygones be bygones. Regardless of the fact that it is not just the mnemonists who feel unceasing sorrow, a society that is constantly forced to confront its past, and to re-live every single moment of it, will also feel perpetually depressed.

From our little world here in the Danube Valley, we watch—with some interest, perhaps, but without much empathy—the former adversaries making peace with each other and cooperating successfully, either as old soldiers who once faced each other from opposite sides of the front line or as travelers on the road to the realization of the European Union through the joint efforts of nations which had long considered each other as ancient adversaries. While none of them seems to forget the essential point, all of them nevertheless forget what it is that stands in the way of shaking hands and establishing peaceful coexistence and even cooperation. We can only hope that our southern neighbors, too, will one day reach this stage. This may remind us of the truth and reconciliation committees in South Africa, East Timor and even in the United States, besides numerous other countries ravaged by war, genocide and other humanitarian catastrophes, where public reconciliation between former enemies, perpetrators as well as victims, is encouraged in the course of a cathartic experience. Such a mixture of rituals, too, sets the stage for remembering, forgetting and forgiving.

And this is the point where historians, or the professional students of our recent history in particular, enter the picture. Some of them are convinced that historians belong to a privileged caste in society, whose mission is to uncover the atrocities of the past. Their raw material is people, i.e. the individuals who populate historical events and documents, and in their missionary zeal, they feel that the end justifies the means. In plain language, they want to be seen as a cut above the rest: they reserve for themselves the right to make decisions about people's life history and personal data. They wish to overrule the subjects' right to self-determination, for example, by denying them the right to forget. Naturally, I am not talking about the actions of persons occupying public office, nor of the crimes committed by politicians, generals and their henchmen and informers: I am referring to the people who simply got caught up in the wheels of history. Through legal provisions characteristic of the democratic transitions, historians, along with any other qualified researchers, do receive some kind of a concession in the study of the past; however, when it comes to the publication of any personal data discovered in the process, the legal limitations kick in. The legislators of the democratic transition were of the opinion—maybe they still are—that historians and professional researchers are people who hold themselves to higher ethical standards and will not abuse their privileged position of being privy to personal information. However, some students of recent history take pride in exploiting their privileged position, not to mention the fact that anyone determined to publish personal data without the consent of the individuals concerned will face little difficulty in obtaining a research certificate.

In 1996, following a lengthy investigation in which I was able to assist, the first Hungarian data protection commissioner produced a detailed analysis of an undertaking in the course of which personal data related to the persecution of Hungarian Jewry during the Nazi era was recorded on microfilm and then sent on to the Yad Vashem Archives in Jerusalem. Holocaust survivors are history's private victims, who are doubly entitled to the right of information self-determination. Those historians and activists who are trying to debunk the Holocaust deniers' claims by publishing the documented life histories of flesh-and-blood people seem to forget that these people, too, have the right to forget and that their moral right to refuse to carry the stigma of Holocaust victims and to bear witness in public is also guaranteed under the law. The Commissioner's recommendation tried to offer at least a symbolic confirmation of this right, shadowed by the hostile feelings of the historians concerned, who regarded the recommendation as a pointless constraint upon their work. They seem to have overlooked the point that by failing to respect the Holocaust survivors' right to forget, they in fact behaved similarly to the persecutors of the Jews: by regarding them as a faceless mass, they actually assist in the virtual, rather than physical, deportation of these people, who are once again being sacrificed on the altar of history.

*

Not so long ago we were still worried about the possibility that the technical problems surrounding the archiving and long-term storage of digital documents could lead to the appearance of a "hole in history." And now we are shocked to discover that we can no longer remember how to forget. We could even say that we have *forgotten how to forget*. It is important to emphasize, however, that it is of course not historical or scientific/technological information that I am concerned with in this essay. What I have been talking about all along is the preserving versus forgetting of human information (concerning individuals and families, small groups and communities); still, this type of information has a crucial bearing on our lives, culture and social environment.

Naturally, we still have the "right"—or I hope that we *still* have the right—to control the data about us, including the framework and time limits of their storage and accessibility, at least according to the letter of the law. The elaborately developed arsenal of the data protection law contains provisions for enforcement, for example through the principle of purpose specificity regulating the time limits of data handling or the data subjects' right to have information regarding them erased. And while we also have the moral right to do all this, the range of *actions* we can actually take is diminishing. Just as the metaphors of Panopticon and Peripticon are being replaced by the metaphor of the Bewitched Palace's House of Mirrors, a hall where our image is multiplied in distorted reflections of different sizes, we can feel that our information reflections, instead of being a real-time process, have now been extended in time and will perhaps continue bouncing back and forth forever.

Can we learn to forget again? Forgetting, similarly to remembering, has many different modes. We may still continue practicing some of them, but others we have started to forget. One of them is infantile amnesia: considered by Freud as highly significant, it still attracts much attention from psychologists who develop various methods either to lock memories up in our brains forever or to bring them out into the open; another one is social amnesia, which the historians studying our recent past attack with great conviction; then there is forgetting for self-defense, deliberate and unconscious; our external memory can break down: our notebook can perish by fire or our Winchester can give up the ghost.

Since digital forgetting is a costly affair, and since it is likely to be even more expensive in the future, there is a danger that it will become the luxury of the wealthy, of the people with the necessary resources, such as money, influence, knowledge and focus. In itself, neither money, nor education, nor critical thinking is sufficient. The liberal intelligentsia, not particularly well-off in terms of money yet much better endowed when it comes to intellectual power, will have to wake up to the reality about the Internet: after bedazzling them and enticing them with its boundless horizons, the Internet also raised their intellectual threshold level, instilling in them a perpetual demand for stimuli. Expecting an information/communication revolution to be delivered to our doorstep every day, we fail to acknowledge that the Internet has become a sort of intellectual fast food of the modern era and that the truly precious things are precisely the individual, human aspects that require focused attention and time. Just as synthetic curtains and windbreakers have lost their appeal as status symbols in Eastern Europe, while hand-woven fabric and hand-embroidery are coming back into fashion, so is a real-life postcard with a handwritten message becoming more valuable than an automatically generated, virtual birthday card. In a sense, forgetting has a similar role in the age of information storage and processing.

The majority of people in modern societies must be convinced to re-cultivate a demand for the right to forget: they must learn how to recognize and use this right. To achieve this, however, a new generation must come forward, whose members attach importance to their personal autonomy while still retaining a strong demand for critical thinking; whose desire for comfort does not necessarily take precedence over their yearning for freedom; and who can still remember how to forget.

References

Below is a list of the more important direct and indirect references, arranged in the actual order of their occurrence:

Lessig, Lawrence. 2000. *Code and other laws of cyberspace*. New York: Basic Books. (Lessig's best-seller, in which he expounds his view on how code becomes law in information societies.)

Beck, Ulrich. 1992. *Risk society: Towards a new modernity*. London: Sage. (The originator of the concept of reflexive or second modernity wrote his classic treatise on risk society as early as 1986.)

Fuchs, Christian. 2011. Critique of the political economy of Web 2.0 surveillance. In *Internet and surveillance: The challenge of Web 2.0 and social media*, ed. Christian Fuchs, Kees Boersma, Anders Albrechtslund, and Marisol Sandoval, 31–70. New York: Routledge. (The author, who applied Marxist ideas to the world of Internet, made an ambitious attempt to develop the political economy of Web 2.0.)

Borges, Jorge Luis. 1962. Funes, the Memorius. In Jorge Louis Borges, *Ficciones*, 107–126. New York: Grove Press. (A famous 1942 short story capturing the short life of the Uruguayan mnemonist; analyzed by Draaisma; see further below!)

Luria, Alexander Romanovich. 1968. *The mind of a Mnemonist: A little book about a vast memory*. New York: Basic Books. (Writing about Solomon Shereshevsky, who is referred to in the book as "S," Luria, the prominent Soviet neurologist, describes the numerous experiments he carried out for many years between the two World Wars.)

Draaisma, Douwe. 2004. *Why life speeds up as you get older*. New York: Cambridge Univ. Press. (Revealing a strong empathy for the various theories of memory, a remarkable book by the famous Dutch psychologist and historian of psychology.)

Mayer-Schönberger, Viktor. 2009. *Delete: The virtue of forgetting in the digital age*. Princeton/Woodstock: Princeton Univ. Press. (While not being the first to address the problem, the author produced the first book with a comprehensive argumentation about the need as well as the difficulties of forgetting.)

Bentham, Jeremy. 1995. *Panopticon, or the inspection-house (1787)*. Its modern edition: *The panopticon writings*, ed. Miran Bozovic, 29–95. London: Verso. (Originally put forward by Foucault, this proposition became the classic metaphor of the surveillance society.)

Székely, Iván. 2010. Kukkoló társadalom—avagy van-e még függöny a virtuális ablakunkon? ["Voyeur Society—Does Our Virtual Window Still Have a Curtain?"] In *Az internet a kockázatok és mellékhatások tekintetében* [The Internet, with regard to possible hazards and side-effects], ed. Judit Talyigás, 93–120. Budapest: Scolar [In Hungarian]. (My study on voyeur society for the above publication.)

Feeley, Michael M., and Jonathan Simon. 1992. The new penology: Notes on the emerging strategy of corrections and its implications. *Criminology* 30:449–474. (A much-cited sourcebook of the new school of penology.)

PRIME and PrimeLife. Privacy and Identity Management for Europe: a project supported by the European Commission's 6th Framework Programme and the Swiss Federal Office for Education and Science, and PrimeLife: Bringing sustainable privacy and identity management to future networks and services, a research project funded by the European Commission's 7th Framework Programme. (These web sites make available documents from the first and second phase of the development of a comprehensive user-centric identity management system.) http://www.prime-project.eu, and http://www.primelife.eu. Accessed 6 December 2011.

Brands, Stefan. 2000. *Private credentials: Zero-knowledge systems, Inc., November 2000*. http://osiris.978.org/~brianr/crypto-research/anon/www.freedom.net/products/whitepapers/credsnew.pdf. Accessed 6 December 2011. (One of the author's earliest expositions of the system of private credentials.)

Nooteboom, Cees. 1983. *Rituals*. Baton Rouge: LSU Press. (The original source of the earlier-quoted metaphor of memory.)

BROAD. Broadening the Range of Awareness in Data protection: a project supported by by the Fundamental Rights and Citizenship Programme of the European Commission. (The homepage of the BROAD project.) http://www.broad-project.eu. Accessed 6 December 2011.

Wells, Herbert George. 1938. *World brain*. London: Meuthuen & Co. Ltd. (A highly visionary piece of early sci-fi about the world brain.)

Bell, Gordon, and Jim Gemmel. 2007. A digital life. *Scientific American* 296:58–65. (A description of the ideology behind MyLifeBits, a tool that records everything.)

Armengo, Roberto, Kent Wayland, and Priscilla Regan. 2010. Facebook funhouse: Notes on personal transparency and peer surveillance. Paper presented at the fourth Biannual Surveillance and Society/SSN conference, April 13–15. London, UK. (A presentation given by the researchers who originally proposed the House of Mirrors metaphor—for the moment still only in highlights.)

Parliamentary Commissioner for Data Protection and Freedom of Information. 1998. Recommendation on the microfilm recording of documents containing personal data relating to the persecution of Jews during the Nazi period, and on their transfer to the Yad Vashem Archives in Jerusalem. In *The first three years of the Parliamentary Commissioner for Data Protection and Freedom of Information*, ed. László Majtényi. Budapest: Office of the Parliamentary Commissioner for Data Protection and Freedom of Information. (An analysis of conflicts between the legitimate purposes of unveiling historical events and the privacy rights of the victims of history.) http://www.osaarchivum.org/publications/accessandprotection/. Accessed 6 December 2011.

A történelemben lesz egy lyuk ["There Will Be a Hole in History"]. 2003. In *E-világi beszélgetések.hu* [E-World Conversations.hu], ed. Judit Talyigás. Budapest: Peszto Kiadó [In Hungarian]. (A conversation with me, originally conducted for a collection of interviews with the main contributors to the establishment of the Hungarian information society.)

MIX
Papier aus verantwortungsvollen Quellen
Paper from responsible sources
FSC® C105338

If you have any concerns about our products,
you can contact us on
ProductSafety@springernature.com

In case Publisher is established outside the EU,
the EU authorized representative is:
Springer Nature Customer Service Center GmbH
Europaplatz 3, 69115 Heidelberg, Germany

Printed by Libri Plureos GmbH
in Hamburg, Germany